STRUCTURE AND DYNAMICS OF MACROMOLECULES: ABSORPTION AND FLUORESCENCE STUDIES

STRUCTURE AND DYNAMICS OF MACROMOLECULES: ABSORPTION AND FLUORESCENCE STUDIES

J.R. ALBANI

Laboratory of Molecular Biophysics
University of Science and Technology
of Lille, Villeneuve d'Ascq, France

ELSEVIER

2004

Amsterdam – Boston – Heidelberg – London – New York – Oxford – Paris
San Diego – San Francisco – Singapore – Sydney – Tokyo

ELSEVIER B.V.
Sara Burgerhartstraat 25
P.O. Box 211, 1000 AE
Amsterdam, The Netherlands

ELSEVIER Inc.
525 B Street
Suite 1900, San Diego
CA 92101-4495, USA

ELSEVIER Ltd.
The Boulevard
Langford Lane, Kidlington,
Oxford OX5 1GB, UK

ELSEVIER Ltd.
84 Theobalds Road
London WC1X 8RR
UK

First edition 2004

Library of Congress Cataloging in Publication Data
A catalog record is available from the Library of Congress.

British Library Cataloguing in Publication Data
A catalogue record is available from the British Library.

ISBN: 0-444-51449-X

♾ The paper used in this publication meets the requirements of ANSI/NISO Z39.48-1992 (Permanence of Paper).
Printed in The Netherlands.

Preface

Absorption and fluorescence spectroscopies are widely used to characterize the structure and dynamics of biological molecules. It is now clearly established that biological function of a molecule depends on its structure and / or dynamics.

This book deals with the fundamentals of the absorption and fluorescence along with applications on the characterization of the structure and dynamics of macromolecules. The example chosen from the literature are discussed, and in many cases we show that it is possible to interpret the same result in two different and contradicted ways.

The examples given in the chapter dealing with the origin of proteins fluorescence clearly rule out the existence of a general model describing this origin. Also, in one of the chapters, we describe fluorescence experiments that led us to obtain a description of the tertiary structure of α_1- acid glycoprotein. In the same chapter, we added the first results we obtained recently with X-ray and electron microscopy on the protein and its carbohydrate residues. Finally, the last chapter describes novel experiments showing that animal and vegetal species and varieties within the same species have characteristic fluorescence fingerprints.

This book will be useful to students, professors and researchers who want to discover, to learn, to teach or to use absorption and fluorescence spectroscopies.

We were not able to cite the works of all the colleagues using absorption and fluorescence spectroscopies in their research. However, we believe that every work carried out in the field of these two techniques is important to find out their performance and their limits.

Finally, the author wishes to thank the designer Mr. Roger Boudiab who redraws most of the graphics already published in the literature and which is used in this book.

Contents

3. Fluorophores : Descriptions and properties

4. Fluorescence quenching

Chapter 1

LIGHT ABSORPTION BY A MOLECULE

l. Jablonski diagram or diagram of electronic transitions

Absorption of light (photons) by a population of molecules induces the passage of electrons from the single ground electronic level S_o to an excited state S_n (n > 1). In the excited state, a molecule is energetically unstable and thus it should return to the ground state S_o . This will be achieved according to two successive steps:
- The molecule at the excited state S_n will dissipate a part of its energy in the surrounding environment reaching by that the lowest excited state S_1.
- From the excited state S_1, the molecule will attain the ground state S_o via different competitive processes:
 a) Emission of a photon with a radiative rate constant k_r. Emission of a photon is called fluorescence.
b) The energy absorbed by the molecules is dissipated in the medium as heat. This type of energy is non radiative and occurs with a rate constant k_i.
c) The excited molecules can give up their energy to molecules located near by. This energy transfer occurs with a rate constant k_q (collisional quenching), or with a rate constant k_t (energy transfer at distance).
d) A transient passage occurs to the excited triplet state T_1 of energy lower than S_1 with a rate constant k_{isc}. The triplet state is energetically unstable. Therefore, deexcitation of the molecule will occur via different competitive phenomena:
- Emission of a photon with a rate constant k_p. This phenomenon is called phosphorescence.
- Dissipation of non radiative energy with a rate constant k'_i.
- Transfer of energy to another molecule at distance (rate constant k'_t) or by collision (rate constant k'_q) (Figure 1.1).
 The S $\rightarrow$ S or the T $\rightarrow$ T transitions are called internal conversion.
 The S_1 -$\rightarrow$ T_1 transition is called intersystem crossing. For each excited state S, an excited state T does exist.
 Molecules that absorb photons are called chromophores and a chromophore that emits a photon is called a fluorophore. Therefore, a fluorophore is inevitably a chromophore, however a chromophore is not necessarily a fluorophore. For example, heme does absorb light however it does not fluoresce. The absence of fluorescence is the result of total energy transfer from the porphyrin ring to the iron. Other metals such as zinc or tin do not quench completely porphyrin fluorescence. Thus, zincporphyrin and tinporphyrin display fluorescence of energy lower than that of porphyrin.
 It is important to indicate that thermal activation of a molecule can induce its passage from the ground to the excited state.

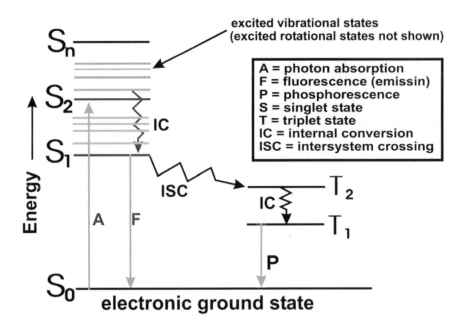

Figure 1.1. Jablonski diagram or the electronic transitions diagram. (Adapted from Jablonski, A, 1935. Z. Phys. 94, 38-64).
Used with permission. ©Thomas G. Chasteen, Sam Houston State University

2. Singlet and triplet states

Transitions described in Jablonski diagram take place within chemical molecules and therefore they concern the atoms that constitute these molecules. The electrons of these molecules are responsible for the different transitions shown in the Jablonski diagram. For this reason, the Jablonski diagram is also called the electronic transitions diagram.

Since localization of an electron is difficult, one referred to four quantum numbers (n, l, m and s) to characterize an electron and to differentiate it from the others. The principal quantum number n determines the energy of any one-electron atom of nuclear charge Z. n can assume any positive integral value, excluding zero.

The angular – momentum quantum number l determines the angular momentum of the electron. It may assume all integral values from 0 to n-1 inclusive.

The magnetic quantum number m characterizes the magnetic field generated by the electric current of the electron, circulating in a loop. m can assume all integral values between –l and +l including zero.

The spin quantum number s is the result of the electron spinning about its own axis. Thus, a local small magnet is generated with a spin s.

Singlet and triplet states depend on the quantum number of spin of the electron (s). According to Pauli Exclusion Principle, two electrons within a defined orbital cannot have equal four quantum numbers and thus will differ by the spin number. Two spin numbers are attributed to an electron, + ½ and – ½. Therefore, two electrons that belong to the same orbital will have opposite spin.

A parameter called the Multiplicity (M) is defined as being equal to:

$$M = |s_1 + s_2| + 1 \tag{1.1}$$

When the spins are parallel, M will be equal to 1 and we shall have the singlet state S.

When the spins are anti-parallel, M will be equal to 3. We are in presence of a triplet state T (Figure 1.2).

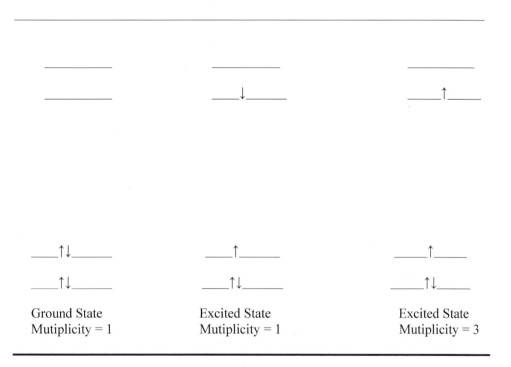

Ground State	Excited State	Excited State
Mutiplicity = 1	Mutiplicity = 1	Mutiplicity = 3

Figure 1.2. Spin configurations of the singlet and triplet states.

Upon excitation, one electron absorbs light energy and goes to an upper vacant orbital. The molecule is in the excited state S_n. After desexcitation, the electron will return to its original orbital and the molecule will be in the ground state S_o.

In the ground state, all molecules except oxygen are in the singlet form. Oxygen is in the triplet state, and when excited it reaches the singlet state. The singlet excited state of oxygen is destructive for cells attained by tumors. Photodynamic therapy is a process in which a light-responsive chemical, when exposed to the appropriate wavelength of light, is activated to undergo either a photophysical process or to initiate photochemistry, producing molecular species which can interact with biological targets (photosensitization).

Such interactions can be exploited for biomedical applications or for basic studies. Photosensitization is exploited for the destruction of tumors and certain non-neoplastic target tissues in an approach termed photodynamic therapy (PDT). Compounds such as porphyrins are localized in target cells and tissues. Upon light activation, energy transfer occurs from porphyrins to oxygen molecules inducing by that a triplet -> singlet transition within the oxygen state. Cells attained by tumor attract excited oxygen molecules and then are being destroyed by these same oxygen molecules. The history of photodetection and photodynamic therapy has been described by Ackroyd et al. (2001). Administrated drugs can also be used instead of porphyrins to induce singlet oxygen formation in cells.

Perotti et al (2004) studied the efficacy of 5-aminolevulinic acid (ALA) derivatives as pro-photosensitising agent. The authors used cell line LM3 for their studies. ALA is a precursor of protoporphyrin IX and thus injection of ALA and / or of its derivatives into cells will induce the formation of protoporphyrin IX, a necessary molecule to induce singlet oxygen and thus cell death.

Figure 1.3 describes porphirin synthesis from ALA and its derivatives. One can notice that synthesis of porphyrin does not follow the same rule for all the coupounds used.

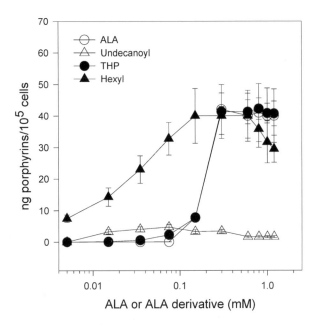

Figure 1.3. Porphyrin synthesis from 5-aminolevulinic acid (ALA) and ALA derivatives, hexyl-ALA (He-ALA), undecanoyl-ALA and R,S-2-(hydroximethyl)tetrahydropyranyl-ALA (THP-ALA). Cells were incubated for 3 hours in the presence of different amounts of ALA or its derivatives. Intracellular porphyrins were determined fluorometrically and relativised per number of cells present at the beginning of the experiment. Source: Perotti C, Fukuda H, DiVenosa G, MacRobert AJ, Batlle A and Casas A, 2004, Br. J. Cancer. 90, 1660-1665. Authorization of reprint accorded by Nature Publishing Group.

Figure 1.4 displays dark and PDT toxicities for ALA and its derivative. Undecanoyl-ALA is intrinsically toxic to cells even in the absence of light (Panel A). Increasing light doses accelerates cell death (Panels B and C).

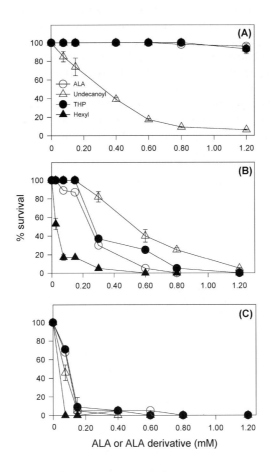

Figure 1.4. Dark and PDT toxicity of ALA and ALA derivatives. Cells were incubated for 3 hours in the presence of different amounts of ALA or its derivatives in the dark (panel A), or exposed to 0.25 (panel B) and 0.4 J. cm^{-2} of light (panel C). Cell survival is expressed as a percentage of the control nonirradiated and exposed to ALA or derivatives. Source: Perotti C, Fukuda H, DiVenosa G, MacRobert AJ, Batlle A and Casas A, 2004, Br. J. Cancer. 90, 1660-1665. Authorization of reprint accorded by Nature Publishing Group.

Finally, the authors calculated tumor porphyrin synthesis after administration of equimolar concentrations of ALA and ALA derivatives to mice. They found that ALA is the best inducer of porhyrin synthesis (15.9 $\pm$ 1.12 nmol/g of tissue), followed by THP-ALA (8.83 $\pm$ 0.95 nmol/g of tissue), He-ALA (3.11 $\pm$ 0.42 nmol/g of tissue) and undecanoyl-ALA (2.64 $\pm$ 0.23 nmol/g of tissue). The values were calculated after three hours of ALA or ALA derivative administration to the cell lines.

The natural non-toxic hypericin is a photosensitizing anti-cancer drug. In fact, neoplastic cells in culture respond to hypericin in a dose-dependent fashion: high doses of light and high concentrations of photosensitizer cause apoptosis or cell necrosis, whereas low levels of activation induce only a `stress response' that involves the synthesis of proteins known as stress-induced or heat shock proteins (HSP) (Varriale et al. 2002 and other cited references). The main function of HSP is to afford additional protection to the cell as the need arises. Figure 1.5 displays hypericin uptake, release and sub-cellular distribution in HEC1-B cells. One can notice that hypericin uptake remained quite steady from 3 to 20 h (insert a). The sub-cellular distribution of hypericin obtained by using ultracentrifuge-based methods was verified by checking for the specific marker proteins c-Myc (nuclear), p-Bip (ER), actin (cytosol) and VDAC (mitochondria) within each sub-fraction (Fig. 1.5. insert b). The amount of hypericin in sub-cellular fractions was estimated fluorimetrically (emission spectra) and expressed after normalization (Fig. 1.5). Sensitization of hypericin in the nuclear fraction induces locally DNA damage. The authors found that hypericin sensitization induces damage that causes apoptosis or cell necrosis according to the energy adsorbed. Also, they show that mild pre-sensitization endows cells with an unexpected high degree of photo-tolerance, enhances HSP70 synthesis, sequentially promotes the expression of specific apoptosis-related proteins and causes cell cycle arrest (Varriale et al. 2002).

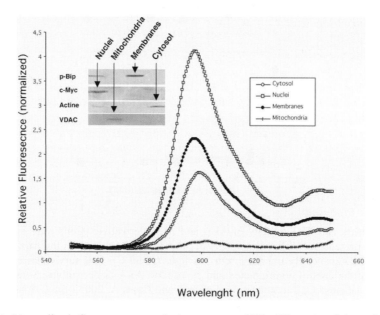

Figure 1.5. Normalized fluorescence emission spectra (550–650 nm) of hypericin from individual sub-cellular compartments of hypericin-loaded HEC1-B cells. a: Relative fluorimetric estimate of cell-trapped drug at indicated times. The arrows indicate the residual hypericin fluorescence at 3 and 20 h. Hypericin was measured fluorimetrically in three experiments and expressed as percent (±S.D.) with respect to hypericin fluorescence obtained at time 0 (λ_{ex} = 490 nm). b: Sub-cellular fractionation was validated by Western blot using protein markers specific for each fraction: actin (cytosol), c-Myc (nucleus), VDAC (mitochondria) and p-Bip (membranes). Source: Varriale, L., Coppola, E., Quarto, M., Veneziani, B. M., and Palumbo, G. 2002, FEBS Letters, 512, 287-290.

3. Forbidden and non forbidden transitions

Upon light absorption, electrons are under the effect of electric and magnetic components. The contributions of these two components in the transition process could be defined as:

$$F = (e\,E) + (e\,v\,H / c)$$ (1.2)

where e, c, v, E and H are respectively the charge of the electron, the velocity of light, the speed of rotation of the electron on itself, the electric and magnetic components of the light wave. The products (E e) and (e v H / c) are respectively the contribution of the electric and magnetic fields to the absorption phenomenon.

The speed of rotation of the electron on itself is very weak compared to the velocity of light, thus the magnetic contribution to the absorption is negligible compared to the electric one. Therefore, upon absorption, electronic transitions are the result of the interaction between the electron and the electric field. For this reason, during absorption, the displaced electron preserves the same spin orientation. This is why, only the $S_o \rightarrow S_n$ transitions are allowed and the $S_o \rightarrow T_n$ transitions are forbidden.

The Jablonski diagram describes very well the phenomena that are occurring after light absorption by a molecule. Understanding these phenomena helps to master absorption and fluorescence methods and to overcome the major problems that a researcher meets during his work.

4. Reading the Jablonski diagram

In the present paragraph, we are simply going to enumerate the most important conclusions we can draw from the Jablonski diagram. Details are given in the different corresponding sections.

a. Absorption allows the passage of the chromophore from the ground state to an excited state, therefore absorption of photons is synonymous with excitation. This feature allows us to compare the absorption and the excitation spectra of a molecule one with each other. This comparison is possible for fluorophores and not for molecules that absorb only.

b. Energy of absorption is higher than energy of emission.

c. Many phenomena other than fluorescence emission contribute to the deexcitation of the fluorophore. These other alternatives to fluorescence are, radiationless loss, phosphorescence, photooxidation and energy transfer. Thus, the weaker will be the competitive phenomena, the highest will be the deexcitation via fluorescence.

d. Dissipation of energy via the non-radiative process has to be controlled if we want to compare absorption or emission spectra at stable temperature.

e. Emission occurs from the excited state S_1, independently of the excitation wavelength. Therefore, the energy of the emission of a fluorophore would be independent of the excitation wavelength.

f. Fluorescence energy is higher than that of phosphorescence.

g. Absorption and fluorescence do not require any reorientation of spin. However, the inter-system crossing and phosphorescence require a spin reorientation. Therefore, absorption and fluorescence are much faster than phosphorescence. Absorption occurs within a time equal to 10^{-15} s and the lifetime of fluorescence goes from 10^{-9} to 10^{-12} s. Phosphorescence is a long transition that can last from the ms to the s.

h. Since phosphorescence is slow, quenching by small molecules such as oxygen, present naturally in solution, makes phosphorescence difficult to be observed. Therefore, fluorescence is the primary emission process of biological relevance.

i. The energy of an electronic transition is equal to the energy difference between the starting energy level and the final one. Therefore, the energy of the transition E is equal to :

$$E = h\nu = h\,c\,/\,\lambda \tag{1.3}$$

where h is the Planck constant (h = 6.63 x 10^{-34} joules.s), ν is the light frequency (s^{-1} or Hertz, Hz), c, the velocity of light and λ, the wavelength expressed in nanometer (nm). The energy is expressed in Joules / mol. Therefore, each transition occurs with a specific energy and thus at a specific and single wavelength.

j. Absorption spectrum occurs from the ground state. Therefore, it will be the mirror of the electronic distribution in this state. Fluorescence and phosphorescence occur from excited states, thus they will be the mirrors of the electronic distribution within the excited states, S for the fluorescence and T for the phosphorescence. Any modification of the electronic distribution in these states, such as in presence of charge transfer, will modify the corresponding spectrum. A common example is the reduction of cytochrome c. Addition of an electron to the ground state for example, will modify the electronic distribution within the molecule and thus will affect the absorption spectrum.

5. Chemical bonds

5a. Atomic and molecular orbitals

In nature, atoms interact one with the others forming molecules that have lower energy than separated atoms. The chemical and physical properties of a molecule are the results of its empirical composition and mainly of the way the atoms are bonded.

Electrons are the main atom components that will play a fundamental role in the structure of the newly formed molecular bond. Quantum mechanics provides us with a mathematical expression describing the probability of finding an electron in every position of space. However, this theory does not explain how an electron moves from one position to another. Therefore, the notion of orbital has been introduced in order to explain the probability of finding an electron at various points in space. Each type of orbitals corresponds to one of the possible combinations of quantum numbers. An orbital can have two electrons of opposite spin signs, one electron or can be vacant. In the ground state, chemical bonds will occur in such a way so that the two electrons will now belong to the formed molecular bond. The two electrons will occupy or share a molecular orbital.

A molecule is defined as a system of atomic nuclei and a common electron distribution. However, describing the electronic structure of molecules is done by approximating the molecular electron distribution by the sum of atomic electron distributions. This approach is known as the *Linear combination of atomic orbitals*

(LCAO) method. The orbitals produced by the LCAO method are called *molecular orbitals* (Mos).

The shape and symmetry of the molecular orbitals differ from those of the atomic orbitals of the isolated atom. The molecular orbitals extend over the entire molecule and their spatial symmetry must conform to that of the molecular framework.

5b. The coordinated bond

Three types of chemical bonds are met, the ionic, the covalent and the coordinated bonds. Once established, the coordinated bond acts as a covalent one. Only the covalent and coordinated bonds intervene in absorption and fluorescence.

Covalent bond occurs between two or more atoms. The formation and stability of these molecules are associated with an equal sharing of valence electrons, i.e., covalent bonding. The electrons participating in the chemical bond will belong to the same molecular bond, i.e., to the same molecule.

There are three types of covalent bonds in the ground state, σ, π and n. In the σ bond, the molecular orbital is symmetric around the axis of the chemical bond and the electronic density is concentrated between the two atoms participating in the bond. The π bond evolves in both sides of the axis of the chemical bond. The electronic density is located between the two atoms participating in the chemical bond. The n bond concerns molecules where the electrons are not shared by the two atoms. The electrons are located in one atom but not in the other.

Transition from the ground state to the excited one is accompanied by a redistribution of the electronic cloud within the molecular orbital. This is an implicit condition so that transitions occur.

The energy of a pair of atoms as a function of the distance between them is given by the Morse curve (Figure 1.6). r_0 is the equilibrium bond distance. At this distance, the molecule is in its most stable position and thus its energy is called molecular equilibrium energy and expressed as E_o or E_e. Stretching or compressing the bond yields an increase in energy.

According to the Franck-Condon principle: electronic transitions are so fast that they occur without change in the position of the nuclei, i.e. the nuclei have no time to move during the electronic transition. In fact, the lifetime of an electronic transition is about 1.4×10^{-15} s, while the lifetime of a nuclei vibration is 22×10^{-15} s. Thus, the lifetime of the nuclei vibration is approximately 16 times longer the lifetime of an electronic transition. This is why the electronic transitions are showed always as vertical lines.

Finally, let us recall that energy within a molecule is the sum of several distinct energies:

E (total) = E(translation) + E (rotation) + E(vibration) + E (electronic) + E (electronic orientation of spin) + E (nuclear orientation of spin).   (1.4)

Rotational energy concerns the rotation of the molecule around its center of gravity, vibrational energy is the one owed to the periodic displacement of atoms of the molecule away from its equilibrium position, and the electronic energy is that generated by electron movement within the molecular bonds. Rotational levels or states are part of the vibrational ones and thus are lower in energy.

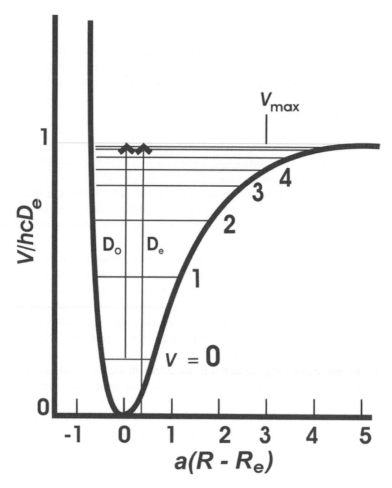

Figure 1.6. Morse curve characterizing the energy of the molecule as a function of the distance R that separates the atoms of a diatomic molecule such as hydrogen. At a distance equal to R_e, that corresponds to point 0, the molecule is in its most stable position and thus its energy is called molecular equilibrium energy and expressed as E_e. Stretching or compressing the bond yields an increase in energy. The number of bound levels is finite. D_o is the dissociation energy and D_e the dissociation minimum energy. The horizontal lines correspond to the vibrational levels.

At the left side of R_e , the two atoms get closer and closer inducing repulsive forces. Thus, we are going to observe an energy increase due to these repulsive forces. At the right side of R_e, the distance between the two atoms increases and thus we are in presence of attractive forces so that the equilibrium distance is reached. Thus, we are going to observe an increase of the energy as the result of the attractive forces. In principle, one should obtain a harmonic oscillation; however, it is not the case. In fact, beyond a certain distance between the two atoms, the attractive forces will exert no more influence and the attraction energy will reach a plateau. Therefore, the Morse curve is anharmonic.

6. Absorption spectroscopy

6a. Origin and properties of the absorption spectra

Looking closely to the Jablonski Diagram, one could see that each electronic transition, whether it concerns absorption, fluorescence or phosphorescence, occurs with a definite energy and thus at one specific wavelength. Therefore, the profiles of these electronic transitions should correspond to a single line observed at the specific wavelength. Instead, one gets a spectrum which is the result of electronic, vibrational and rotational transitions. Thus, the energy absorbed by a molecule is equal to the sum of energies absorbed by the above three mentioned states. For this reason, the molecular electronic levels modifications are accompanied by changes in the vibrational and rotational energy levels. The absorption spectrum will be the result of all these transitions, the maximum of the spectrum (the peak) will correspond to the electronic transition line and the rest of the spectrum will be formed by a series of lines that correspond to the rotational and vibrational transitions. Therefore, absorption spectra are sensitive to temperature. Rising the temperature increases the rotational and vibrational states of the molecules and induces the broadening of the recorded spectrum.

The profile of the absorption spectrum will depend extensively on the relative position of the value of E_r which depends on the different vibrational states. The intensity of absorption spectrum will depend, among others, on the population of molecules reaching the excited state. More this population is important, more the intensity of the corresponding absorption spectrum is high. Therefore, recording an absorption spectrum of the same molecule at different temperatures should in principle yield altered or modified absorption spectra.

An absorption spectrum is characterized by the position of its peak (the maximum), and the full width at half maximum (FWHM) which is equal to the difference

$$\delta \nu = \nu_2 - \nu_1 \qquad (1.5)$$

ν_1 and ν_2 correspond to the frequencies that are equal to the half of the maximal intensity.

If the studied molecules do not display any motions, the spectral distribution will display a Lorenzien type profile. In this case, the probability of the electronic transition $E_i ---> E_f$ is identical for all the molecules that belongs to the E_i level.

Thermal motion of the molecules will induce different speed of displacement of the molecules and thus different transition probabilities. These facts change from one molecule to another and from a population of molecules to another. In this case, the spectral distribution will be a Gaussian type. The full width at half maximum of a Gaussian spectrum is larger than that of a Lorenzian one.

It is important to mention that effect of temperature is less important on absorption than on fluorescence spectra. However, it is better to keep the spectrophotometer thermostated along the experiment. Also, one should avoid placing the spectrophotometers in front of the sun. The sun will heat the electronics of the spectrophotometer inducing important errors in the optical density measurements. Also, exposing the spectrophotometer to the sunlight can induce continuous fluctuations in the optical density making impossible any serious measurement.

In absorption such as in fluorescence, light intensity is plotted as a function of the wavelengths. Figure 1.7 displays the absorption spectra of tryptophan, tyrosine and phenylalanine in water. One can notice the presence of a strong band at 210-220 nm and a weaker one at 260-280 nm.

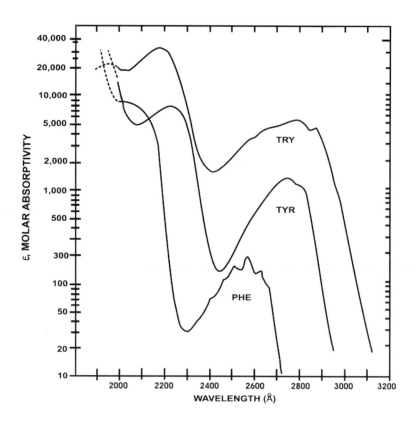

Figure 1.7. Absorption spectra of tryptophan, tyrosine and phenylalanine in water. Source: Wetlaufer, D. B., 1962, Ultraviolet absorption spectra of proteins and amino acids, Advan. Protein Chem. 17, 303-390.

Also, one can record the excitation spectrum of a molecule. The excitation spectrum of an isolated and pure molecule is equivalent to its absorption spectrum. Figure 1.8 displays the fluorescence excitation spectrum of Phenylalanine in water recorded at λ_{em} = 310 nm while figure 1.9 displays the fluorescence excitation spectrum of cytochrome b_2 core extracted from the yeast *Hansenula anomala* recorded at λ_{em} = 330 nm. The value of the spectrum bandwidth (34 nm) is characteristic of tryptophan residues.

A molecule can have one, two or several absorption peaks or bands (Figure 1.10). The band located at the highest wavelength and therefore having the weakest energy is called the first absorption band.

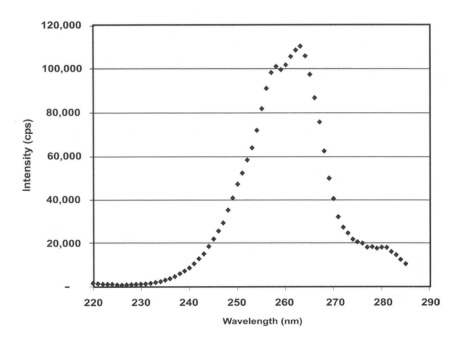

Figure 1.8. Fluorescence excitation spectrum of Phenylalanine in water recorded at λ_{em} = 310 nm. Curtosey from Professor Rachel E. Klevit, Department of Biochemistry, University of Washington, Seattle.

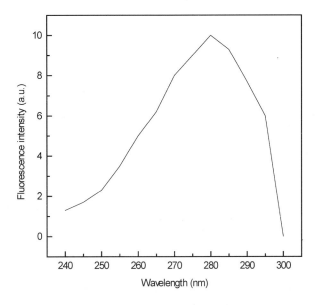

Figure 1.9. Fluorescence excitation spectrum of cytochrome b_2 core extracted from the yeast *Hansenula anomala.* λ_{em} = 330 nm.

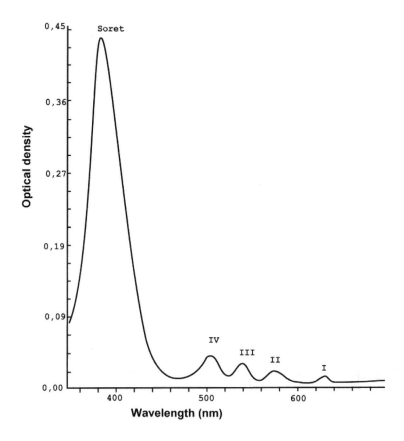

Figure 1.10. Absorption spectrum of protoporphyrin IX dissolved in DMSO.

The positions of the absorption peaks of protoporphyrin IX vary with the nature of the solvent. When dissolved in water, aggregation of porphyrins occurs very easily (more details on porphyrin spectral properties are given in chapter 9).

6b. Beer-Lambert-Bouguer law

An absorption spectrum is characterized by two parameters, the position of the maximum (λ_{max}) and the molar extinction coefficient calculated usually at the maximum (ε). The relation between ε, the sample concentration (c) and the thickness (l) of the absorbing medium is characterized by the Beer-Lambert-Bouguer law. Since the studied solution is put in a cuvette and the monochromatic light beam is passing through the cuvette, the thickness of the sample is called optical pathlength or simply the pathlength.

While passing through the sample, the light is partly absorbed and the spectrophotometer will record theoretically the non-absorbed or the transmitted light.

Plotting the transmittance which is the ratio of the transmitted light I_T over the incident light I_o

$$T = I_T / I_o \qquad (1.6)$$

as a function of the optical pathlength and the sample concentration yields an exponential decrease (Fig. 1.11).

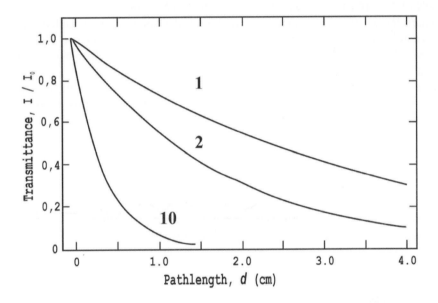

Figure 1.11. Transmittance variation with optical pathlength for three sample concentrations.

Therefore T is proportional to exponential (-cl) (Eq.1.7).

$$T \; \alpha \; e^{-cl} \qquad (1.7)$$

Equation 1.7 can be written also as

$$Ln \; T = Ln \; (I_T / I_o) \; \alpha \; -c \; l \qquad (1.8)$$

A constant of proportionality k can be introduced and we get

$$Ln \; T = Ln \; (I_T / I_o) \; = - \; kcl \qquad (1.9)$$

$$- \; Ln \; T = Ln \; (I_o / I_T) \; = \; kcl \qquad (1.10)$$

Transforming Eq.1.10 to decimal logarithm, we obtain

$$\log T = \log (I_o / I_T) = \frac{kcl}{2.3} = \varepsilon \, cl = O.D. = A. \tag{1.11}$$

Equation 1.11 is the Beer-Lamber-Bouguer law. At each wavelength, we have a precise optical density. Since the absorption or the optical density is equal to a logarithm, it does not possess any unity. The concentration c is expressed in molar (M) or mole / l, the optical pathlength in cm and thus ε in M^{-1} cm^{-1}.

ε characterizes the absorption of 1 mol./l of the solution. It is called extinction coefficient because when the incident light goes through the solution, part of this light is absorbed. Thus, the light intensity is "quenched" and thus it will be attenuated while passing through the solution.

If the concentration is expressed in mg/ml or in g/l, the unity of ε will be in mg^{-1} ml. cm^{-1} or in g^{-1} l cm^{-1}. The conversion to M^{-1} cm^{-1} is possible by multiplying the value of ε expressed in g^{-1} l cm^{-1} by the molar mass of the chromophore.

In general, ε is determined at the highest absorption wavelength(s). However, it is possible to calculate an ε at each wavelength of the absorption spectrum. Figure 1.12 displays the absorption spectrum of the fluorescent probe 2, p-toluidinylnaphthalene-6-sulfonate (TNS). TNS possesses 4 electronic transitions at 223, 263, 317 and 366 nm. The values of the ε at these wavelengths are already known. Plotting the absorption spectrum of a known concentration of TNS allows obtaining the values of ε at different wavelengths. Knowing the values of ε at different wavelengths is important for energy transfer studies. Table 1.1 shows the values of ε at different wavelengths for TNS dissolved in water.

Table 1.1. Values of the molar extinction coefficient ε of TNS dissolved in water at different absorption wavelengths.

λ_{abs} (nm)	ε (M^{-1} cm^{-1})	λ_{abs} (nm)	ε (M^{-1} cm^{-1})
223	4.7×10^4	335	8.42×10^3
263	2.45×10^4	340	6.6×10^3
280	1.32×10^4	350	5.26×10^3
295	1.21×10^4	360	4.74×10^3
317	1.89×10^4	366	4.08×10^3
325	1.42×10^4	370	3.15×10^3
330	1.16×10^4	380	1.58×10^3

One should understand that we can measure or calculate a molar extinction coefficient for a molecule at each wavelength of its absorption spectrum. In general, the molar extinction coefficient is determined at the absorption peaks simply because the molecule absorbs most at these wavelengths.

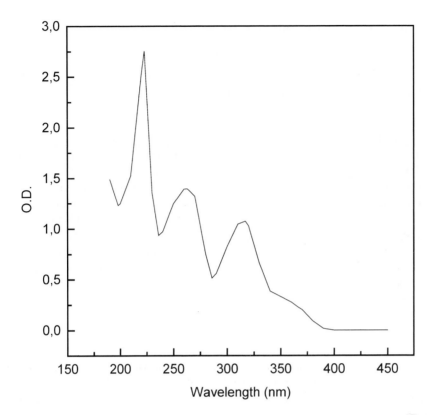

Figure 1.12. Absorption spectrum of the fluorescent probe 2, p-toluidinylnaphthalene-6-sulfonate (TNS). The molecule possesses 4 electronic transitions at 223, 263, 317 and 366 nm.

In proteins, three amino acids, tryptophan, tyrosine and phenylalanine, are responsible for the U.V. absorption. ε in proteins is determined at the maximum (278 nm) (Fig. 1.13) and thus, concentrations of proteins are calculated by measuring the absorbance at this wavelength. Cystine and the ionized sulfhydryl groups of cysteine absorb also in this region but much weakly than the three aromatic amino acids. Ionization of the sulfhydryl group induces an increase in the absorption and an appearance of a new peak around 240 nm. The imidazole group of histidine absorbs in the 185-220 nm region. Finally, important absorption of the peptide bonds occurs at 190 nm.

When a protein possesses a prosthetic group such as heme, its concentration is usually determined at the absorption wavelengths of the heme. The most important absorption band of heme is called the Soret band and it is localized around 408-425 nm. The position of the peak of the Soret band depends on the heme structure, and in cytochromes, it will depend on whether the cytochrome is oxidized or reduced. For example, cytochrome c extracted from the yeast *Hansenula anomala* shows a Soret band at 408 nm with an ε of 100 mM^{-1} cm^{-1} in the oxidized form and at 415 nm with an ε of 183 mM^{-1} cm^{-1} in the reduced form (see also Table 1.2).

Table 1.2. Values of ε of different biological products.

Product	ε
DNA	6.6 mM^{-1} cm^{-1} at 258 nm
RNA	7.4 mM^{-1} cm^{-1} at 258 nm
Oxidized FMN	12.5 mM^{-1} cm^{-1} at 445 nm
Oxidized cyt.c	100 mM^{-1} cm^{-1} at 408 nm
Reduced cyt.c	183 mM^{-1} cm^{-1} at 415 nm
Free hemin	60.3 mM^{-1} cm^{-1} at 390 nm
HbO$_2$	15.15 mM^{-1} cm^{-1} at 575 nm
MetHb	11.5 mM^{-1} cm^{-1} at 540 nm
Chain α of hemoglobin	15.05 mM^{-1} cm^{-1} at 576 nm
Chain β of hemoglobin	14.5 mM^{-1} cm^{-1} at 577 nm
Trypsine	1.5 [mg / ml]$^{-1}$ cm^{-1} at 280 nm

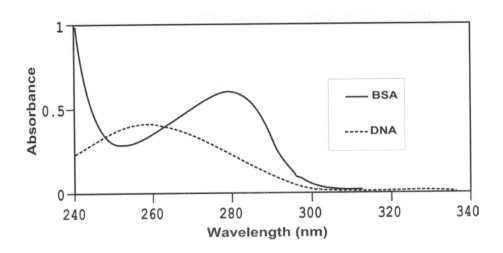

Figure 1.13. Absorption spectra of DNA and of bovin serum albumin.

6c. Determination of the molar extinction coefficient of proteins

The Beer-Lambert-Bouguer law clearly indicates that at each wavelength, the optical density is linearly dependent on ε. Therefore, the simplest method to determine the value of ε at a certain wavelength is to dissolve a precise quantity of the protein in 1 ml of buffer and then to measure the optical density of the corresponding solution.

Measurements are done for 7 to 10 protein concentration. Plotting the optical density as a function of protein concentration expressed in mg/ml yields a linear plot with a slope equal to the product (l ε). For a cuvette pathlength l equal to 1 cm, the slope will be equal to ε, expressed in mg^{-1} l. cm^{-1}. Multiplying the value of ε by the molecular weight of the protein yields an ε expressed in mM^{-1} cm^{-1}.

This type of experiment gives an accurate value of ε if cautions are taken. Experiments should be repeated two or three times to be sure that the data are reproducible; also one should work at low protein concentrations so that to avoid aggregation.

A second method exists, it consists on counting the number of tryptophans, tyrosines and cysteines in the protein sequence and then determine the value of ε with equation 1.12 (Gill and von Hippel, 1989; 1990) :

$$\varepsilon_{(M\text{-}1\ cm\text{-}1)} = (5560 \times \text{nb. of Trp}) + (1200 \times \text{nb. of Tyr}) + (60 \times \text{nb. of cys}). \qquad (1.12)$$

We determined the value of ε at 278 nm with the two above described methods for the α_1-acid glycoprotein extracted from human plasma. Plotting the optical density as a function of protein concentration yields a straight line with an ε equal to 0,723 mg^{-1} ml cm^{-1}. The molecular weight of the protein is equal to 41000, ε was found equal to 29.7 mM^{-1} cm^{-1} (Albani, 1992).

α_1-Acid glycoprotein contains three Tryptophans, 11 tyrosines and 5 cysteines. The value of ε determined from equation 1.12 is 30.410 mM^{-1} cm^{-1}. One can notice that the experimental and theoretical values are very close.

The values of ε of a chromophore can change whether it is free in solution or bound to a protein. This change can be important or non-significant. For example we determined the value of ε of calcofluor white free in solution and bound to α_1-acid glycoprotein which contains 40% carbohydrate by weight. Calcofluor White binds specifically to the hydrophilic carbohydrate residues (Albani and Plancke, 1998; 1999), although interaction with hydrophobic pockets of some proteins such as serum albumin occurs also (Albani and Plancke, 1998; 1999).

Figure 1.14 shows the optical density plots of calcofluor white at 352.7 nm as a function of its concentration in absence (a) and in presence (b) of 10 μM α_1-acid glycoprotein. We notice that the value of ε is not the same, the value of ε for bound calcofluor white is around 9 % lower than that of free calcofluor in solution.

The decrease in the value of ε was also observed for the fluorescent probe 1-anilino-8-naphthalene sulfonate (ANS) in presence of apomyoglobin. In fact, the absorption peak shift from 265 nm when free in solution to 271 nm in presence of apomyoglobin while the extinction coefficient decreases from 19200 to 13000 cm^2 / mmole L, (Fig. 1.15).

Modification in the values of ε and the shift in the absorption peak of the chromophore when it is bound to a macromolecule is common to many chromophores. For example, free hemin absorbs at 390 nm. However, in the cytochrome b$_2$ core extracted from the yeast *Hansenula anomala*, absorption maximum of heme is located at 412 nm with a molar extinction coefficient equal to 120 mM^{-1} cm^{-1} (Albani, 1985). In the same way, protoporphyrin IX dissolved in 0.1 N NaOH absorbs at 510 nm while when bound to apohemoglobin it absorbs in the Soret band at around 400 nm.

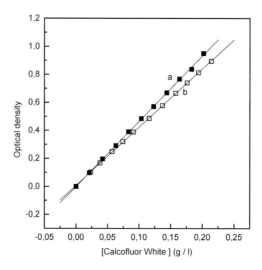

Figure 1.14. Optical density plots of calcofluor white at 352.7 nm as a function of its concentration in absence (a) and in presence (b) of 10 μM α_1-acid glycoprotein. In the absence of protein, $\varepsilon = 4.65443$ g^{-1} l cm^{-1} $= 4387.76$ M^{-1} cm^{-1}. In presence of the protein, $\varepsilon = 4.15921$ g^{1} l cm^{-1} $= 3920$ M^{-1} cm^{-1}. We notice that the values of ε are not the same, the value of ε for bound calcofluor white is around 9 % lower than that of free calcofluor in solution.

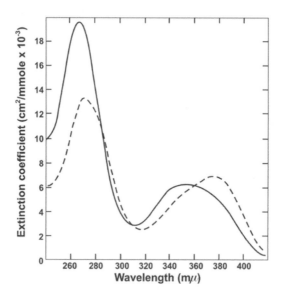

Figure 1.15. Absorption spectra of 5 x 10^{-5} M ANS in 0.1 M phosphate buffer, pH 5.8 (full line) and of ANS complexed to apomyoglobin (dotted line). The spectrum of the complex was obtained by recording the spectrum of ANS-apomyoglobin complex versus apomyoglobin spectrum. Source: Stryer, L, 1965, J. Mol. Biol. 13, 482-495.

6d. Effect of high optical densities on the Beer-Lambert-Bouguer law

The Beer-Lambert-Bouguer law is no more respected for two reasons, one is due to the instrument and the other depends on the chromophore itself.

The linearity between the optical density and the concentration of a chromophore is no more respected at high optical densities. This can be the result of the saturation of the photomultiplier which will be not able to count more than a precise number of photons. When this number is reached, the optical density given by the spectrophotometer will be the highest and we are going to say that the photomultiplier is saturated. Beyond the "concentration of saturation" the spectrophotometer will always give the same optical density that corresponds to that of saturation.

The second reason is owed to the chromophore itself. At high concentrations, aggregation could occur. In this case, we are going to have two values of ε, corresponding to low and high concentrations. Plotting the optical density as a function of the macromolecule concentration will yield two plots with two different values of ε.

6e. Effect of the environment on the absorption spectra

The environment here can be the temperature, the solvent, the interaction of the chromophore with another molecule, etc...

In general, interaction between the solvent and the chromophore occurs via electrostatic interactions and hydrogen bonds. In the presence of a highly polar solvent, dipole-dipole interaction requires high energy and thus the position of the absorption peak will be located at low wavelengths. In the contrary, when the dipole-dipole interaction is weak such as when the chromophore is dissolved in a solvent of low polarity, the energy required for the absorption is weak, i.e., the position of the absorption band is located at high wavelengths. A blue shift in the absorption band is called hypsochromic shift and a red shift is called bathochromic shift.

When a chromophore binds to a protein, the binding site is in general more hydrophobic than the solution. Therefore, one should expect to observe a shift in the absorption band to the highest wavelength compared to the peak observed when the chrompohore is free in solution. The shift in the absorption peak of ANS in presence of apomyoglobin from 265 to 271 nm is a typical example of this phenomenon (Stryer, 1965). Another example is the binding of flavin adenine dinucleotide (FAD) to electron transfer flavoprotein (ETF). Figure 1.16 displays absorption spectra of ETF and of FAD-saturated ETF and titration of ETF with FAD. Distortion in the absorption spectrum of ETF is observed in presence of FAD and titration curve indicates that saturation occurs at a stoichiometry of 1 FAD for 1 ETF. However, since ETF contains already one mol FAD, FAD-saturated ETF contains 2 mol FAD. Thus, spectral modifications were attributed by the authors to the binding of the second FAD molecule. Figure 1.17 shows the absorption spectra of free FAD and of the FAD bound to ETF. The last spectrum presents an absorption band at 401 nm instead of 370 nm for free FAD. The extinction coefficients at the peaks are 9.8 mM^{-1} cm^{-1} at 450 nm for free FAD and 14.3 mM^{-1} cm^{-1} at 401 nm for bound FAD. The unusual absorption spectrum of FAD observed at saturation has been attributed to the flavin environment in the protein and not to chemical modifications of the flavin ring.

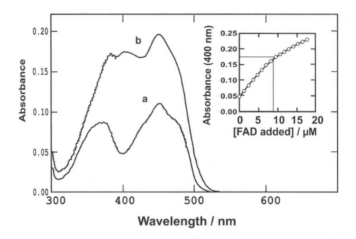

Figure 1.16. Absorption spectrum of ETF in the absence (a) and presence of saturated concentration of FAD (b). The spectrum of the FAD-saturated ETF was obtained by analyzing the set of spectra recorded during the titration. The inset shows the titration plot of absorbance at 400 nm and a theoretical curve with a dissociation constant of 0.07 µM and an added FAD concentration at 1:1 molar ratio of 8.76 µM (solid curve). The perpendicularly connected lines indicate the 1:1 point. Source: Sato, K., Nishina, Y. and Shiga, K., 2003, J. Biochem, 134, 719-729. Authorization of reprint accorded by the Japanese Biochemical Society.

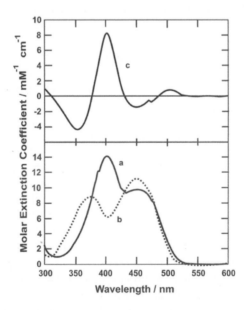

Figure 1.17. Spectral change of FAD upon binding to ETF. Curve a shows the difference spectrum between a and b in Fig. 1.14 divided by the added FAD concentration at a 1:1 ratio to convert to the molar extinction coefficient. Curve b shows the spectrum of free FAD and curve c shows the difference spectrum, a – b. Source: Sato, K., Nishina, Y. and Shiga, K., 2003, J. Biochem, 134, 719-729.

The presence of a solvent may perturb the optical spectroscopy of a protein, in this case, the difference spectrum should be obtained between (protein-perturbent) vs protein. Figure 1.18 displays the perturbation difference spectra of amino acids produced by 20% (v/v) ethylene glycol (Donovan, 1969).

Temperature increases the vibrational transitions, i.e., the vibrational levels of the electronic ground state is more and more populated. These transitions occur from the highest vibrational levels. The physiological properties of a macromolecule are modified with the variation of temperature. If temperature variations are sufficient to induce important variations in the vibrational and electronic transitions in the molecule, the absorption spectrum will be deeply modified. The molecule will loose its physiological properties only when we reach a temperature where structural denaturation is reached. However, one should be cautious when studying folding and / or unfolding of proteins since the different regions of the proteins do not behave similarly with the temperature. In fact, in an enzyme, denaturation of the active site with temperature could be more difficult than other regions of the protein.

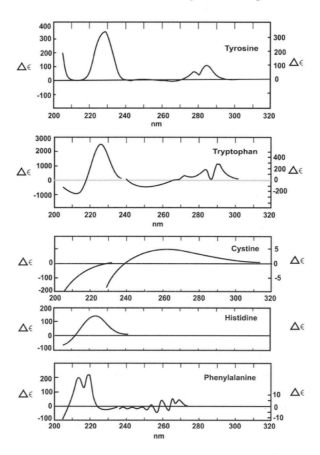

Figure 1.18. Perturbation difference spectra of amino acids produced by 20% (v/v) ethlyene glycol. Source: Donovan, J. W., 1969, J. Biol. Chem. 244, 1961-1967. Authorization of reprint accorded by the Biochemical Society for Biochemistry and Molecular Biology.

Structural modification can be followed also by monitoring the variation of the optical density at a fixed wavelength. A typical example is the optical density variation of DNA with temperature (Fig. 1.19). One can notice that DNA denaturation can be followed by the increase of the optical density at one wavelength, 260 nm. When complete denaturation is reached, the optical density does no more increase.

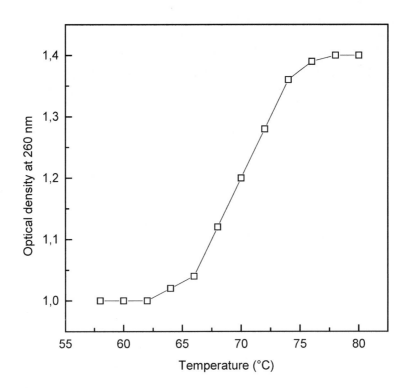

Figure 1.19. Denaturation of DNA with temperature followed by the optical density increase at 260 nm.

Protein denaturation induces a loss of its secondary and tertiary structures. Since the tertiary structure is responsible for the physiological properties of the proteins, its loss will induce a total inactivation of the protein. Denaturation can occur by different ways, thermal, pH, urea, guanidine salts mainly guanidine chloride, organic solvents and detergents. We are not going to detail all these types of denaturation since it is not the purpose of this book. However, it is important to understand that denaturation does not occur instantaneously and it depends not only on the nature of the denaturing factor or agent but also on the concentration of the molecule and on its structure. The most effective denaturants is guanidine chloride. Total loss of the molecules structure occurs at 6 M GndCl. The salt is a high electrolyte, which means it will interact on the protein via electrostatic charges.

Protein charges that maintain the folded structure will be neutralized by the electrostatic interaction with guanidine chloride inducing by that an unfolded protein. Also, interaction occurs with the amino acids of the inner core of the protein inducing by that an irreversible unfolded state in presence of the denaturant. Equation 1.13 gives the relation that exists between the intrinsic viscosity of proteins (η) in presence of 6 M guanidine chloride and the number (n) of amino acid residues per protein chain (Tanford et al. 1967):

$$\eta = 0.71\ n^{a} = 0.71\ n^{0.66} \tag{1.13}$$

where a characterizes the efficiency of the solute-solvent interaction. The values of a goes from 0.5 to 1 for random coil. Plotting log η as a function of log n yields a straight line (Fig. 1.20). Also, one can plot log η as a function of the logarithm of the molecular weight, a linear relation is also obtained.

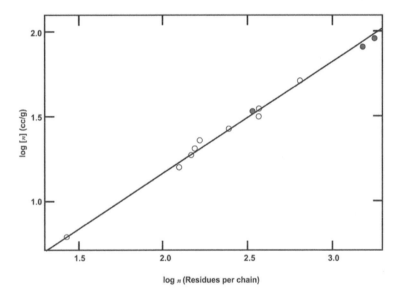

Figure 1.20. Intrinsic viscosity as a function of chain length, for protein polypeptide chains in 5M (closed circles) and 6M (open circles) guanidinium chlodide at 25°C. The straight line represents Eq. 1.13. Source: Tanford, C., Kawahara, K. and Lapanje, S. 1967. J. Am. Chem. Soc. 89, 729-736. Authorization of reprint accorded by the American Chemical Society.

β-Galactosidase is a tertameric enzyme that consists of identical subunits with a molecular mass of 116 kDa. Each monomer is composed of five compact domains and a further 50 additional residues at the N-terminal end. Within the cell, the enzyme cleaves lactose to form glucose and galactose. The latter is exchanged with lactose via a lactose-galactose antiport system. The composition and structure of the cell wall change continuously during plant development. Thus, it is not a static structure but a dynamic one. Plant cell walls consist of cellulose microfibrils coated by xyloglucans and embedded in a complex matrix of pectic polysaccharides (Talbot and Ray, 1992 ;

Carpita and Gibeaut, 1993). Plant development involves a coordinated series of biochemical processes that, among other things, result in the biosynthesis and degradation of cell wall components. Enzymes such as β-Galactosidase and α-arabinosidase play a role in the cross-linking of pectins and cell wall proteins by catalyzing the formation of phenolic-coupling activity. The enzymes hydrolyze corresponding phospho-para-nitrophenylderivatives (α-L-arabinofuranoside and β-D-galactopyranoside, respectively), into para-nitrophenyl that absorbs at 410 nm. Thus, it is possible to follow the evolution of the plant development by quantification of the PNP formed. This can be done by following the absorption of para-nitrophenol (PNP) at 420 nm ($4.8 \times 10^3 \ M^{-1} \ cm^{-1}$) (Stolle-Smits and Gerard Beekhuizen, 1999).

In the following example, we have studied denaturation of β galactosidase by guanidine chloride following the optical density of PNP at 410 nm. Denaturation of the enzyme will induce a loss in the enzyme activity, which will be observed by the decrease of the optical density at 410 nm.

Figure 1.21 displays the loss of activity of the β galactosidase with increased concentrations of guanidine chloride. We notice that the loss of the total activity is reached at 12.5 mM of guanidine chloride. The optical densities shown are those recorded after 25 mn of reaction. The reaction was stopped by addition of Na_2CO_3. The hyperbolic shape of the curve allows determining dissociation constant for the enzyme-Guanidine complex equal to 3 mM. Enzyme concentration was small, which explains the total loss of activity at concentrations of guanidine chloride much lower than 6 M. Also, it would be interesting for each denaturant concentration to record the fluorescence excitation spectrum of the enzyme so that one can see the structural modifications of the protein.

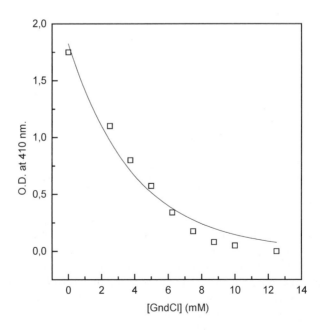

Figure 1.21. Effect of Guanidine chloride concentration on the activity of β-galactosidase.

Figure 1.22 displays the activity of β galactosidase in presence of 7 mM of guanidine chloride as a function of time. Analysis of the decay curve yields two decay times equal to 2.9 and 14.5 min. These times reveal mainly that denaturation process of a protein does not occur in one simple step. The structure and the dynamics of the protein should play important role in the accessibility of the guanidine chloride to the amino acids and thus to the unfolding process.

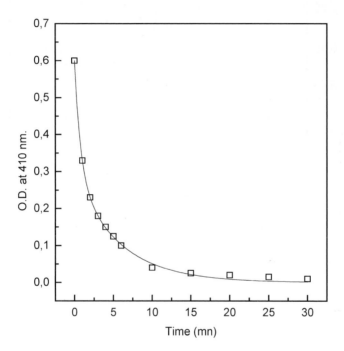

Figure 1.22. Activity of β-galactosidase with time in presence of 7 mM guanidine chloride. Analysis of the exponential decrease as the sum of two exponentials yields value of the decay times equal to 2.9 min and 14.5 min.

6f. Absorption spectroscopy and electron transfer mechanism in proteins

Electron transfer between electron donor and acceptor located in two different proteins or within the same protein induces a spectral modification in the absorption of both redox centers and thus kinetics parameters and electron transfer data can be studied with absorption. Figure 1.23 displays the absorption spectra of oxidized (a) and reduced (b) forms of cytochrome b_2 core extracted from the yeast *Hansenula anomala*. Reduction of cytochrome c with cytochrome b_2 core is followed at the isobestic point of cytochrome b_2 equal to 416.5 nm. At this wavelength, only absorption of cytochrome c increases upon reduction.

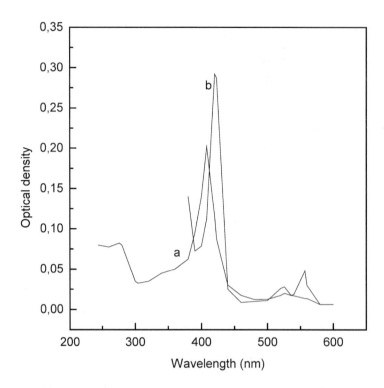

Figure 1.23. Absorption spectra of cytochrome b_2 core in the oxidized (a) and in the reduced forms (b). Reduction occurs with dithionite which absorbs below 400 nm. Cytochrome b_2 core concentration is calculated at 423 nm using an ε equal to 178 $mM^{-1} cm^{-1}$.

In general, the redox centers are separated by distances that average between 10 to 25 Å, although in almost all productive reactions, the edge to edge distances of long range physiological electron transfer was found to be less than 14 Å (Fig. 1.24).

Electron will be transferred via the protein matrix and the protein-protein interface. Therefore, one should ask whether the structure and the dynamics of the proteins play an important role in the determination of the rate of this electron transfer or not? Also, the relative orientation and distance of the two redox potential centers will have another determinant role in the electron transfer efficiency.

It has been considered that protein dynamics are too slow to affect electron transfer. However, the different kinetic constants measured for the electron transfer for proteins indicates that the electron transfer can occur within the range of the protein dynamics timescale. For example, the highest kinetic constant for the electron transfer between Zn porphyrin to Ruthenium both bound to myoglobin is found equal to 7.2×10^7 s^{-1} (Casimiro et al. 1993.) This means a time equal to 14 ns. This time is in the same range of the protein rotation and even slower from the time of the local rotation which averages 1 ns or less.

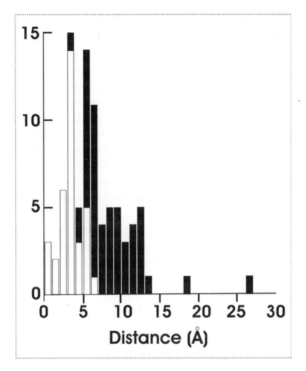

Figure 1.24. A histogram of edge-to-edge distances of long-range physiological electron transfer for 31 different redox proteins in the PDB data base shows that almost all productive reactions are less than 14 Å. The two examples at longer distances may represent physiologically inappropriate structures Solid bars, distances between cofactors within chains; open bars, distances between closely spaced substrates and redox centers within catalytic clusters. Source: Page, C.C., Moser, C. C., Chen, X and Dutton, P. L. 1999. Nature. 402, 47 – 52. Authorization of reprint accorded by Nature Publishing Group.

Another example, a range of 1.5×10^7 s^{-1} was found for the kinetic constant electron transfer between cytochrome b_2 core and cytochrome c (Capeillère-Blandin and Albani, 1987), i.e. a time equal to 66 ns. This time is higher than the rotational correlation time ($\phi = 6$ ns at 20°C) of cytochrome b_2 core (MW = 14000). Therefore, if the electron transfer is considered a very fast process, does the experimental kinetic constants reflect the real electron transfer or the effect of the electron transfer on local structural or / and dynamical modification within the proteins?

Unlike adiabatic chemical reactions, electron transfer mechanism does not affect the molecule bonds of the concerned materials. The kinetic constant of electron transfer can be written as (Marcus and Sutin,.1985):

$$K_{Et} = \frac{4 \pi^2 H_{AB}^2}{h (4 \pi R T)^{1/2}} \exp \left[- \Delta G^\circ + \lambda \right)^2 / 4 \lambda k_B T] \qquad (1.14)$$

$\Delta G^{\circ} + \lambda)^2 / 4\lambda$ is the activation free energy, ΔG° is the driving force of the redox redaction and λ is the reorganization energy. H_{AB} is the probability of the reaction, k_B the Boltzmann constant and h the Planck's constant.

λ is the energy needed to modify the nuclear configuration of the whole system that concerns the surrounding medium, the redox centers, the bond length, protein-protein interface reorganization if any and the protein(s) matrix local modifications. Therefore, electron transfer will be the result of the effect of each of these parameters in the overall reaction. The electrons will follow a route within the system to go from a center to another, a route that will be less or more condensed with atoms (protein, solvent, etc…).

From this definition of the electron transfer, two mechanisms were proposed to explain the electron transfer mechanism, one is based on the better pathway(s) the electron can follow within the system and the second is based on the packing density of the medium between the two redox centers.

Electron transfer can take place via one or multiple pathways. When different pathways occur, usually electron transfer occurs via the strongest path. However, the different pathways can inhibit one each other and electron transfer will occur via an unexpected pathway controlled mainly by the structure and the dynamics of the overall system.

This means also that any conformational change within the system would induce a modification in the electron transfer process.

Also, electron transfer can be the result of the packing density between the two redox centers. In other terms, the medium within the proteins between the two redox centers will affect the electron transfer rate. For example, if electron transfer occurs between two redox centers separated by a vacuum, theoretically it can take place with high-energy barrier, thus a phenomenon difficult to observe.

Different questions have to be raised here too concerning this model: What is the contribution of the solvent enveloping and passing through the proteins in the electron tunneling through the proteins? Do the structure and / or the dynamics of the proteins play any role in the electron-tunneling phenomenon? Taking the fact that electrons will follow precise pathways (one or multiple) in the proteins, do we know if protein domains, not concerned by the electron transfer pathways, affect in a way or another the rate and the probability of the electron transfer?

Anyway, regardless the different questions we have raised for the moment, and in order to take into account the medium and the paths the electron are taking, equation 1.14 will be simplified and written as:

$$K_{Et} = k_o \exp \left[-\beta (r - r_o) \right] \qquad (1.15)$$

k_o ($= 10^{13}$ s^{-1}) is the rate constant of electron transfer at the van der Waals contact distant ($r_o = 3$ Å), r is the distance between the donor and the acceptor and β is a parameter that will evaluate the effectiveness of the medium in the electron transfer. β is called also the electronic decay factor. In proteins, values of β were found from 0.7 to 1.4 Å^{-1}. More the value of β is low better will be the electron transfer rate.

First, it has been estimated that more the medium between the two redox centers is highly packed with atoms, more electron tunneling will be fast and probable. However, from the intramolecular electron transfer and structure deposited in the protein Data bank, Page et al. (1999) found the packing density is statistically identical whether measured between redox centres of physiologically productive reactions, between redox centers of unphysiological and counterproductive electron transfer (Fig. 1.25). The authors concluded "nature does not use protein structural heterogeneity to physiological benefit".

Also, the same authors found the absence of any correlation between the quality of electron transfer path and the physiological benefit of the reaction.

Figure 1.26 displays the logarithm of the rate constant as a function of edge-to-edge distance between the two redox centers.

Comparative studies between the pathways and the packing density models have shown that both models are equivalent in describing the electron transfer process. Both models combine through-bond and through-space electronic interactions in proteins. Also, it appears that the strongest single pathway is generally a member of a family of many pathways with similar coupling strengths. Although theoretically the two models ignore the possible multiple pathways interference and local fluctuations, the results obtained from the two methods are from experiments and in a way or in another they are taken into consideration in the analysis (Liang et al. 2000).

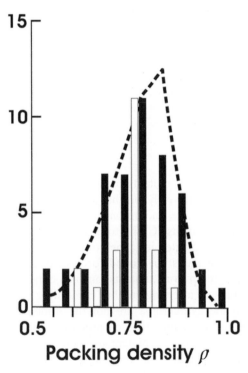

Figure 1.25. Distributions of packing density for physiologically productive (solid bars) and unphysiological (open bars) electron transfer reactions for proteins in the PDB database. The dashed line shows a normalized distribution of the presumably arbitrary interior packing found between standard donor cofactors and the 6-atom aromatic ring of tyrosine and phenylalanine side chains as mock cofactors. Source: Page, C. C., Moser, C. C., Chen, X. and Dutton, L. P. 1999, Nature, 402, 47-52. Authorization of reprint accorded by Nature Publishing Group.

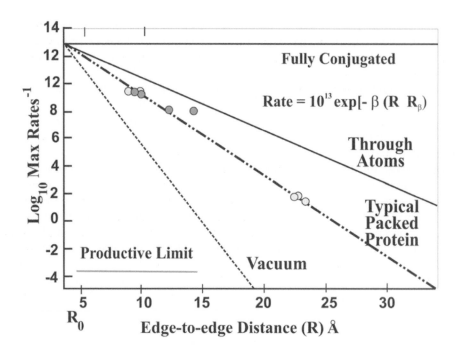

Figure 1.26. Free- energy-optimized rate of electron tunneling. Source : Professor Dutton, L. P. See also figure 1 in: Page, C. C., Moser, C. C., Chen, X. and Dutton, L. P. 1999, Nature, 402, 47-52.

Electron transfer studies within mutants of $Mb^{Fe \to Zn}$ and of cytochrome c between zinc porphyrin and Ruthenium have shown that the number of paths is very important for myoglobin (from 79 to 186, depending on the mutant) while it goes from 1 to 22 paths for the cytochrome c mutants (Casimiro et al. 1993). Therefore, it appears that electron transfer is better coordinated in cytochrome c to the difference of myoglobin. Also, since small number of paths was found for cytochrome c and almost 10 times more were found for myoglobin in all mutants, one can conclude that each protein has its own specificity. The factors affecting and / or controlling electron tunneling through myoglobin and through the cytochrome are not the same. If now we compare the number of paths between the mutants of each protein, cytochrome or myoglobin, we found that it is not the same from a mutant to another. Therefore, one amino acid mutation was sufficient to modify the number of pathways significantly in both proteins. This could simply be the result of modification in the local structures within each protein. In other terms, global but also local structures could be an important factor in the electron transfer tunneling mechanism.

Studies performed on electron transfer between zinc – myoglobin ($Mb^{Fe \to Zn}$) and cytochrome b_5 have revealed that electron transfer between the two proteins do not depend on their affinity (Liang et al. 2000; 2002). In fact, within the region of cyt.c-Mb interaction, the two heme propionates of myoglobin possess a pair of negative charges that would tend to repel cyt.b$_5$. This will induce a very low affinity complex between the two proteins. Therefore, the idea was to neutralize the heme propionates so that affinity between the two protein increases inducing in the same time a burst in the electron transfers efficiency and rate. However, the result was an increase in the electron transfer between the two proteins without any modification in their affinity (K_a = 10^3 M^{-1}). For example, at pH 7, the bimolecular constant rate k_2 increases from 6.1 x 10^6 M^{-1} s^{-1} to 5.5 x 10^8 M^{-1} s^{-1} in the neutralized state. Therefore, the authors concluded that binding is not substantially enhanced by charge neutralization and reaction rates are decoupled from binding affinities. Pathways calculations indicate that the neutralization does not change either the area of reactive surface on myoglobin or the reactivity of this particular surface. The authors proposed a model called the dynamic docking model. In this model they considered the existence of a large ensemble of conformations, each being associated with its own rate constant for electron transfer (Fig. 1.27.)

Electron transfer between Mb$^{Fe \rightarrow Zn}$ and cyt.b$_5$ occurs from a specific conformation. Neutralization of heme propionates induces a more productive protein docking increasing by that electron transfer efficiency. Therefore, although the affinity of the complex is still the same because the area of the two proteins interaction did not change, the orientation of the two proteins within the complex is more favorable to electron transfer.

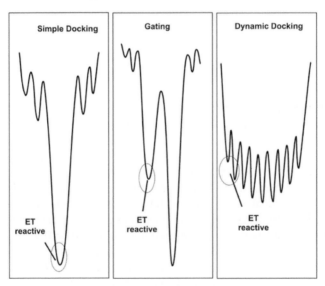

Figure 1.27. Schematic energy landscapes, which plot the conformational energy as a function of a "conformational coordinate" that connects the bound configurations of the complex, for "simple docking", gated (or conformationally coupled) docking which involves two major conformational substrates, and dynamic docking, which involves multiple conformations of the eomplex. Source: Liang, Z-X., Nocek, J. M., Huang, K., Kurnikov, I. V., Beratan, D. N. and Hoffman, B. M, 2002, J. Am. Chem. Soc. 124, 6849-6859. Authorization of reprint accorded by the American Chemical Society .

Studies between different cytochromes extracted from the yeast *Hansenula anomala* have shown that electron transfer is dependent on the stability of the reaction complex formed between the proteins. Results have shown that modification of one of the redox center affects drastically the rate of the energy transfer.

Flavocytochrome b_2, the L(+)-lactate : cytochrome c oxidoreductase, is a tetrameric enzyme (Mr 235,000) containing one FMN and one protoheme IX per protomer. The enzyme, located in the mitochondrial inter-membrane space, is able to bind one cytochrome c per protomer. The complex is stable at ionic strengths equal to or lowers than 40 mM.

Flavocytochrome b_2 can be cleaved by controlled proteolysis. Each protomer is folded into two domains having different functions, the L-lactate dehydrogenase, the flavodehydrogenase $(FDH)_4$, a tetramer of molecular mass of 160 kDa, and its electron acceptor, the cytochrome b_2 core , a monomer of molecular mass of 13 kDa (Gervais et al. 1980), which then acts as a one-electron donor to cytochrome c .

Several steps are involved in the overall electron transfer from L-lactate to ferricytochrome c. These occur from L-lactate to the flavodehydrogenase, then to the cytochrome b_2 core and finally to the cytochrome c:

$$\text{Lactate} \rightarrow \text{flavodehydrogenase FMN} \rightarrow \text{cyt.}b_2 \text{ heme} \rightarrow \text{cyt.c heme}$$

Figures 1.28 and 1.29 display respectively the variation of the second-order rate constant with ionic strength for flavocytochrome b_2 → cyt.c and cytochrome b_2 core → cyt.c electron transfer, illustrated by the classical Debye-Hückel plot defined by log k_+ versus the square root of ionic strength:

$$\text{Log } k_+ = \log k_{+\,0} + 2A\ Z_1\ Z_2\ \sqrt{I} \tag{1.16}$$

where k_+ is the apparent rate constant.

For the cytochrome b_2 core → cyt.c electron transfer, a linear dependence is observed with a negative slope of 1.9 and an extrapolated value at zero ionic strength $k_{+\,0} = 6$ x $10^7\,M^{-1}\,s^{-1}$ at 5°C.

The plot obtained for the flavocytochrome b_2 → cyt.c is completely different. For ionic strength higher than 0.07 M, a linear dependence is observed with a negative slope equal to 5.6. However, at ionic strength lower than 0.07 M, the electron transfer is intramolecular, i.e., it occurs inside the cytochrome c – flavocytochrome b_2 complex and is ionic strength independent. At ionic strength higher than 0.07 M in the linear part of the graph, the extrapolated second order rate constant, at I = 0, corresponds to $k_{+\,0} = 2$ x $10^{10}\,M^{-1}\,s^{-1}$. This value is equal to that calculated theoretically for a diffusion controlled reaction occurring when collisions are at high frequencies. This is not the case for the cytochrome b_2 core – cytochrome c complex since the affinity of the latter complex is found equal to 2.5 μM while the dissociation constant of the flavocytochrome b_2 – cytochrome c complex is equal to 0.1 μM.

A conclusion that can be drawn from these results is that the reactivity of cytochrome b_2 core towards cytochrome c is not the same. The reason for this could be the less stable complex. Therefore, one role of the flavodehydrogenase is to stabilize the flavocytochrome b_2 – cyt.c complex. It is important to mention that the K_d of the flavodehydrogenase - cyt.c complex is equal to 0.1 μM (Albani, 1993).

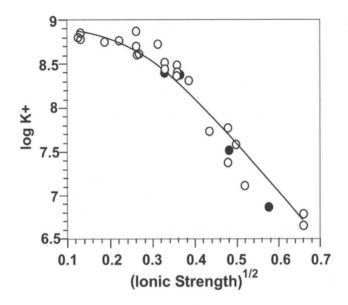

Figure 1.28. Ionic strength dependence of the bimolecular rate constant, for the reduction of cytochrome c by flavocytochrome b_2. Temperature = 5°C. Source: Capeillere-Blandin, C. 1982. Eur. J. Biochem. 128, 533-542. Authorization of reprint accorded by Blackwell Publishing

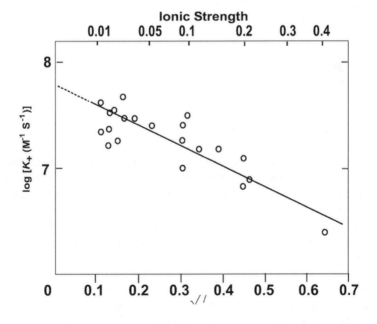

Figure 1.29. Ionic strength dependence of the bimolecular rate constant, for the reduction of cytochrome c by cytochrome b_2 core. Temperature = 5°C. Source: Capeillère-Blandin, C, and J. Albani. 1987. Biochem. J. 245,159-165. Authorization of reprint accorded by Portland Press.

We can describe the following scheme for the cytochrome b_2 core-cyt.c interaction:

$$\text{Cyt.b}_2{}^{2+} + \text{cyt.c}^{3+} \xrightarrow[k_{-1}]{k_{+1}} (\text{cyt.b}_2{}^{2+} - \text{cyt.c}^{3+}) \xrightarrow{k_{+2}} \text{cyt.b}_2{}^{3+} + \text{cyt.c}^{2+}$$

In such a two-step mechanism, the value of k_+ will be equal to:

$$k_+ = \frac{k_{+2}}{[\text{cyt.b}_2{}^{2+}]} = \frac{k_{+2}}{[\text{cyt.b}_2{}^{2+}] + K_d} \qquad (1.17)$$

Determining the electron-exchange constant (k_{+2}) for flavocytochrome b_2 – cyt.c and cyt.b_2 core – cyt.c complexes at different ionic strengths yields very close values for both complexes. For example, at ionic strength of 0.06 and 0.36 M, k_{+2} was found equal to 296 and 395 s^{-1} for the cyt.b_2 core – cyt.c electron transfer and 315 and 200 for the flavocytochrome b_2 -$\rightarrow$ cyt.c electron transfer.

This leads us to the following conclusion: the flavodehydrogenase domain in the native flavocytochrome b_2 exerts very little influence on the efficiency of the electron transfer between cytochrome b_2 and cytochrome c within the complex. This result implies that similar transition states must be achieved in the reactions of cytochrome *c* with both free and bound cytochrome b_2. Consequently, the electron transfer occurs via the same mechanism involving the same pathway through the intra-molecular reaction complex. Thus, in both flavocytochrome b_2 – cyt.c and cyt.b_2 core – cyt.c complexes, the relative distance and orientation of the heme b_2 and heme c planes are optimal to achieve the electron-transfer process.

In an attempt to calculate the distance R that separates the two cytochrome hemes (cytochrome b_2 core and cytochrome c), we applied Eq.1.15 taking for β a value equal to 1.4 Å^{-1}. With k_{ET} equal to 300 s^{-1}, this yields a value of 20 Å for R. This value is in the same range of that, 19 Å, found for the hemes of cytochrome c and flavocytochrome b_2. In conclusion, the relative orientation of the two heme planes is the key to electron transfer between the two cytochromes. The affinity exerts almost no influence on the kinetic constant of the electron exchange. Also, the protein matrix does not seem to play a major role in the electron transfer process. In fact, the absence or the presence of the flavodehydrogenase does not modify the data. This also allows us to conclude that the numbers of pathways are very limited. A highly efficient pathway has been considered for electron transfer between cytochrome b_2 and cytochrome c. This pathway was suggested to be Ile-50, Lys-51 and Phe-52 of cytochrome b_2 (13 covalent bonds making a distance of 12.6 Å and two through – space jumps at the methyl edge of pyrrole c ring of heme b_2 and heme c corresponding to distances of 4.5 Å and 6.3 Å, respectively) (Tegoni et al. 1993).

In another study, we investigate electron transfer between flavocytochrome b_2 and cytochrome c in presence of TNS. In principle, in absence of FMN or heme, electron transfer cannot occur between flavocytochrome b_2 and cytochrome c. 2-p-toluidinylnaphthalene-6-sulfonate (TNS) is a fluorescent probe used to study the interaction between proteins, proteins and their ligands and to follow the global and local motions in proteins. TNS binds to the flavocytochrome b_2 and to the flavodehydrogenase.

Figure 1.30a shows the reduction of cytochrome c in presence of L-lactate, flavodehydrogenase and TNS. At 415 nm, the value of ε of cytochrome c is equal to 129 mM^{-1} cm^{-1}. From the value of the absorption (0.261), we reach the conclusion that all the cytochrome c (2 μM) has been reduced. TNS alone, in presence of lactate and in absence of flavodehydrogenase, does not reduce the cytochrome c (Fig. 1.30c).

In presence of flavocytochrome b_2 (0.1 μM), we observe an instantaneous reduction of cytochrome c (Fig. 1.30d). Since the time course of reduction of cytochrome c by reduced flavocytochrome b_2 occurs in the millisecond range, it is happening too fast to be totally observed. Thus, only the final phase of reduction was recorded with the spectrophotometer.

In presence of TNS, reduction of cytochrome c occurs in the range of minutes instead of seconds, as it is the case for the reduction with flavocytochrome b_2 or isolated cytochrome b_2 core. Thus, the relative distance and orientation of the FMN, TNS and heme planes induces an electron transfer pathway different from that known for the cytochrome b_2 - cytochrome c electron transfer. This clearly shows the importance of cytochrome b_2 core in the electron transfer to cytochrome c.

Binding of TNS to the flavodehydrogenase - cytochrome c complex induces a conformational change around FMM (Albani, 1993). In presence of TNS, the flavin plane orientation is different from that observed in its absence.

In many proteins, it has been shown that electron transfer between ruthenium and heme is faster when the ruthenium is physically located further from the heme (Beratan et al. 1991; Casimiro et al. 1993). Thus, it is possible that the TNS and the heme plane of cytochrome c are very close one to each other.

The dissociation constant of the cyt.c-flavodehydrogenase complex can be calculated from equation 1.18, using the data of figure 1.30a:

$$L_b = 0.5 \ [(L_o + P_o + K_d) + \{(L_o + P_o + K_d)^2 - 4 \ L_o \}^{1/2}] \qquad (1.18)$$

where L_b and L_o are the concentrations of bound and total cyt.c, respectively, P_o is the flavoprotein concentration and K_d is the dissociation constant of the cyt.c-flavoprotein complex.

When the binding site of cyt.c on the flavoprotein is saturated, L_o is equal to 2 μM. L_b considered for the calculation is equal to 1 μM and P_o the flavoprotein concentration is equal to 2 μM. We found a value of 0.15 μM for the dissociation constant. This value is equal to that (0.1 μM) already determined by fluorescence titration (Albani, 1993). The velocities for cytochrome c reduction by flavodehydrogenase at different ionic strengths and in the presence of a constant concentration of TNS (1.5 μM), are plotted (Fig. 1.31) with the classical Debye-Huckel equation:

$$\log v = \log vo + 2A \ Z_1 \ Z_2 \ \sqrt{I} \qquad (1.19)$$

where v is the velocity at ionic strength I, vo is the velocity at zero ionic strength, 2A = 1 and $Z_1 \ Z_2$ is the product of the charges implicated in the interaction. The slope $Z_1 \ Z_2$ obtained is equal to -1.25, a value different from that (-5.7) found for the flavocytochrome b_2 - cytochrome c (Blandin, 1982). At ionic strengths equal to and lower than 54 mM, the electron transfer is intramolecular, i.e., it occurs inside the cyt.c - flavodehydrogenase - TNS complex and is ionic strength independent.

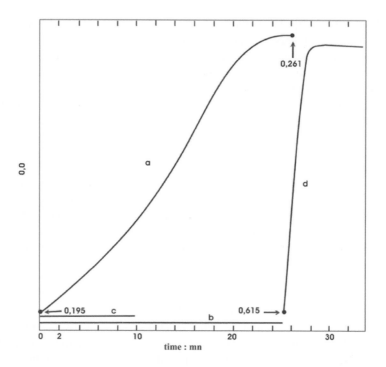

Figure 1.30. Reduction of 2 μM of cytochrome c at 415 nm in presence of 3 mM lactate, 1.5 μM of flavodehydrogenase and 1.5 μM of TNS (plot a). Neither the flavoprotein alone (plot c) nor TNS alone (plot b) can reduce cytochrome c. Plot d shows the final phase of the reduction of 5 μM of cytochrome c in presence of 3 mM lactate, 2 μM flavodehydrogenase and 0.1 μM flavocytochrome b_2. The optical density of cytochrome c before addition of the flavodehydrogenase-flavocytochrome b_2 solution is equal to 0.473. We notice from the optical density at the beginning of the recording that 77% of the cytochrome c has been already reduced by the flavocytochrome b_2. Since this reduction occurs in the millisecond range, it is too fast to be rcorded entirely with the spectrophotometer, thus only the final phase was obtained.

The product of charges (-1.25) would be the result of the interaction between the negative charge of TNS and a positive charge near the heme crevice of cytochrome c. These charges are in no way responsible for the high affinity between the two proteins. Electron transfer does not occur between flavodehydrogenase and cytochrome b_2 core. Thus, the structure organization of flavocytochrome b_2 facilitates the electron transfer between the flavin and the heme. TNS alone has no potential for electron transfer (Fig. 1.30b). However, when bound to the flavodehydrogenase, the organization energy in the system allows the electron transfer. TNS properties are highly dependent on the microenvironment of the probe. Relaxation rate of its surrounding dipole, i.e., binding site and the solvent, is a key factor of its electron transfer efficiency and rate and also of its spectral properties such as fluorescence. TNS binds to flavovodehydrogenase at a highly relaxed state, therefore, dipole relaxation is important, i.e., electron transfer between TNS and cytochrome c will depend on the dynamics of the binding site of the probe. In presence of TNS, electron transfer properties between flavodehydrogenase and cytochrome c differ from those observed between flavocytochrome b_2 and cytochrome c, i.e. the organization energies in the two systems are not identical. However, in both systems, electron transfer occurs in highly organized states. In the absence of these states, electron transfer does not exist. Data obtained with TNS reveals that flavodehydrogenase is necessary for the electron transfer toward cytochrome c, but it does not play a major role in the velocity of this transfer. The predominant factor appears to be the structure and the relative positions of the donor (here, the TNS) and the acceptor (cyt.c heme).

However, another possibility can be drawn, which is the fact that in the absence of cytochrome c, a second tunneling route appears via flavodehydrogenase. This route could be inhibited in the presence of cytochrome b_2 and will be reactivated in its absence.

All these studies indicate that electron transfer within the flavocytochrome b_2 - cytochrome c complex is dependent upon a number of factors such as the distance between donor (cytochrome b2 core or TNS) and acceptor (cytochrome c), their relative orientation, their chemical nature and the structure of the protein medium involved in the electron transfer.

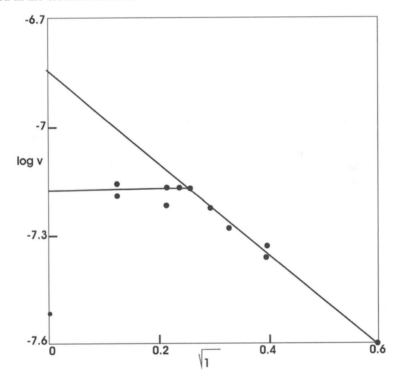

Figure 1.31. Ionic strength dependence of electron transfer between flavodehydrogenase and cytochrome c, in presence of lactate (3 mM) and of TNS (1.15 μM). The velocity is expressed in optical density of reduced cytochrome c per minute.

6g. Second – derivative absorption spectroscopy

6g1. Theory

Absorption in proteins occurs from tryptophan, tyrosine and phenylalanine. However, absorption is dominated by tryptophan and tyrosine. Although in solution the three amino acids fluoresce, in proteins Trp residues dominate the fluorescence in general. In other terms, in presence of tyrosine and tryptophan, phenylalanine has no chance to be observed either by classic fluorescence or absorption.

When the structure of a protein is perturbed as the result of addition of a denaturant or solvent of high viscosity such as glycerol or sometimes even just by the binding of a ligand, the absorption spectra of the protein is perturbed and in general its maximum shifts to higher or lower wavelengths. Still, the spectral modifications of the absorption spectrum are small and do not allow always effective and serious quantitative or / and qualitative studies. Derivative absorption spectra of proteins allowed to study the environment of each of the three amino acids and are very sensitive to the modification occurring within the protein and mainly in their close microenvironments.

Our interest will go to the second derivative spectrum since in the literature there are many relevant papers describing how we can monitor local structural perturbation within the environment of the three concerned amino acids and mainly phenylalanine and tyrosine.

Figures 1.32 to 1.34 display the absorption, the first derivative and second derivative spectra of free L-phenylalanine, L-tyrosine and L-tryptophan in water, respectively. We notice important differences mainly in the second derivative spectra since the peaks and troughs positions differ in certain spectral ranges. For studies in proteins, since the three amino acids are bonded, comparative studies will be done not with L-amino acids in solution but with the N-acetyl amino acids derivative. However, before we go further in these comparative studies, let us see very briefly the origin of these derivative spectra.

The limiting value of a variable x is the value a that x continuously approaches closer and closer so that the differences between them will be less than any assignable magnitude. If a is a finite number, the limiting value of the variable will be a finite limiting value. If a tends to infinity, the limiting value will be infinity.

An absorption spectrum is described by a function and the limiting value of a function is the value approached the most by the function when $x \dashrightarrow a$. For example, the function $f(x) = 2x^2 + 4x - 5 \dashrightarrow -1$ when $x \dashrightarrow 1$.

The derivative of a function f(x) at $x = x_0$ is the limiting value of the ratio

$$\frac{f(x) - f(x_0)}{x - x_0} \tag{1.20}$$

The derivative $f'(x) = $ limiting value of $\quad \dfrac{f(x) - f(x_0)}{x - x_0} \tag{1.21}$
$\qquad\qquad\qquad\qquad\qquad x \dashrightarrow x_0$

What is the geometrical significance of the derivative?

The equation of a linear plot is

$$y = a x + b \qquad (1.22)$$

In order to determine the slope a of the plot, we choose two points on the plot with coordinate $(x_1 ; y_1)$ and $(x_2 ; y_2)$. Thus, we can write:

$$a = \frac{y_2 - y_1}{x_2 - x_1} \qquad (1.23)$$

Comparing Eq. 1.21 and 1.23 indicates that the derivative of a function taken for a precise value *n* is nothing else than the slope of the tangent to the curve, plotted at *n*.

Therefore, one can obtain the first derivative spectrum by plotting the tangent at the maximum and at the two curvatures on both sides of the maximum of the absorption spectrum. The second derivative is then obtained in the same way.

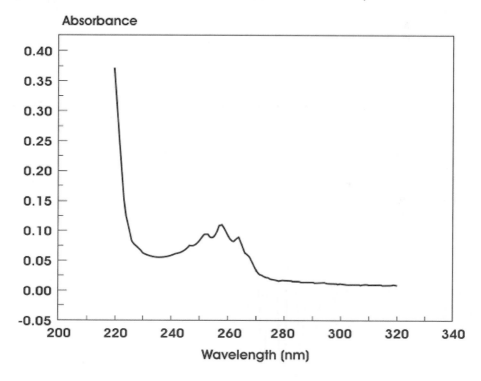

Figure 1.32a. Absorption spectrum of free L-phenylalanine in water.

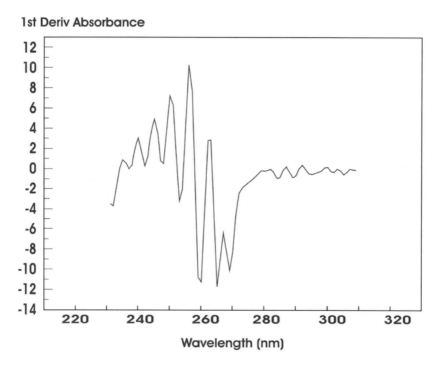

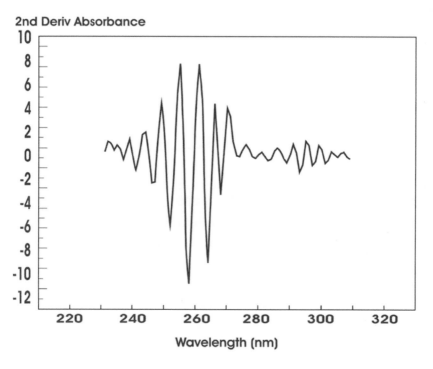

Figure 1.32b,c. First and second derivative absorption spectra of free L-phenylalanine in water.

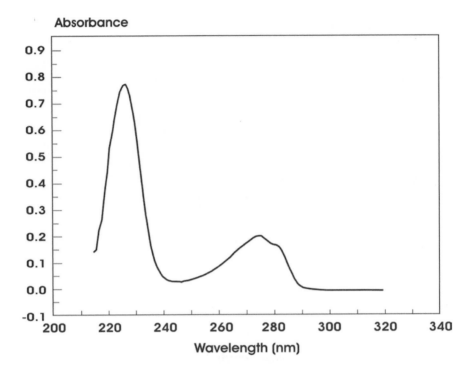

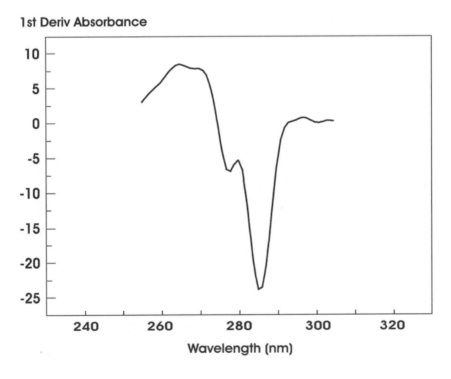

Figure 1.33a,b. Absorption and first derivative spectra of free L-tyrosine in water.

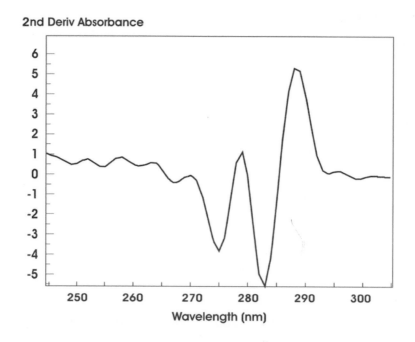

Figure 1.33c. Second derivative absorption spectrum of free L-tyrosine in water.

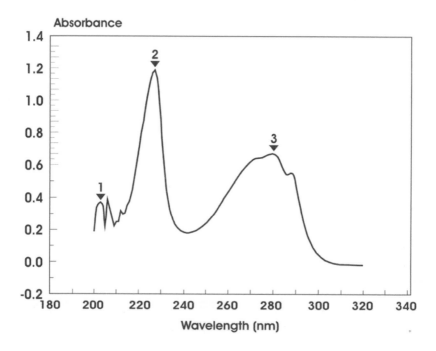

Figure 1.34a. Absorption spectrum of free L-tryptophan in water.

1st Deriv Absorbance

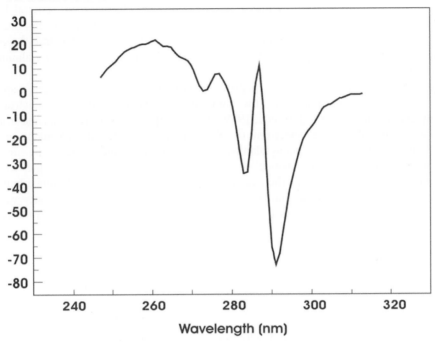

Wavelength (nm)

2nd Deriv Absorbance

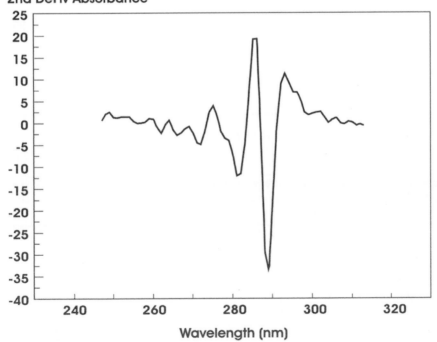

Wavelength (nm)

Figure 1.34b,c. First and second derivative spectra of free L-tryptophan in water.

The second derivative spectra of tyrosine and tryptophan are flat between 245 and 270 nm, while the spectrum of phenylalanine shows its characteristic spectral bands (λ_{max}: 249, 256, 260, 262, 266 and 270 nm and λ_{trough}: 247, 252, 258, 261, 264 and 268 nm between 245 and 270 nm). Ichikawa and Terada (1977) used the spectral properties of the second derivative spectrum between 245 and 270 nm to study phenylalanine in proteins.

In fact, the authors found that in different proteins, the second derivative spectra below 270 nm were essentially the same as the second derivative spectrum of phenylalanine (Ichikawa and Terada, 1979). The absorption at certain peaks and troughs of the second derivative spectrum of phenylalanine are found dependent on the microenvironment of the phenylalanine. Denaturation of the proteins by urea or guanidine modifies the intensities of the spectral bands of phenylalanine without inducing a significant shift in their positions. Table 1.3 shows the difference between peaks and troughs in the second derivative spectrum of phenylalanine under various conditions and Table 1.4 shows the difference observed for four proteins in the native and denatured states.

Table 1.3. Differences between peaks and troughs $\Delta(d^2 \varepsilon / d\lambda^2)$, in the second derivative spectrum of phenylalanine under various conditions.

Medium	$\Delta(d^2 \varepsilon / d\lambda^2) * (10^{15} \; M^{-1} \; cm^{-3})$			
	B – A (266 – 268 nm)	B – C (266 – 264 nm)	D – C (262 – 264 nm)	D – E (262 – 258 nm)
Phosphate buffer	3.04	4.15	4.29	3.89
6 M guanidine chloride	2.61	3.58	3.72	3.41
1- propanol	4.29	6.26	7.16	5.95
Ethlene glycol	4.01	5.43	5.78	4.91

Source: Ichikawa, T. and Terada, H. 1979, Biochim. Biophys. Acta. 580, 120-128.

Table 1.4. Average absorbance difference of second derivative specta of proteins and average number of phenylalanine residues in proteins calculated from second derivative spectra at the wavelength pair 266 – 268 nm.

	Average absorbance difference, $\Delta(d^2 A / d\lambda^2) / C_P (10^{15} \; M^{-1} \; cm^{-2})$	
	Native protein	Denatured protein
Serum albumin	88.16 ± 0.15	26.0 ± 0.05
Insulin	9.73 ± 0.12	3.0 ± 0.03
Ribonuclease	9.42 ± 0.12	8.13 ± 0.11

Source: Ichikawa, T. and Terada, H. 1979, Biochim. Biophys. Acta. 580, 120-128.

The values of the absorbance difference, $\Delta\, (d^2 A / d\lambda^2) / C_P$ of the denatured proteins allowed to calculate the average number n of phenylalanine residues in the proteins (Table 1.4) by using equation 1.24 :

$$n = \frac{1}{C_P.l} \; x \; \frac{\Delta\, (d^2 A / d\lambda^2)}{\Delta\, (d^2 \varepsilon / d\lambda^2)} \qquad (1.24)$$

The authors found that the best wavelength pair that allows obtaining qualitative and quantitative information on the state of the phenylalanine is 266 – 268 nm. Looking closely to the intensity that separates the peak (266 nm) and the trough (268 nm), one can notice that more the medium is hydrophobic more the intensity is small.

Other studies performed on the second-derivative spectrum (Ragone et al. 1984) have shown that the ratio r between two peak to trough distances (287-283 and 295-290) was related to the tyrosine / tryptophan ratio and was found dependent on the surrounding medium polarity of tyrosine. Table 1.5 shows the values of r for different proteins measured in the native (r_n) and denatured (r_u) states. We can notice that r increases when the proteins are denatured, i.e., tyrosines are much more exposed to the solvent in the denatured state. The increase of r with the protein denaturation is in principle dependent on the tyrosines locations in the proteins in the native state.

Table 1.5. Effect of protein unfolding on the ratio between the peak-to-peak distances 287-283 and 295-290.5 nm.

Protein	r_n	r_u
Lysozyme	0.74	0.79
α-Chymotrypsine	0.76	0.79
Horse apomyoglobin	0.58	0.94
Sperm whale apometmyoglobin	0.64	1.09
Sperm whale metmyoglobin	0.62	1.09
Bovine aspartate aminotransferase	0.93	1.15
Tuna apomyoglobin	1.1	1.30
Pepsin	1.43	1.79

Source: Ragone, R., Colonna, G., Balestrieri, C., Servilo, L. and Irace, B. 1984. Biochem. 23, 1871-1875. Authorization of reprint accorded by the American Society of Chemistry.

6g2. Binding of progesterone to α₁-acid glycoprotein

Progesterone binds to α₁-acid glycoprotein on a hydrophobic region of a pocket present in the protein. Figure 1.35 shows the absorption spectrum of 13 µM α₁-acid glycoprotein in buffer at pH 7, its first and second derivative spectra

Figure 1.36 displays more detailed parts of the second derivative spectrum from 245 and 270 nm (a) and from 270 to 300 nm (b). One can notice that the general feature of the derivative spectra corresponds very well with those of the three amino acids (Trp, Tyr and Phe), responsible for the protein absorption.

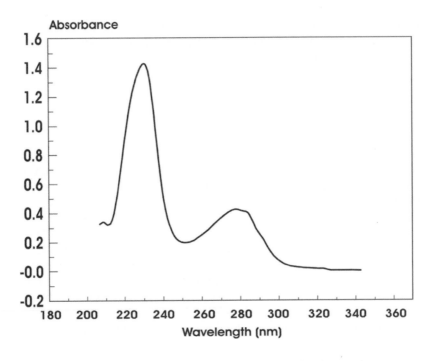

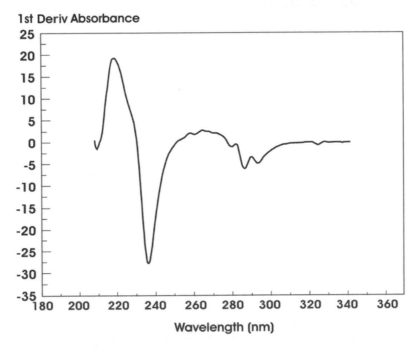

Figure 1.35a,b. Absorption and first derivative spectra of 13 µM α_1-acid glycoprotein in buffer at pH 7.

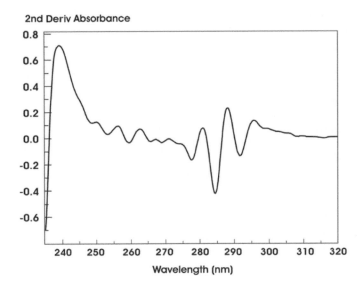

Figure 1.35c. Second derivative absorption spectrum of 13 μM α_1-acid glycoprotein in buffer at pH 7.

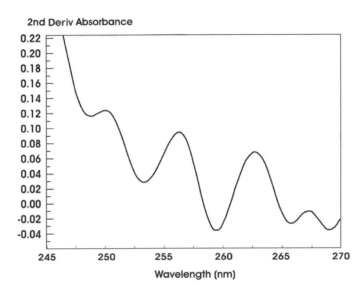

Figure 1.36a. Second derivative absorption spectrum from 245 and 270 nm of 13 μM α_1-acid glycoprotein in buffer at pH 7

λ peaks	λ trough
B 267.2 nm	A 269 nm
D 262.6 nm	C 265.2 nm
F 256.2 nm	E 259.4 nm
H 250 nm	G 253.2 nm

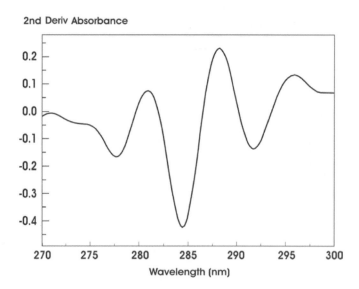

Figure 1.36b. Second derivative absorption spectrum from 245 and 270 nm of 13 μM α_1-acid glycoprotein in buffer at pH 7.

λ peaks	λ trough
A 295 nm	B 291 nm
C 288.2 nm	D 284.4 nm
E 281 nm	F 277.8 nm
G 275.4 nm	H 273.6 nm

Figure 1.37 displays the absorption (a), the first (b) and the second (c) derivative spectra of progesterone in buffer pH 7.

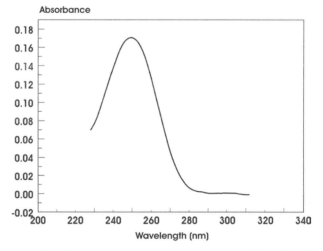

Figure 1.37a. Absorption spectrum of progesterone in buffer, pH 7.

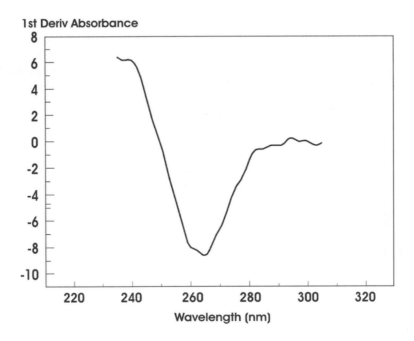

Figure 1.37b. First derivative absorption spectrum of progesterone in buffer, pH 7.

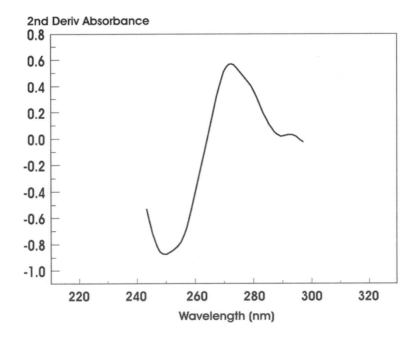

Figure 1.37c. Second derivative absorption spectrum of progesterone in buffer, pH 7.

Since progesterone absorbs at the same wavelengths of α_1-acid glycoprotein, it is important to record the difference spectra (α_1-acid glycoprotein – progesterone complex) vs progesterone, if we want to see the effect of ligand binding on the protein. The result is shown in figure 1.38, where we can see the absorption (a), the first (b) and the second derivative (c) spectra. Tables 1.6.and 1.7 detail the positions of the different peaks and troughs of the second derivative spectra with their respective absorption.

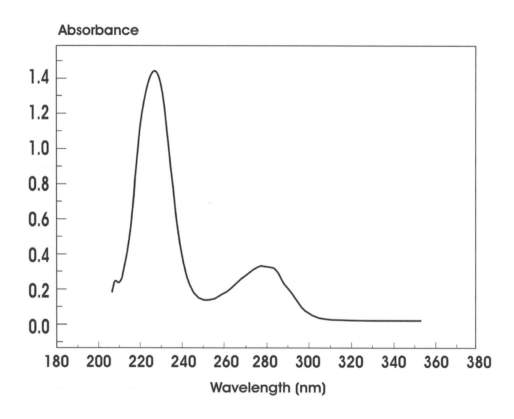

Figure 1.38a. Absorption difference spectrum of (α_1-acid glycoprotein – progesterone complex) vs progesterone.

Comparing the difference absorption spectrum (Fig. 1.38a) to the absorption spectrum of α_1-acid glycoprotein alone (Fig. 135a) does not reveal any significant modifications between the two spectra. Nevertheless, in presence of progesterone, a decrease in the absorption of the protein at 280 nm can be observed. Despite this absorption decrease, we prefer analyzing the results with the derivative spectra.

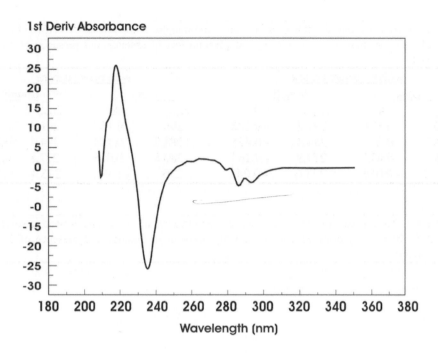

Figure 1.38b. First-derivative difference spectrum of (α_1-acid glycoprotein – progesterone complex) vs progesterone.

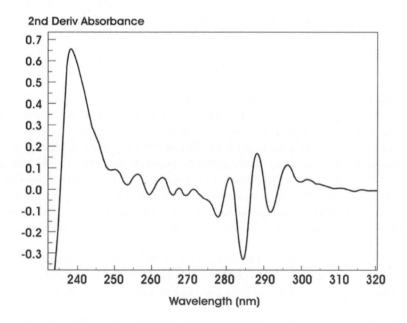

Figure 1.38c. Second-derivative difference spectrum of (α_1-acid glycoprotein – progesterone complex) vs progesterone.

Table 1.6. Position and absorption of the peaks and troughs of the second derivative spectra in the 270-300 nm region of 13 μM α_1-acid glycoprotein in absence and presence of 50 μM progesterone.

without progesterone				with progesterone			
peak		trough		peak		trough	
$\lambda_{(nm)}$	A	$\lambda_{(nm)}$	A	$\lambda_{(nm)}$	A	$\lambda_{(nm)}$	A
295	0.134	291.8	- 0.138	296	0.187	291.6	-0.104
288.2	0.232	284.4	- 0.425	288.2	0.170	284.4	-0.329
281	0.075	277.8	- 0.167	280.8	0.056	277.8	-0.128
275.4	- 0.058	273.6	- 0.043	275	- 0.049	273.4	-0.033

Table 1.7. Position and absorption of the peaks and troughs of the second derivative spectra in the 245-270 nm region of 13 μM α_1-acid glycoprotein in absence and presence of 50 μM progesterone.

without progesterone				with progesterone			
peak		trough		peak		trough	
$\lambda_{(nm)}$	A	$\lambda_{(nm)}$	A	$\lambda_{(nm)}$	A	$\lambda_{(nm)}$	A
267.2	-0.01	269	- 0.035	267.2	0.008	269.2	-0.029
262.8	0.069	265.2	- 0.026	262.8	0.057	265.6	-0.025
256.2	0.095	259.4	- 0.036	256.2	0.079	259.4	-0.023
250	0.124	253.2	0.029	250	0.095	253.4	0.022

First let us see the effect of progesterone binding to α_1-acid glycoprotein on the of tyrosines microenvironment. Let us call *a* the distance between the trough at 284.4 nm and the peak at 288.2 nm and *b* the distance between the trough at 291 nm and the peak at 295 nm. In absence of progesterone, *a* and *b* are equal to 0.656 and 0.272, respectively. Thus the ratio *r = a / b* is equal to 2.41. In presence of progesterone, *a* and *b* are equal to 0.499 and 0.291, and thus *r* is equal to 1.715.

We notice that *r* decreases upon binding of progesterone to α_1-acid glycoprotein. Thus, tyrosines microenvironment is more hydrophobic in presence of progesterone. Therefore, binding site of progesterone contains at least one tyrosine.

In order to find out the effect on phenylalanines of progesterone binding to α_1-acid glycoprotein, we have compared the distance between the peak at 267 nm and the trough at 269 nm in absence and presence of progesterone. This distance increases from 0.025 to 0.037 in presence of progesterone. Thus, binding of progesterone to α_1-acid glycoprotein modifies phenylalanines microenvironment, i.e., they are more hydrophobic. This means that binding site of progesterone includes at least one phenylalanine.

Progesterone is a hydrophobic molecule and cannot bind to the polar surface of proteins. In α_1-acid glycoprotein, progesterone binds to the hydrophobic region of a pocket. Homology studies have shown that binding site of progesterone contains three tyrosines and one phenylalanine (Kopecky et al. 2003). Thus, the effect we have observed with the second derivative spectra concerns the phenylalanine and the tyrosines located in the pocket of α_1-acid glycoprotein.

Chapter 2

FLUORESCENCE: PRINCIPLES AND OBSERVABLES

1. Introduction

Fluorescence is the emission by a fluorophore of a photon from the lowest excited state S_1 to the ground state S_o. Emission of the photon occurs with a precise energy E and thus it will be observed at a precise wavelength (λ_{em}). Emission occurs from a population of n excited fluorophores, therefore it will occur with an intensity I

$$I = n E \qquad (2.1)$$

Transition from different vibrational levels will occur simultaneously with the emission, with energies lower than that of the emitted photon, generating a fluorescence spectrum with a maximum corresponding to the emission transition.

Emission lifetime is within the pico-to the nanosecond range. Thus, emission is a very fast process. Therefore, in order to observe fluorescence emission, the fluorophore should be excited continuously.

Variation of temperature provokes variation of global and local motions of the environment of the fluorophore and thus modifies its fluorescence emission feature. Also, temperature can affect the local dynamics of the fluorophore itself inducing an additive effect to that of the surrounding medium.

Intensity, position of the emission wavelength, lifetime are some observables that are going to characterize a fluorophore. Modification of the temperature and / or the viscosity of the medium will affect the values of these observables. Each fluorophore has its own fluorescence properties and observables. These properties are intrinsic to the fluorophore and are modified with the environment. We shall see in the next chapter that these modifications do not follow the same rules for all fluorophores. Therefore, it is important to understand the nature of the fluorophore environment before taking conclusions that could be misleading in the interpretation of the studied phenomenon. Also, we are going to see that fluorophore structure can influence its fluorescence lifetimes. This will vary from a fluorophore to another.

A fluorescence spectrum is the plot of the fluorescence intensity as a function of wavelength (Fig. 2.1). Since we have to excite the sample and to record the emitted intensity at different wavelengths, the layout of a fluorometer consists of an excitation source (a lamp or a laser), an excitation monochromator or a filter (if a lamp is used as the source of excitation), a cuvette holder where we can put the sample, an emission monochromator or a filter (if we do not want to record the whole spectrum but just the fluorescence intensity), a photon detector and a recorder (Fig. 2.2).

When we want to record the fluorescence emission spectrum, the excitation wavelength is kept fixed and the emission monochromator is run (Fig. 2.1b). Excitation spectrum is obtained by fixing the emission wavelength and running the excitation monochromator (Fig. 2.1a).

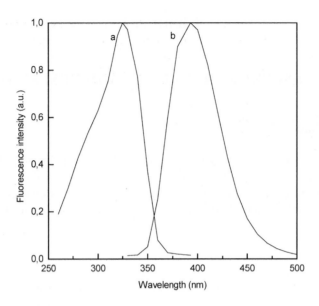

Figure 2.1. Normalized fluorescence excitation (a) and emission (b) spectra of 7-amino-4-methylcoumarin-DEVD complex. DEVD = Asp-Glu-Val-Asp.

(a) $\lambda_{em} = 470$ nm and $\lambda_{max} = 323$ nm.

(b) $\lambda_{ex} = 260$ nm and $\lambda_{max} = 393$ nm.

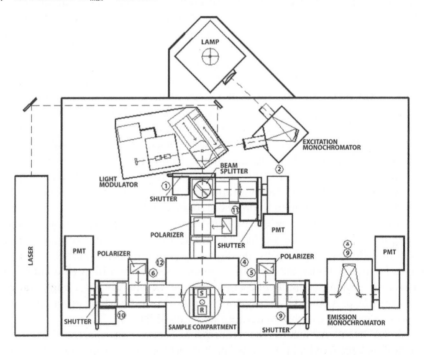

Figure 2.2. Layout of a spectrofluorometer. Curtosey from ISS instruments.

The efficiency of photon detection of the emission monochromator and the photomultiplier is not 100% at all wavelengths. Therefore, the recorded emission spectrum can be distorted at many wavelengths. Corrections can be done by comparing the recorded spectrum of the sample to the recorded spectrum of a standard whose corrected fluorescence emission spectrum is known. Also one can use standard lamps whose real outputs is known. Comparison of the lamp output obtained with the laboratory material with the real one allows calculating the sensitivity of the fluorometer at every wavelength.

Some fluorometers can be equipped with correction devices, others not. Therefore, one should be careful when comparison is done between two different instruments. Figure 2.3 displays the uncorrected and corrected fluorescence emission spectra of zincporphyrin IX dissolved in dioxane, recorded with an SLM instrument. One can notice that the difference between the two spectra exists at wavelengths higher than 600 nm. One can notice also, that correction of the spectrum yields a small red shift of the peak of 2 to 3 nm.

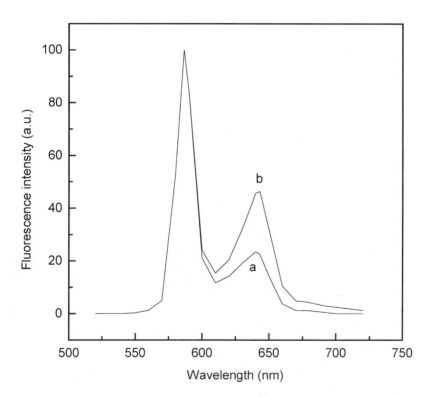

Figure 2.3. Uncorrected (a) and corrected (b) fluorescence emission spectra of znppIX in dioxane at 20°C. λ_{ex} = 415 nm.

2. Fluorescence properties

2a. Stokes shift

The energy absorbed by a fluorophore at the ground state to reach the excited state is more important than the energy of the emitted photon. In fact, as we have seen in the Jablonski diagram, deexcitation of the molecule occurs via different competitive steps. The energy E is equal to

$$E = h \nu = h c / \lambda \tag{2.2}$$

where h, ν, c and λ are respectively the Planck constant, the frequency of the electromagnetic beam, light velocity and the wavelength.
Absorption energy is equal to

$$E_a = h c / \lambda_\alpha \tag{2.3}$$

and emission energy is equal to

$$E_{em} = h c / \lambda_{em} \tag{2.4}$$

Since

$$E_{em} < E_a \tag{2.5}$$

We are going to have

$$\lambda_{em} > \lambda_a \tag{2.6}$$

One should remember that the wavelengths we are talking about are those of the peaks of the absorption and emission spectra. The real absorption and emission are located at the maximum of the spectra.

Eq.2.6 means that emission wavelength is higher than the absorption one. The emission spectrum has its maximum shifted to the red compared to the maximum of the absorption spectrum (Fig. 2.4). Stokes observed this shift for the first time in 1852 and since this time it is called the Stokes shift.

When the absorption and / or the emission spectra of a fluorophore possess two or more bands, the Stokes shift will be equal to the difference that separates the two most intense bands of the two spectra (Fig. 2.5). Stokes shift is dependent on the structure of the fluorophore itself and on its environment (the solvent where it is dissolved and / or the molecule to which it is bound. For example, the cyanine dye, thiazole orange (TO) absorbs at 500 nm and emits at 535 nm when it is dissolved in solution at low concentrations. These spectral properties are characteristic of that of a monomer. Increasing the concentrations yields to the formation of aggregates that absorb at 473 nm and emit at 635 nm. Thus, aggregates how larger Stokes shift compared to the monomer. The same phenomenon is also observed when TO monomers interact with DNA at high concentration of TO compared to that of DNA. In this case, the Stokes shift is found equal to at least 100 nm (Ogul'chansky et al. 2001).

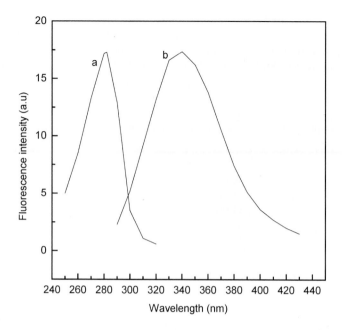

Figure 2.4. Fluorescence excitation (a) (λ_{em}= 340 nm) and emission (b) (λ_{em} = 280 nm) spectra of bovin serum albumin.

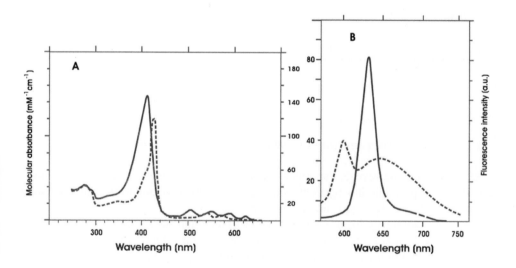

Figure 2.5. Absorbance (A) and emission (B) spectra of Mb^{desFe} (—) and $Mb^{Fe \rightarrow Zn}$ (---). For both fluorescence emission spectra, the absorbance at the excitation wavelength (514 nm) was 0.05. Source: Albani, J. and Alpert, B. 1987, Eur. J. Biochem. 162, 175-178. Authorization of reprint accorded by Blackwell Publishing.

Tables 2.1 and 2.2 give two examples concerning the positions of the absorption and emission of the fluorescence probes 1-anilino-8-naphthalene sulfonate and 6-propionyl-2-(dimethylamino)naphthalene or PRODAN with the solvent. We can notice that for both fluorophores, the Stokes shift decreases with the decrease of the polarity of the solvent.

Table 2.1. Absorption and emission positions of 1-anilino-8-naphthalene sulfonate in different solvents

Solvent	λ_{abs}	λ_{em}	$\Delta\lambda$
H_2O	335 nm	515 nm	180 nm
D_2O	334 nm	514 nm	180 nm
MeOH	341 nm	485 nm	144 nm
DMF	351 nm	461 nm	110 nm

Source: Drew, J., Thistletwhaite, P. and Woolfe, G. 1983. Chem. Phys. Letters. 96, 296 - 301.

Table 2.2. Absorption and emission positions of PRODAN in different solvents.

Solvent	λ_{abs}	λ_{em}	$\Delta\lambda$
H_2O	364 nm	531 nm	167 nm
MeOH	362 nm	505 nm	143 nm
Propylene glycol	370 nm	510 nm	140 nm
DMF	355 nm	461 nm	106 nm
Cyclohexane	342 nm	401 nm	59 nm

Source: Weber, G. and Farris, F. J. 1979, Biochemistry, 18, 3075-3078. Authorization of reprint accorded from The American Chemical Society.

2b. Relation between emission spectrum and excitation wavelength

Jablonski diagram reveals that emission occurs from the excited state S_1. Therefore, emission is in principle independent from the excitation wavelength. In another chapter we are going to see that this is not always the case.

Emission occurs with lower energy than absorption, and not all the molecules present at the excited states are going to participate in the fluorescence. Therefore, a fluorescence quantum yield Φ_F exists and is equal to:

$$\Phi_F = \frac{I_F}{I_A} \tag{2.7}$$

$\Phi_F < 1$.

Since

$$I_A = I_o - I_F \tag{2.8}$$

and

$$I_T = I_o \, 10^{-\varepsilon c l} = I_o \, e^{-2,3\,\varepsilon\,c\,l} \tag{2.9}$$

$$I_A = I_o - I_o \, e^{-2,3\,\varepsilon\,c\,l} = I_o \, (1 - e^{-2,3\,\varepsilon\,c\,l}) \tag{2.10}$$

At very low optical densities (<< 0.05), Eq. 2.10 can be written as

$$I_o \, (1 - (1 - 2,3\,\varepsilon_{(\lambda)}\,c\,l) = 2,3\,I_o\,\varepsilon_{(\lambda)}\,c\,l \tag{2.11}$$

Thus,

$$I_{F(\lambda)} = 2,3\,I_o\,\varepsilon_{(\lambda)}\,c\,l\,\Phi_F \tag{2.12}$$

Equation 2.12 describes the relation that exists between the fluorescence intensity at a precise emission wavelength and the intensity of the incident light (I_o), the quantum yield of the fluorophore and the optical density of the fluorophore at the excitation wavelength.

Apparently, the fluorescence intensity increases linearly with all these parameters. However, we are going to see the limits of this linearity especially when experiments are performed at high optical density.

I_o is the intensity of the incident beam that is going to excite the sample. Thus, excitation with a laser will allow getting better fluorescence intensity than if excitation occurs with a lamp. This is true only if at the excitation wavelength, the laser beam does not destroy the sample.

The fluorescence intensity is proportional to the quantum yield, i.e., a high quantum yield will lead automatically high fluorescence intensity.

The product $\varepsilon\,\Phi_F$ is called fluorophore sensitivity. More the sensitivity is high, more important will be the fluorescence intensity.

The fluorescence intensity is proportional to the optical density. This means that it is proportional to ε, c and l. Therefore, modifying one of these parameters allows modifying the fluorescence intensity. When a fluorophore is very sensitive to light, it is usually recommended not to excite it at high values of ε, for example at or near the absorption maximum. Since ε is wavelength dependent, the fluorescence intensity will be also wavelength dependent (Fig. 2.6). The fluorescence intensity increases with the increase of the molar extinction coefficient or / and with the increase of the optical density. Normalizing the fluorescence spectra with the same optical density at the excitation wavelength yields identical spectra. This means that although the fluorescence spectrum intensity increases with the optical density, the quantum yield is the same.

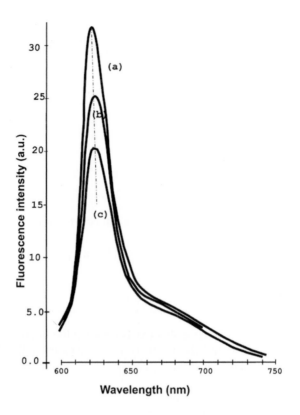

Figure 2.6. Fluorescence emission spectra of Mb^{desFe} at 20°C recorded at three excitation wavelengths, 514 (a), 525 (b) and 540 nm (c). λ_{em} = 626 nm.

2c. Relation between the fluorescence intensity and the optical density

Equation 2.12 states that the fluorescence intensity is linearly proportional to the optical density. However, this is not always correct. For simplicity let us take a usual fluorometer where the emission beam is recorded perpendicularly to the excitation beam. In general, the emission is detected at the cuvette center. Thus, the incident beam will go through half the cuvette pathlength before the recording of the emission. Thus, from the surface of the cuvette to its center, the intensity of the incident beam will be absorbed following the Beer – Lambert - Bouguer law (Eq. 1.11). For example, let us assume the optical density at the excitation wavelength as equal to 0.05. Eq. 1.11 yields at the center of the cuvette a value for I_{exc} equal to 0.95 I_o. Since fluorescence intensity is proportional to the excitation intensity, at the center of the cuvette the emitted light will be 5% less intense than the intensity of a very diluted solution.

The decrease in the fluorescence intensity is not only the result of the optical density at the excitation wavelengh, but also of the absorption at the emission wavelength. In fact, we do not record a fluorescence spectrum in the center of the cuvette, but from outside the cuvette. Thus, the beam emitted at the center of the cuvette should go through half of the cuvette pathlength before being detected. Therefore, the intensity of the emitted beam will decrease according to Eq.1.11.

In conclusion, the recorded emission intensity will be underestimated as the result of the optical density at both excitation and emission wavelengths. This autoabsorption by the solution is called the inner filter effect. This effect can modify the whole fluorescence spectrum inducing a shift in the emission maximum. Figure 2.7 shows the effect of cytochrome c binding on the fluorescence spectrum of TNS bound to apocytochrome b_2 core. Since cytochrome c absorbs at both excitation and emission wavelengths, the inner filter effect is important and thus we observe a shift in the position maximum of TNS emission.

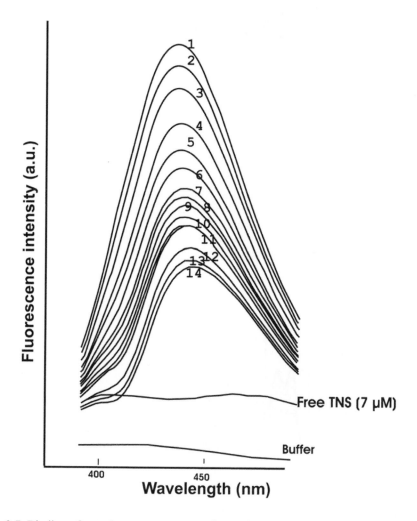

Figure 2.7. Binding of cytochrome c to apocytochrome b_2 core – TNS complex. The decrease of the fluorescence intensity of TNS is accompanied by a red shift of the peak due to the autoabsorption by cytochrome c at both excitation (320 nm) and emission wavelengths (433 nm).

Figure 2.8b describes the interaction between a complementary target DNA and excimer-monomer switching molecular beacons (EMS-MBs), dually labeled with pyrene at both 3' and 5' ends of single-stranded oligonucleotides (2.8a). The emission peak of the excimer does not change upon increasing DNA 5 concentrations. The reason for that is the very weak absorption (0.0147) at the excitation wavelength (345 nm).

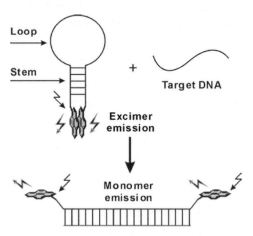

Figure 2.8a. Schematic representation of excimer-monomer switching molecular beacons (EMS-MBs). EMS-MBs were dually labeled with pyrene at both 3' and 5' ends of single-stranded oligonucleotides with a stem-and-loop structure. In the absence of target DNAs, the stem-close-shaped EMS-MBs predominantly emit the excimer fluorescence (yellow-green). Upon hybridization with target DNAs, the EMS-MBs undergo the dynamic conformational change to emit the monomer fluorescence (pale blue).

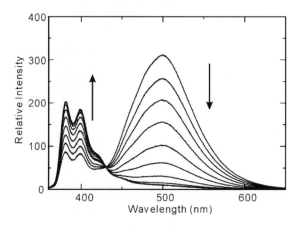

Figure 2.8b. Fluorescence titration spectra of **3** ([**3**], 200 nM; [MgCl$_2$], 5 mM; [KCl], 50 mM; [Tris-HCl], 20 mM, at pH 8.0) in the presence of a fully complementary target DNA **5** (0.1-1.0 equiv). Upon DNA 5 binding, the decrease in the fluorescence intensity of the excimer is accompanied with an increase of the fluorescence intensity of the monomer. Source of figure 2.8: Fujimoto, K., Shimizu, H. and Inouye, M. 2004, J. Org. Chem., 69, 3271 –3275. Authorization of reprint accorded by the American Chemical Society.

Correction of the fluorescence intensity for the inner filter effect can be done as follows:

The optical density at the excitation wavelength is equal to:

$$O.D_{(ex)} = \log I_o / I_T \text{ or } I_o / I_T = 10^{D.O.(ex)} \tag{2.13}$$

The intensity I_o of the incident beam will decrease as the result of the inner filter effect and will be equal to I_T at the center of the cuvette. The optical density of the solution measured with a spectrophotometer is equal to εcl, l being the pathlength of the cuvette.

When fluorescence measurements are performed, the pathlength that should be considered is l/2 and thus the optical density measured with the spectrophotometer is divided by two. This will lead to

$$I_o / I_T = 10^{O.D.(ex)/2} \quad \text{or} \quad I_o = I_T * 10^{O.D(ex)/2} \tag{2.14}$$

The intensity of the beam emitted from the center of the cuvette will decrease upon going through the solution to the surface of the cuvette leading to an apparent fluorescence intensity F_{rec}.

Thus, if the inner filter effect does not occur at the emission wavelength, the recorded fluorescence intensity would be equal to that emitted at the center of the cuvette F_c :

$$F_c = F_{rec} * 10^{O.D.(em)/2} \tag{2.15}$$

F_c was obtained with excitation intensity I_T lower than that of the incident beam I_o as the result of the inner filter effect at the excitation wavelength. Thus, F_c should be multiplied by a correction factor we are going to call Y.

Determination of Y.

$$F = F_{rec} * 10^{O.D.(em/2)} \quad \text{proportional to } I_T. \tag{2.16}$$

$$F_{corr} = F_c * Y \quad \text{proportional to } I_o = I_T * 10^{O.D.(ex)/2} \tag{2.17}$$

$$F_{corr} = F_{rec} * 10^{O.D.(em)/2} * Y \quad \text{proportional to } I_o \tag{2.18}$$

$$\frac{F_{corr}}{F_c} = \frac{F_{rec} * 10^{O.D.(em)/2} * Y}{F_{rec} * 10^{O.D.(em)/2}} = \frac{I_T * 10^{O.D..(ex)/2}}{I_T} \tag{2.19}$$

$$\frac{F_{corr}}{F_c} = Y = 10^{O.D.(ex)/2} \tag{2.20}$$

$$F_{corr} = F_{rec} * 10^{\,O.D..(em)\,/\,2} \quad * \quad 10^{\,O.D.(ex)\,/\,2} \tag{2.21}$$

$$F_{corr} = F_{rec} * 10^{\,[\,O.D..(em)\,+\,O.D..(ex)\,]\,/\,2} \tag{2.22}$$

2d. Fluorescence excitation spectrum

The fluorescence excitation spectrum characterizes the electron distribution of the molecule in the ground state. Excitation is equivalent to absorption since upon absorption, the molecule reaches the excited state S_n. The fluorescence excitation spectrum is obtained by fixing the emission wavelength and by running the excitation monochromator. Figure 2.9 displays the fluorescence excitation spectrum of *Lens culinaris* agglutinin. Fluorescence occurs from the Trp residues of the protein.

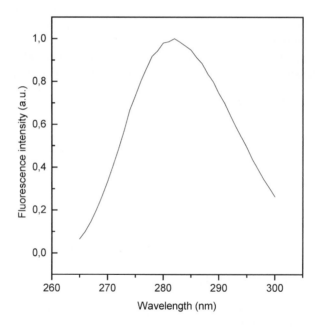

Figure 2.9. Fluorescence excitation spectrum of LCA. $\lambda_{em} = 330$ nm.

For a pure product and in the absence of any interference with other molecules in the solution, the excitation and the absorption spectra of a fluorophore should be identical. Figure 2.10 displays the absorption (b), the fluorescence excitation (a) and emission (c) spectra of quinine present in a Tonic Soda. One can notice how the excitation and the absorption peaks fit well one with each other. However, in the absorption spectrum one can observe clearly the presence of two peaks. In the excitation spectrum, the peak at short wavelength is not very evident, indicating that fluorescence occurs mainly from one specific structure or component in quinine.

Since the sensitivity of fluorescence is higher than absorption, the fluorescence excitation spectrum of a pure molecule can be obtained at low concentrations compared to the absorption spectrum.

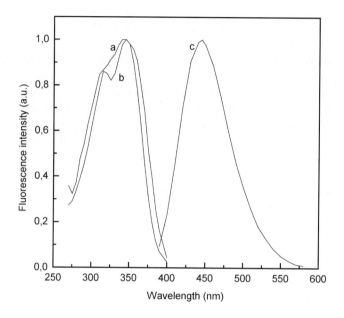

Figure 2.10. Normalized fluorescence excitation (λ_{max} = 345 nm) (a), absorption (b) and emission (λ_{max} = 445 nm) (b) spectra of quinine present in Tonic Soda.
Emission and excitation spectra were recorded at λ_{ex} = 300 nm and λem = 420 nm, respectively.

The excitation spectrum is technically perturbed by two problems: the light intensity of the excitation lamp, which varies with the wavelength, and the intensity upon detection, which is also wavelength-dependent. Using rhodamine B dissolved in glycerol as a reference could perform corrections. Radiation from rhodamine is proportional to the excitation intensity independently of the excitation wavelengths. Therefore, excitation of rhodamine will yield a fluorescence excitation spectrum that characterizes the spectrum of the excitation lamp. In order to obtain the real fluorescence excitation spectrum of the studied fluorophore, the recorded excitation spectrum will be divided by the excitation spectrum obtained from the rhodamine. This procedure is done automatically within the fluorometer.

In general, when one wants to find out if structural modifications have occurred within a protein, circular dichroism experiments are performed. This will help to find out whether the global structure of the protein or / and its local structure have been altered or not. However, one can record also the fluorescence excitation spectrum of the protein. If perturbations occur within the protein, one should observe excitation spectra that differ from a state to another. One should not forget to correct the recorded spectra for the inner filter effect.

68   *Structure and Dynamics of Macromolecules*

For example, we studied the interaction between the fluorescent probe calcofluor and α_1-acid glycoprotein at two concentrations of calcofluor. Also, we recorded the fluorescence excitation spectrum of the protein so that to find out whether binding of calcofluor to the protein modifies its structure or not.

At λ_{em} = 330 nm, calcofluor does not emit and thus, only the excitation spectrum of the Trp residues would be recorded. Therefore, any modification of the fluorescence excitation spectrum in presence of calcofluor would be the result of a structural modification of the protein in the ground state.

Figure 2.11 displays the fluorescence excitation spectrum of the Trp residues of α_1-acid glycoprotein in absence (a) and presence of 10 µM (b) and 120 µM (c) of calcofluor. We notice that in presence of high concentrations of calcofluor, the peak is located at 280 nm instead of 278 nm and a shoulder is observed at 295 nm. Thus, the global shape of the spectrum is modified in presence of high calcofluor concentrations, i.e., binding of calcofluor at high concentrations has modified the local structure of the Trp residues of α_1-acid glycoprotein and of the protein structure around the Trp residues. The fact that binding of low concentrations of calcofluor (10 µM) does not affect this local structure indicates that a minimum concentration of fluorophore is necessary to induce any structural modification of the protein.

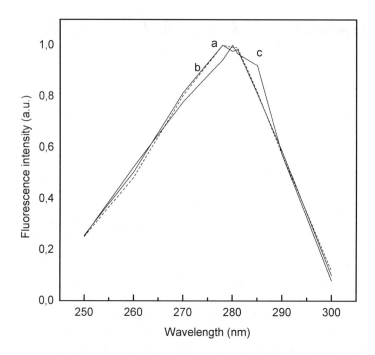

Figure 2.11. Fluorescence excitation spectra of 3 µM of α_1-acid glycoprotein in absence (a) and in presence of 12 µM of calcofluor (b) and of 120 µM of calcofluor (c). λ_{em} = 330 nm. Source: Albani, J. R. 2001. Carbohydrate Research, 334, 141-151.

2e. The mirror-image rule

The emission spectrum of a fluorophore is the image of its absorption spectrum when the probability of the $S_1 \rightarrow S_0$ transition is identical to that of the $S_o \rightarrow S_1$ transition. If however, excitation of the fluorophore leads to a $S_o \rightarrow S_n$ transition, with $n > 1$, internal relaxation that will occur, so that the molecule reaches the first excited singlet state before emission, induces an emission transition different from the absorption one. The mirror-image relationship is in general observed when interaction of the excited state of the fluorophore with the solvent is weak. Figure 2.12 displays the absorption and emission spectra of the fluorophore 7-nitrobenz-2-oxa-1,3-diazole (NBD) covalently bound to the N-terminal of apoC-II$_{19-39}$ of sequence (KESLSSYWESAKTAAQDLYEK-NH$_2$).

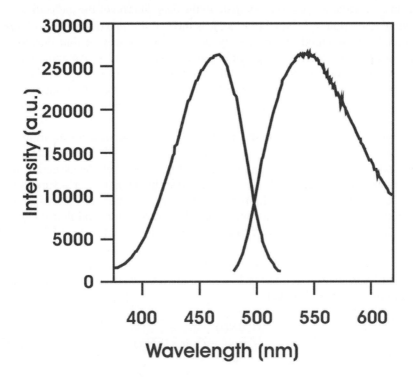

Figure 2.12. Normalized absorption (left) and emission (right) spectra of NBD-apoC-II$_{19-39}$ in aqueous solution. Source: MacPhee, C. A., Howlett, G. J., Sawyer, W. H. and Clayton, A. H. A. 1999, Biochemistry, 38, 10878-10884. Authorization of reprint accorded by the American Chemical Society.

The fluorescence emission spectrum of NBD-labeled peptide shows a good image-mirror with respect to the excitation spectrum.

In the contrary, quinine excitation and emission spectra do not follow the mirror-image rule (Figure 2.10).

2f. Fluorescence lifetime

2f1. Definition of the fluorescence lifetime

After excitation, molecules remain in the excited state a short time before returning to the ground state. The lifetime of the excited state is equal to the mean time during which molecules stay in the excited state. This time is considered as the fluorescence lifetime. This time goes from the nanosecond (10^{-9} s) to the picosecond (10^{-12} s).
The fluorescence lifetime τ_f is equal to:

$$\tau_f = 1/k = 1/(k_r + k_{isc} + k_i) \tag{2.22}$$

The mathematical definition of the fluorescence lifetime arises from the fact that non radiative and radiative processes participate in the deexcitation of the molecule.
The radiative lifetime τ_r is equal to $1/k_r$. It is the real lifetime of emission of a photon that it should be measured independently of the other processes that deactivate the molecule. However, since these processes occur in parallel to the radiative one, it appears that it is impossible to eliminate them during the measurement of the radiative lifetime. Therefore we are going to measure a time characteristic of all the deexcitation processes. This time is called the fluorescence lifetime and is lower than the radiative lifetime.
A fluorophore can have one or several fluorescence lifetimes. So in proteins, it is difficult to assign one specific fluorescence lifetime to a specific Trp residue. Cautions should be taken simply because from one protein to another the structure and the dynamics of the environment of the Trp residue differ inducing by that a variation in the fluorescence lifetime(s) of the fluorophore. Protein denaturation yields a uniform fluorescence lifetime equal to 2.5 ns, indicating that the structure and thus the dynamics of the protein in the tertiary structure play an important role in the determination of the fluorescence lifetime
Let's see how can we describe the intensity decay with time. Let us consider N_o the population of fluorophores having reached the excited state. The velocity of decrease of this population with time t is equal to:

$$-\frac{dN(t)}{dt} = (k_r + k')\, N(t) \tag{2.23}$$

where k_r is the radiative rate constant and k' the sum of rate constant of all other competing processes, and $N(t)$ the population at the excited state at time t.

$$\frac{dN(t)}{N(t)} = -(k_r + k')\, dt \tag{2.24}$$

$$\text{Log } N = -(k_r + k')\, t + \text{constant} \tag{2.25}$$

At t = 0, N = N_o and Log N = Log N_o

$$Log\ N = - (k_r + k\,') t + Log\ N_o \qquad (2.26)$$

$$Log\ N/N_o = - (k_r - + k\,') t \qquad (2.27)$$

$$N / N_o = e^{-(kr + k\,')} \qquad (2.28)$$

$$N = N_o\ e^{-(kr + k\,')} \qquad (2.29)$$

$$\tau = 1 / (k_r + k\,') \qquad (2.30)$$

$$N = N_o\ e^{-t/\tau} \qquad (2.31)$$

Therefore, fluorescence intensity decreases exponentially.

When t = τ

$$N = N_o\ e^{-1} = N_o / e \qquad (2.32)$$

Thus, we can define the fluorescence lifetime as being the time required for the excited-state population to be reduced by 1 / e of its initial population, immediately after excitation (Fig. 2.13)

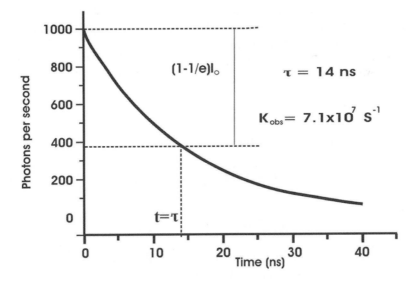

Figure 2.13. Fluorescence intensity decay of quinine sulfate. Curtosey from Professor Professor Jean-Claude Bünzli, Swiss Federal Institute of Technology, Lausanne, Institute of Molecular and Biological Chemistry, Laboratory of Lanthanide Supramolecular Chemistry.

2f2. Mean fluorescence lifetime

The fluorescence intensity decrease of any fluorescent probes whether free in solution, bound to a protein or to a membrane such as extrinsic fluorophores or being parts of a protein such as intrinsic fluorophores can take place with one, two or several fluorescence lifetimes.

Among the naturally occuring opioid peptides, the D-aminoacid containing peptide dermorphin (Montecucchi et al. 1981), Tyr-DAla-Phe-Gly-Tyr-Pro-Ser-NH$_2$, (DRM) isolated from frog skin, combines the highest affinity and selectivity for the μ-opioid receptor (Erspamer, 1992). With respect to its analgesic activity, it is now established as the most potent natural peptide actually known being 1,000 times more active than morphin in inducing long-pasting analgesia after intra-cranial injection in rodents (Erspamer, 1992). Conversely, the related peptide dermenkephalin (Sagan et al. 1989), Tyr-DMet-Phe-His-Leu-Met-Asp-NHz (DREK) (also referred to as deltorphin or dermorphin gene-associated peptide) is a highly potent agonist at the φ-opioid receptor.

The fluorescence intensity of the unique Tyrosine residue of DREK obtained at λ_{ex} of 280 nm can be adequately represented by the sum of two exponentials,

$$I\ (\lambda\ ,t) = 0.54\ e^{\ -t/0.60} + 0.46\ e^{\ -t/2.12} \tag{2.33}$$

where 0.54 and 0.46 are the preexponential factors, 0.60, and 2.12 ns are the fluorescence lifetimes. The emission wavelength (λ) is 303 nm. $\chi^2 = 1.2$.

Also, fluorescence intensity of the two Tyrosine residues of the peptide (Tyr-Dala-Phe-Gly-Tyr-Pro-Ser-NH$_2$ (DRM) measured in the same experimental conditions of DREK, can be described as the sum of two exponentials decays :

$$I\ (\lambda\ ,t) = 0.698\ e^{\ -t/0.96} + 0.302\ e^{\ -t/1.91} \tag{2.34}$$

where 0.698 and 0.302 are the preexponential factors, 0.96, and 1.91 ns are the fluorescence lifetimes. $\chi^2 = 1.1$.

The mean fluorescence lifetime is the second order mean:

$$\tau_o = \Sigma\ f_i\,\tau_i \tag{2.35}$$

and

$$f_i = \beta_i\,\tau i\ /\ \Sigma\ \beta_i\,\tau_i \tag{2.36}$$

where β_i are the preexponential terms, τ_i are the fluorescence lifetime and f_i the fractional intensities. $\lambda_{em} = 330$ nm. τ_o of DREK and DRM is equal to 1.74 and 1.48 ns, respectively. These values are in the same range of that (1.4 ns) found for the peptide [Leu^5] enkephalin (Tyr-Gly-Gly-Phe-Leu) (Lakowicz and Maliwal, 1983).

Since the intensity average lifetime or the mean lifetime is the average amount of time a fluorophore spends in the excited state, it is normal that this time should be applied to determine the rotational correlation time of a fluorophore, the bimolecular diffusion constant of small molecules such as oxygen, iodide and cesium ions in macromolecules and in energy transfer studies.

One can notice that for both peptides (DREK and DRM) we found two fluorescence lifetimes although the number of tyrosines is not the same. We can notice also that fluorescence lifetimes and the largest preexponential factors are not identical. Therefore, the origin of the two fluorescence lifetimes differs from one peptide to another. The structure and the dynamics of the peptides would exert an important influence on the fluorescence decay times. The local environment of the tyrosines plays an important role in the attribution of these lifetimes. We mean by the local environment, the amino acids that are in close contact with the tyrosines.

2f3. Fluorescence lifetime measurement

Three techniques are actually available for measuring fluorescence lifetime. The Strobe and the Time Correlated Single Photon Counting (TCSPC) are based on measurement in the time domain and the third method measures fluorescence lifetimes in the frequency domain. Time domain allows a direct observation of the fluorescence decay, while frequency domain is a more indirect approach in which the information regarding the fluorescence decay is implicit.

In the time-correlated single-photon counting (TCSPC) technique, the sample is excited with a pulsed light source. The light source, optics, and detector are adjusted so that, for a given sample, no more than one photon is detected. When the source is pulsed, a timer is started. When a photon strikes the detector, the time is measured. Over the course of the experiment, the fluorescence decay curve is constructed by measuring the "photon events," or accumulated counts, versus time (Fig. 2.14).

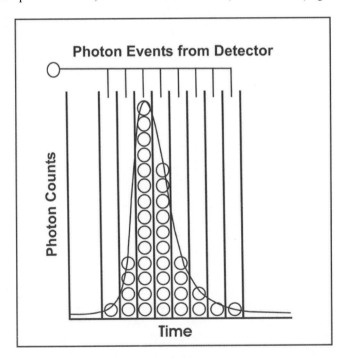

Figure 2.14. Schematic profile of the time-correlated single-photon counting technique. Curtosey from Photon Technology International (PTI).

Fluorescence decays are very sensitive to the modifications that occur within a protein. For example, human hemoglobin possesses 6 tryptophan residues, one in each α subunit (α_{14}) and two in each β subunit (β_{15} and β_{37}). The fluorescence intensity decay of the Trp residue in the α subunit occurs with two fluorescence lifetime equal to 80 ps and 2 ns. The fluorescence intensity decay of the Trp residues of the β chain can be described with three fluorescence lifetimes around 90 ps, 2.5 ns and 6.4 ns. Upon oxidation of the cysteine residues (α_{104}, β_{93} and β_{112}), a small absorption band appears near 625 nm (Fig. 2.15) accompanied with a significant change in the fluorescence decay curves and an increase in the long emission lifetime (Fig. 2.16 and 2.17) (Albani et al. 1985). These spectral modifications are observed regardless of the state of ligation. One conclusion that may be drawn here is the fact that a single tryptophan residue may give rise to two fluorescent components so that in cases where there is more than one tryptophan in a protein it is difficult and in many cases impossible to assign a specific decay component to a specific tryptophan residue. We can assign each fluorescence lifetime to an average conformation of the protein in which the single tryptophan residue is oriented differently with respect to the heme residue. In one of these orientations energy transfer would be highly efficient and thus fluorescence lifetime weak, and in the second orientation energy transfer would be very weak inducing a long fluorescence lifetime.

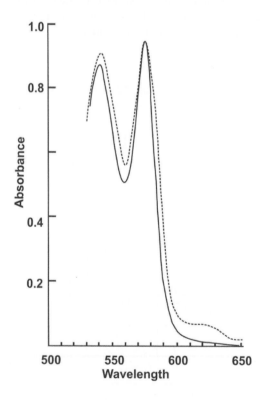

Figure 2.15. Absorption spectra of intact (continuous line) and oxidized (dotted line) hemoglobin alpha chain.

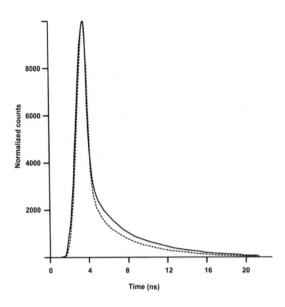

Figure 2.16. Fluorescence decay curve for the intact (continuous line) and thiol oxidized (broken line) hemoglobin alpha chain obtained with the single photon counting technique. Source: Albani J, Alpert B, Krajcarski D.T. and Szabo, A.G. 1985, FEBS Letters, 182, 302-304.

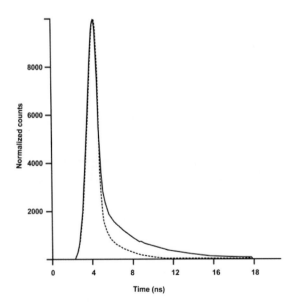

Figure 2.17. Fluorescence decay curve for the intact (continuous line) and thiol oxidized (broken line) hemoglobin beta chain obtained with the single photon counting technique. Source: Albani, J., Alpert, B., Krajcarski, D . T. and Szabo, A. G. 1985, FEBS Letters, 182, 302-304.

In the strobe or pulse sampling technique, the sample is excited with a pulsed light source. The intensity of the fluorescence emission is measured in a very narrow time window on each pulse and saved in the computer. The time window is moved after each pulse. When the data have been sampled over the appropriate range of time, a decay curve of emission intensity versus time can be constructed.

The name "strobe technique" comes about because the photomultiplier PMT is gated - or strobed - by a voltage pulse that is synchronized with the pulsed light source. The strobe has the effect of "turning on" the PMT and measuring the emission intensity over a very short time window (Fig. 2.18) (James et al. 1992).

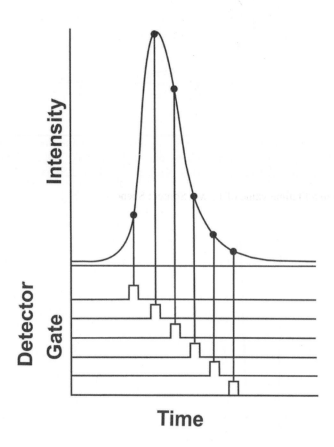

Figure 2.18. Schematic profile of the fluorescence lifetime weasured with the Strobe method. Curtosey from Photon Technology International (PTI)

In frequency domain instruments, the intensity of the excitation beam is sinusoidally modulated at a variable frequency. The fluorescence emission is phase-shifted and amplitude demodulated relative to the excitation. The fluorescence lifetime can be calculated from the phase shift and demodulated amplitude (Fig. 2.19) (Spencer, 1970).

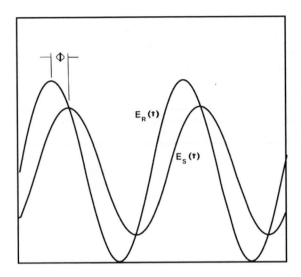

Figure 2.19. Comparison of phase and modulation differences between sinusoidally modulated exciting light. $E_R(t) = A + B \sin wt$, and the resulting fluorescent emission $E_S(t) = A + B \cos \phi \sin(wt - \phi)$. In this illustration a phase shift ϕ of 45 degrees and a demodulation factor, $\cos \phi$, of 0.707 indicate a lifetime value of $1/w$. Source: Spencer, R. D. 1970. Ph.D. Thesis

Table 2.3 shows the values of the fluorescence lifetimes of the Trp residues of the sialylated α_1-acid glycoprotein measured with the multifrequency and TCSPC techniques.

Table 2.3. Values of the fluorescence lifetimes of the 3 Trp residues of the sialylated α_1-acid glycoprotein measured with the multifrequency and TCSPC techniques.

Frequency domain						
τ_1	f_1	τ_2	f_2	τ_3	f_3	τ_o
0.354	0.101	1.664	0.66	4.638	0.238	2.285
± 0.034	± 0.05	± 0.072	± 0.03	± 0.342	± 0.01	
TCSPC method						
τ_1	f_1	τ_2	f_2	τ_3	f_3	τ_o
0.53	0.16	1.88	0.66	5.27	0.18	2.285
± 0.045	± 0.06	± 0.045	± 0.05	± 0.565	± 0.01	

Source of the multifequency data: Albani, J., Vos, R., Willaert, K. and Engelborghs, Y. 1995. Photochem. Photobiol. 62, 30-34.

One can notice that the two methods give very close results. Therefore, both frequency and time domains methods are reliable.

Table 2.4 yields the fluorescence lifetimes recovered by stroboscopic, TCSPC and phase techniques of some fluorophores.

Table 2.4. Fluorescence lifetime of fluorophores measured with Strobe, the phase and TCSPC methods.

Sample	Technique	τ ns	Standard deviation
p-terphenyl	Strobe	1.001	0.031
	TCSPC	1.005	0.032
	Phase	1.002	-
Perylene	Strobe	5.09	0.050
	TCSPC	5.12	0.041
Pyrene	Strobe	411	7.8
		403	4.6

The three techniques give identical results.

The main problem with lifetime measurements is the stability of the studied sample. Too long irradiation of the sample can destroy it yielding erroneous fluorescence lifetimes.

Table 2.5 yields the different values of the mean fluorescence lifetime and of intensity of fluorescein in presence of increased concentrations of KI. Lifetimes were measured with both frequency domain and Time correlated single photon counting methods. Figure 2.20 displays the normalized values at different KI concentrations.

Table 2.5. Fluorescence intensity and fluorescence lifetimes of fluorescein as a function of KI

[KI] (M)	Frequency domain (ns)	TCSPC (ns)	Intensity
0	3.99	3.99	7.87
0.1	2.13	2.20	4.32
0.2	1.30	1.34	2.67
0.3	0.92	nd	1.87
0.4	0.67	0.72	1.41
0.5	0.47	nd	1.08

Data are from Amersham Science.

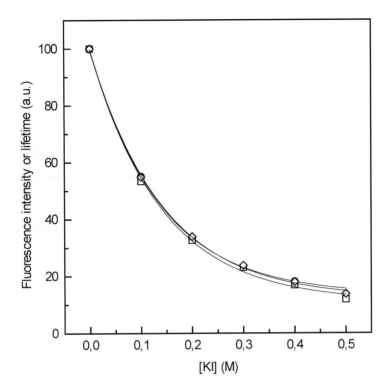

Figure 2.20. Fluorescence intensity / lifetime quenching of fluorescein in presence of increased concentrations of KI. Curtosey from Amersham Science.

2f4. Time Correlated Single Photon counting

It consists of exciting the sample with package of short 20 pulses at maximum. After each excitation, only the first emitted photon is detected and counted. The excitation pulses could have slightly different intensities but they reach the sample after a defined time interval. The time that elapses between excitation and the detection of the first emitted photon is recorded. The distribution of these times gives a statistical picture of the fluorescence decay.

The description of the diagram of the single photon counting decay fluorometer is given in figure 2.20 (Badae and Brand, 1979). The heart of the instrument is the time to pulse height or time to amplitude converter (TAC). The electronic device measures the time between the pulse start and the arrival of a photon to the detector. Each detected photon is then stored in a multichannel pulse height analyzer (MCPHA). For each detection time of a photon, the single count is stored in the appropriate time-location of the MCPHA. Figure 2.21 describes the principle of function of the TAC and of the single photon counting method (Lakowicz, 1983, 1999).

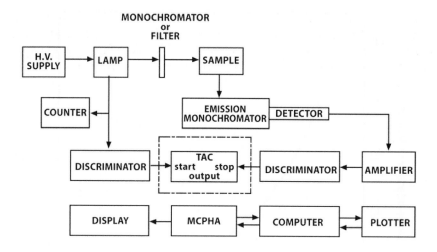

Figure 2.21. Block diagram of a monophoton counting nanosecond decay fluorometer.

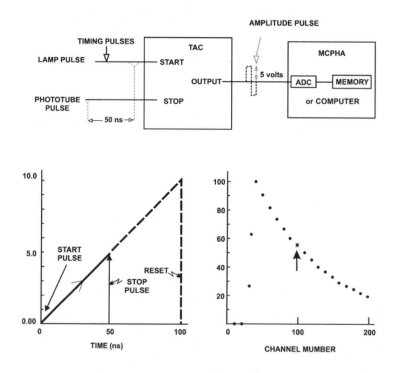

Figure 2.22. Schematic illustration of the operation of a time to amplitude converter (TAC). ADC, analog to digital converter. MCPHA, multichannel pulse height analyzer. Source: Badea, M.G. and Brand, L. 1979, Methods in Enzymology. 61, 378-425.

The data obtained representing the time – resolved decay is distorted by the width of the excitation pulse. Different methods are used to determine the impulse response function which is the decay that would be observed for an infinitely sharp excitation pulse or for a pulse whose width is much shorter than the decay time of the sample (Figure 2.23).

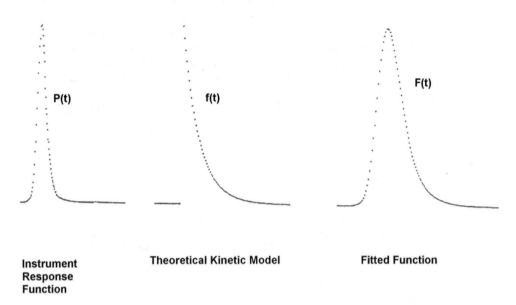

Instrument Response Function　　**Theoretical Kinetic Model**　　**Fitted Function**

Figure 2.23. Reconvolution analysis of the fluorescence decay curve. Curtosey from Dr. John R. Gilchrist, Edinburgh Instruments Ltd.

Equation 2.37 describes the mathematical description of a reconvolution analysis

$$F(t) = \int_{-\infty}^{+\infty} P(t-t')\, f(t')\, dt' \tag{2.37}$$

where F(t), P(t) and f(t) are respectively the expected decay curve, the instrumental response curve and the theoretical decay law describing the kinetic model.

An example between others, fluorescence intensity decay of the Trp residues of the lectin *Lens culinaris* agglutinin was obtained by the single photon counting technique. Lectins are proteins or glycoproteins capable of recognizing and binding specific carbohydrate residues (Liener, 1976). In fact, the term lectin is attributed to proteins that bind specifically and reversibly carbohydrate residues. They play an important role in immunology and hematology and are used as specific probes for membrane glycoprotein structures. Many lectins agglutinate red blood cells. Legume lectins have an important role in seed maturation, cell wall assembly, and the defense mechanism. Some lectins such as those from castor bean are dangerous since they are highly toxic and can kill if ingested in even small amounts.

Composed of two α and β chains (MW = 5710 and 20572, respectively), *Lens culinaris* agglutinin (LCA) is a tetramer $\alpha_2 \beta_2$ with a molecular weight equal to 52570 (Loris et al. 1993). LCA contains five Trp residues, three embedded in the protein matrix (Trp 152β, Trp 19α and Trp 40α) and two near the protein surface (Trp 53β and Trp 128β).

Fluorescence measurements were performed on the experimental setup installed on the SB_1 window of the Synchrotron Radiation machine super-ACO (Anneau de collision d'Orsay). The storage ring provides a light pulse with a full width at half maximum of about 500 ps at a frequency of 8.33 MHz for a double-bunch mode. Data for I_{vv} (t) and I_{vH} (t) were stored in separate memories of a plug-in multichannnel analyzer card in a microcomputer. Accumulation was stopped when 10^5 counts were stored in the peak channel for the total fluorescence intensity decay. The instrument response function was automatically monitored in alternation with the parallel and perpendicular components of the polarized fluorescence decay by measuring the sample-scattering light at the emission wavelength. The automatic sampling of the data was driven with the microcomputer (Vincent et al. 1993).

In the single photon counting, the detection system measures the time between the excited pulse and the arrival of the first photon. The distribution of arrival times represents the decay curve.

Analysis of the fluorescence intensity decay data as a sum of 150 exponentials was performed by the Maximum Entropy Method (MEM) (Livesey and Brochon, 1987), using the commercially available library of subroutines MEMSYS 5 (MEDC Ltd., UK) as a library of subroutines.

With vertically polarized light, the parallel I_{vv} (t) and perpendicular I_{vh} (t) components of the fluorescence intensity at time after the start of the excitation are

$$I_{vv} (t) = (1/3) E (t) * \int_0^\infty \int_0^\infty \int_{-0.2}^{0.4} \gamma (\tau, \phi, A) e^{-t/\tau} (1 + 2 A e^{-t/\phi}), d\tau \, d\phi \, dA \qquad (2.38)$$

and

$$I_{vh}(t) = (1/3) E(t) * \int_0^\infty \int_0^\infty \int_{-0.2}^{0.4} \gamma (\tau, \phi, A) e^{-t/\tau} (1 - A e^{-t/\phi}), d\tau \, d\phi \, dA \qquad (2.39)$$

where E (t) is the temporal shape of the excitation flash, * denotes a convolution product and $\gamma (\tau, \phi, A)$ represents the number of fluorophores with fluorescence lifetime, τ, rotational correlation time, (ϕ), and initial anisotropy, A.

The fluorescence intensity decay is obtained by summing the parallel and perpendicular components:

$$T\{t\} = E(t) * \int_0^\infty \alpha(\tau) \, \exp(- t / \tau dt. \qquad (2.40)$$

$\alpha(\tau)$ is the lifetime distribution given by

$$\alpha(\tau) = \int_0^\infty \int_{-0.2}^{0.4} \gamma (\tau, \phi, A) \, d\phi \, dA \qquad (2.41)$$

In order to ensure that the recovery distribution agrees with the data, the entropy S is maximized:

$$S = \int_0^\infty \alpha(\tau) - m(\tau) - \alpha(\tau) \log (\alpha(\tau) / m(\tau))] \; dt \qquad (2.42)$$

where $m(\tau)$ is the starting lifetime distribution flat in $\log\tau$ space and $\alpha(\tau)$ is the resulting distribution. The entropy S is maximized under the condition that X^2 is minimized to unity,

$$\chi^2 = 1/N \sum_{k=1}^N \frac{I_c(k) - I_o(k}{I_o(k)} \qquad (2.43)$$

where N is the number of channels, and $I_c(k)$ and $I_o(k)$ are the calculated and observed number of photons in channel k, respectively.

When fluorescence anisotropy decay is measured, and assuming that there is no correlation between τ and ϕ, $\alpha(\tau)$ and $\beta(\tau)$ are independent. $I_{vv}(t)$ and $I_{vh}(t)$ can be written as

$$I_{vv}(t) = (1/3) \int_0^\infty \alpha(\tau) \, e^{-t/\tau} \, d\tau \int_0^\infty [1 + 2 \beta(\phi) \, e^{-t/\phi}] \, d\phi \qquad (2.44)$$

$$I_{vh}(t) = (1/3) \int_0^\infty \alpha(\tau) \, e^{-t/\tau} \, d\tau \int_0^\infty [1 - \beta(\phi) \, e^{-t/\phi}] \, d\phi \qquad (2.45)$$

The integrated amplitude $\beta(\phi)$ corresponds to the fundamental anisotropy A_o, and the time dependence of the anisotropy can be described with the integral

$$A(t) = \int \beta(\phi) \, e^{-t/\phi} \, d\phi \qquad (2.46)$$

In this analysis of the fluorescence and fluorescence anisotropy decay, the entropy of the cross-product $\alpha(\tau) . \beta(\phi)$ is maximized under the constraint of minimum χ^2,

$$S = - \int_0^\infty \alpha(\tau) . \beta(\phi) \; \log \frac{\alpha(\tau) . \beta(\phi)}{m(\tau)} \; d\tau \qquad (2.47)$$

where $m(\tau)$ is the initial guess of the $\alpha(\tau) . \beta(\phi)$ distribution.

The maximum entropy method can handle Laplace transforms such as those found in pulse fluorometry, without restricting the validity of the solution or suffering from any instability. It allows the recovery of the distribution of exponentials describing the decay of the fluorescence (i.e., inverting the Laplace transform), which is, in turn, convolved by the shape of the excitation flash. Also, it can determine the background level and amount of parasitically scattered radiation.

Fluorescence intensity, $I(\lambda,t)$ of Trp residues of LCA obtained at λ_{ex} equal to 295 nm can be adequately represented by a sum of four exponentials (Albani, 1996),

$$I(\lambda,t) = 0.7065\ e^{-0.016} + 0.0796\ e^{-0.484} + 0.1285\ e^{-1.6} + 0.0854\ e^{-4.057} \qquad (2.48)$$

where 0.7065, 0.0796, 0.1285, and 0.0854 are the preexponential factors, 0.016 ± 0.007, 0.484 ± 0.113, 1.600 ± 0.442, and 4.057 ± 0.690 are the decay times (ns), and λ the emission wavelength (335 nm). The mean fluorescence lifetime τ_o is equal to 2.912 ns. The value of χ^2 is 1.12.

2f5. The Strobe technique

Bennett developed the Strobe technique and described it in 1960 (Bennet, 1960). However, the method has not been developed until 1990. The first paper describing the method, the instrument developed and some results was published in 1992 (James et al. 1992.) The methods needs less electronic than the two other ones. Everything is run and controlled by the computer where data are analyzed. The following description was given by PTI, Inc, and / or described by (James et al. 1992). A nanosecond flash lamp of wavelength range from 180 to 750 nm (gas dependent) is used as an excitation source. The pulse width is lower than 1.6 ns. The lamp thanks to its flash rate provides a high intensity illumination equal to 25 kHz. Emitted light is collected by quartz at 90 degrees to the incident beam optics and passed through a 200 mm monochromator to a stroboscopic detector. The strobe or the gate is created directly in the PMT through the use of a strip line amplification process. All system functions are under computer control. In fact, a master clock generates a pulse train that is routed simultaneously to the flash lamp trigger circuit and a computer controlled digital delay unit (DGG). The DGG will provide an accurate delay to the photomultiplier tube gating circuit. Since the pulse train from the clock provides a common source to both the flash lamp and the delay unit, the PMT gating circuit will be triggered at a time precisely delayed by δt with respect to the flash lamp. The PMT gating circuit functions by transiently pulsing (strobing) the PMT thus providing a gate during which the PMT is active. Only photons emitted from the sample that arrive at the PMT photocathode during the gate time will be detected. The emission intensity as a function of time can be acquired by sweeping the gate delay time from before the lamp flash to any suitable termination point of the data acquisition; the computer controls both the sweeping of the DGG and the acquisition of the PMT output. Data are collected and analyzed by proprietary windows-based advanced fluorescence software with exclusive NTDA (non-linear timescale data acquisition) capability.

The stroboscopic detector is composed by a standard PMT. With the flash lamp, the minimum lifetime measured is equal to 100 ps. Most interesting compared to the other two methods, is the data collection rate equal to less than 30 seconds.

As in the TCSPC, the data obtained are distorted by the width of the excitation pulse. Deconvolution methods are used to obtain the real decay intensity as function of time.

The absence of electronic allows selling this fluorometer at much less price than the TCSPC or the multifrequency methods. Also, acquisition of the data is very fast which makes this type of fluorometer adequate for many studies that use fluorophores very sensitive to light illumination.

Because the strobe technique is intensity-dependent, strobe instruments are much faster than TCSPC, and even faster than Phase. Strobe systems can make accurate measurements in minutes or even seconds (Fig. 2.24).

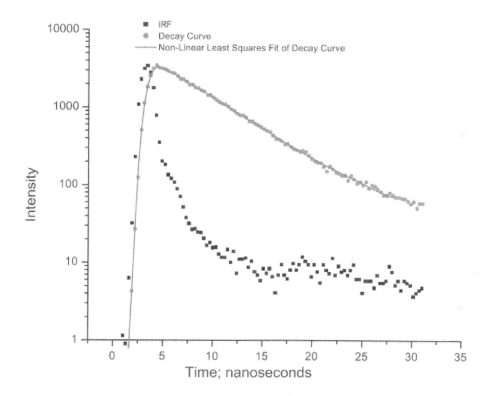

Figure 2.24. Decay curve acquired with a PTI TimeMaster TM-3/2003 N_2 -dye laser fluorescence lifetime spectrometer. Curtosey from Photon Technology International.

2f6. Excitation with continuous light : the phase and demodulation method

Principles

The sample is excited by a light source whose intensity is sinusoidally modulated (w). The emitted light by the sample will be also sinusoidally modulated at the same frequency of the excited light but will be delayed by a phase angle, ϕ, due to the finite persistence of the excited state. The phase angle ϕ is equal to (w θ), where θ is the phase delay time. The intensity of the emitted light is lower than that of the excitation one (Fig. 2.19 and 2.25).

The fluorescence lifetime of the sample can be calculated from phase delay and/or from light demodulation. The longer the sample lifetime, the more the emission is delayed and greater the phase shift between the two lights.

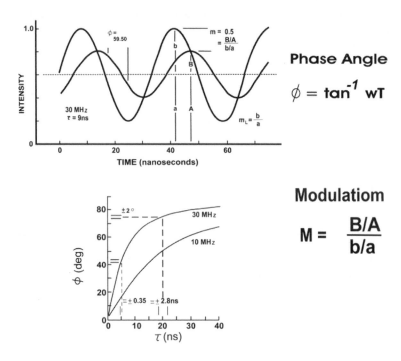

Figure 2.25. Determination of the fluorescence lifetime from the phase angle and intensity modulation

A glycogen solution placed in the emission compartment will scatter light and be used as reference ($\tau_f = 0$) to determine phase delay and fluorescence demodulation. For each measurement, the intensity of the reference is adjusted in order to have it equivalent to the intensity of the fluorescence signal of the sample. Phases and modulations of the fluorescence and scattered light are obtained relative to the reference photomultiplier or instrumental (internal) reference signal. In fact, two identical detection electronic systems are used to analyze the outputs of the reference and sample photomultipliers. Each channel consists of an alternative and continuous currents (AC and AD, respectively).

$$m = (AC/DC)_{EM} \: / \: (AC/DC)_{EX} \: = \: (B/A) \, / \, (b/a) \qquad (2.49)$$

where $_{EM}$ and $_{EX}$ refer to fluorescence and scattered light.

Lifetimes can be obtained by the phase method and it will be denoted τ_P or by modulation and it will be denoted τ_M.

A monoexponential decrease of fluorescence (one lifetime) yields $\tau_P = \tau_M$. A multiexponential decrease gives $\tau_P < \tau_M$ (Table 2.6.)

Table 2.6. Values of τ_P and τ_M of zinc protoporphyrin bound at the heme pocket of apomyoglobin ($Mb^{Fe \to Zn}$) and measured at 6 frequencies with an SLM fluorometer.

Frequency (MHz)	τ_P (ns)	τ_M (ns)
50	2.03	2.26
70	1.96	2.24
90	1.87	2.22
110	1.77	2.20
130	1.68	2.17
150	1.58	2.14

We notice that for the 6 applied frequencies, $\tau_P < \tau_M$. Thus, the decay is heterogeneous and at least two fluorescence lifetimes exist. In fact, the data yield two fluorescence lifetimes 2.320 and 0.333 ns with fractional intensities of 0.9 and 0.1, respectively. The mean fluorescence lifetime is equal to 2.213 ns.

Actually, the possibility of measuring fluorescence lifetimes in the multifrequency method allows in principle to obtain much more accurate results than with small numbers of frequencies. However, below one nanosecond, the higher modulation frequencies necessary to make accurate measurements require an expensive laser-based excitation source.

Figure 2.26 displays the phase shift and frequency modulation measurements of rhodamine in water.

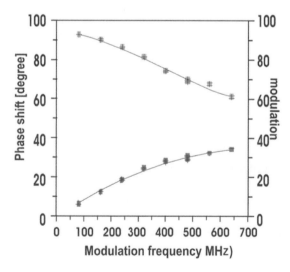

Figure 2.26. Lifetime measurements of rhodamine in water performed in the frequency domain (multifrequency phase fluorometry) by means of a photomultiplier on the camera port of a microscope. The range of frequencies currently in use is 80 - 720 MHz and lifetimes of the order of 100 ps up to few nanoseconds can be measured. Curtosey: Diaspora, A., Robello, M., Chirico, G. and Baldini, G. National Institute of the Physics of Matter (INFM).

Multifrequency and cross-correlation

Multifrequency is achieved by using frequency modulators. The modulator currently used is an electro-optic modulator that is crossed transversely by an electric current (Pockels cell). In the phase detection system, high frequencies signals are used. In the cross-correlation technique, the high frequency signals are converted to low-frequency ones. The excitation light is first modulated and then sent to the cell compartment equipped with a revolving turret so that the excitation beam can once excite the studied sample then the reference that is fluorophore of known fluorescence lifetime. This fluorophore will replace the glycogene solution.

A part of the excitation light is sent toward a reference photomultiplier so that the intensity and the phase of the excitation signal are recorded.

A frequency synthesizer connected to the Pockels cell imposes the wanted frequency. Rather than to apply a continuous current on detection and reference photomultipliers, the two photomultipliers are submitted to a sinusoidal electric current R(t) of frequency (w + Δw) as close as possible of the frequency w of the fluorescence light. Δw is the cross-correlation frequency. The signal recorded by the detection photomultiplier is thus the product of the emitted light F(t) and of the R(t). Therefore, the resulting signal evolves with a weak frequency Δw possessing all information "contained" in the fluorescent light. One measures the evolution of the value < R(t) * F(t) > of the current delivered by the photomultiplier. This method called cross-correlation compares the excitation and the emission signals. This double procedure increases considerably the signal to noise ratio.

What is the advantage of the cross-correlation method in fluorescence lifetime measurements? Fluorescence lifetimes are in the order of the ns or / and ps. Thus, high frequencies from the MHz to the GHz are needed to perform fluorescence lifetimes measurements. Still, the accuracy of the measured values is not reached at high frequencies. Therefore, one can translate the high frequency and phase modulation information to lower frequency carrier signal, Δw. In this case, the measured fluorescence lifetimes are highly accurate.

The mathematical treatment of the cross correlation method is as follows:

$$F(t) = A + B \sin (wt - \phi) \qquad (2.50)$$

where ϕ is the fluorescence phase delay

$$R(t) = a + b \sin (wt - \phi) \qquad (2.51)$$

where ϕ is the phase delay of the sinusoidal current that multiplies the fluorescence emission.

$F(t)*R(t) = Aa + Ab \sin (wt - \phi) + aB \sin (wt - \phi) + bB \sin (wt - \phi) * \sin (wt - \phi)$
$\qquad = Aa + Ab \sin (wt - \phi) + aB \sin (wt - \phi) + [bB \cos (\phi - \phi] /2 - [bB \cos (2 wt - \phi - \phi)] /2 \qquad (2.52)$

The integration of the signal over one time $2 \pi / w$ yields as a mean value:

$$< F(t) * R(t) > = w / 2 \pi \int_0^{2\pi} < F(t) * R(t) > dt = aA + bB \cos (\phi - \phi] \qquad (2.53)$$

The mean value is: $< F(t) * R(t) > = aA + [bB \cos (\Delta wt - (\phi] / 2 \qquad (2.54)$

The resulting signal of the photomultiplier preserves the phase difference ϕ and the relative modulation B /A of the fluorescence modulated at the original frequency. By this technique, fluorescence lifetimes are obtained with a high precision from the phase angle and of the relative demodulation for range of frequencies going from 0 to 250 MHz without any problem. The weak values of Δw (< 100 Hz) permit to measure a temporal shift Δt that allows determining the phase difference between excitation and emission light. The phase angle ϕ is equal to:

$$\phi = 360° \times t / T \qquad (2.55)$$

where T is the signal period.

It is possible to reach frequencies in the range of the GHz when the excitation source is a mode-locked laser.

Figure 2.27 shows the phase variation according to the frequency for the 3 Trp residues of α_1 - acid glycoprotein complexed to progesterone.

Development, description and some applications of the multifrequency are described by different resarchers (Alacala et al. 1985, Lakowicz et al. 1985, Gratton, 1989, De Beuckeler et al. 1999, Fisher et al. 2000, Malicka et al. 2003).

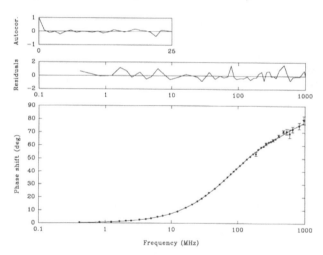

Figure 2.27. Phase measurements of 5 μM α_1 - acid glycoprotein in presence of 4 μM progesterone at 20°C. The average values of five phase lmeasurements are given together with their standard deviation. The continuous line is the best fit to a sum of three exponentials. Plotted are the measured phases and weighted residuals as a function of the frequency and the autocorrelation function of the weighted residuals. The reduced chi-square $\chi_R^2 = 1.03816$. The measured fluorescence lifetimes are equal to : $\tau_1 = 0.343 \pm 0.040$, $\tau_2 = 1.585 \pm 0.084$ and $\tau_3 = 4.258 \pm 0.373$ ns, with preexponential factors equal to $\alpha_1 = 0.103 \pm 0.014$, $\alpha_2 = 0.6488 \pm 0.0342$ and $\alpha_3 = 0.248$. The measurements were performed on an automated multifrequency phase fluorometer built in the Laboratory for Chemical and Biological Dynamics of Professor Yves Engelborghs, Leuven, Belgium

Structure and Dynamics of Macromolecules

Fluorescence parameters such as lifetimes and emission maximum can be monitored to follow structural modifications in a macromolecule. For example, ion binding to a protein can induce conformational changes leading to modifications in all of some of the fluorescence parameters. Calcium binding protein isolated from the sacroplasm of the muscles of the sand worm *Nereis diversicolor* (NSCP) can be used as an example for studying conformational changes upon calcium or magnesium binding. The protein has 174 amino acids (Collins et al. 1988) and a molecular mass of 19485. NSCP possesses four domains with a typical EF-hand Ca^{2+} binding sequence, but only sites I, III and IV can bind calcium or magnesium ions. The EF-hand calcium binding proteins use a distinctive helix-loop-helix structural motif to bind calcium ions. NSCP contains three tryptophan residues at positions 4, 57 and 170, respectively. Tryptophan 4 is located at the beginning of helix A and is not involved in an EF-hand. Tryptophan 57 is positioned at the second (inactive) domain with the indole group close to the binding site of the first domain (active). Tryptophan 170 is found at the end of the last helix H, but the indole group is located in the hydrophobic core in the center of the protein. Fluorescence decay measurements and steady-state emission spectra of the wild-type and of double mutants have been applied to monitor structural modifications in NSCP upon calcium and magnesium binding (Sillen et al; 2003) (Table 2.7).

Table 2.7. Position of the emission peak (in nm), fluorescence lifetimes (τ_i) with their fractional intensities f_i (in %) and the mean fluorescence lifetime (τ_o) of wild-type NSCP and of three double protein mutants, in the apo form and in presence of calcium and magnesium ions.

	λ_{max}	τ_1	f_1	τ_2	f_2	τ_3	f_3	τ_4	f_4	τ_o
					WT					
Ca	338	0.21	2.2	1.3	6.25	4.8	90.7	12	0.8	4.535
Mg	339	0.32	3.7	0.99	3.60	3.2	32.9	5.6	60	5.450
Apo	338	0.42	2.95	1.9	24	4.6	68	11	5	4.146
					W57F/W170F					
Ca	339	0.42	0.25	3.2	13	5.2	86			4.880
Mg	339	0.32	3.7	0.99	3.60	3.2	80			5.022
Apo	338	0.42	2.95	1.9	24	4.6	76.5			5.107
					W4F/W57F					
Ca	338	0.61	2.5	2.7	25	5.1	72			4.362
Mg	341	0.767	2.2	2.745	21	5.227	77			4.618
Apo	350	0.63	5.2	2.77	51	5.63	44			3.922
					W4F/W170F					
Ca	317	0.19	20.6	0.66	25	2.4	35	5.8	19	2.146
Mg	324	0.14	13.7	0.52	25	2.1	39	5.1	21.6	2.070
Apo	330	0.27	1.8	1.4	11.5	4.0	63	7.2	24	4.414

Adapted from Sillen, A.,Verheyden, S., Delfosse, L., Braem, T., Robben. J., Volckaert, G. and Engelborghs, Y., 2003, Biophysical Journal 85, 1882-1893. Authorization of reprint accorded by the American Biophysical Society.

One can notice that while a shift is observed only for two mutants, the fractional intensities obtained from the lifetime measurements are modified in the wild type and in the three mutants (Table 2.7). Emission peak of Tryptophan 4 is not modified with the state of ligation, however fractional intensities f_1 and f_2 increase while f_3 is not modified. Emission of tryptophan 170 is located at 338 nm characterizing an emission from a slightly hydrophobic environment. Modification of the state of ligation, Mg^{2+} instead of Ca^{2+}, and in the apo form, a red shift to 341 and 350 nm is observed. This means that conformational changes are observed in the vicinity of Trp 170 that is more exposed to the solvent in presence of Mg^{2+} than in presence of Ca^{2+}. In the apo form, important structural modifications occur within the protein.

Fluorescence lifetime measurements show that the fractional intensities f_1 and f_2 corresponding to the shortest fluorescence lifetimes increase in parallel to the decrease of the fractional intensity f_3.

The λ_{max} of tryptophan 57 in the Ca^{2+} state is very blue-shifted (317 nm). In presence of Mg^{2+} and in the apoform, the maximum shifts to 324 and 330 nm, respectively. Although this red shift indicates a modification in the microenvironment of the Trp residue, the positions of the maximum reveal that Trp 57 is buried in the hydrophobic core of the protein. We observe also that the fractional intensities of the fluorescence lifetimes are not the same with the state of ligation. The mean fluorescence lifetime is drastically modified when the protein goes from the apo to the ligated forms.

In the wild-type protein, while no shift is observed in the emission peak, the fractional intensities of the fluorescence lifetimes vary with the state of ligation.

Many conclusions can be drawn from these data:
- Fluorescence lifetimes are more sensitive than emission maximum to the modifications occurring in the vicinity of the Trp residues.
- Although a Trp residue does not belong necessarily to the domain where the interaction between the protein and the ligand occurs, measuring fluorescence lifetime parameters of this particular Trp residue can monitor effects of these interactions.
- The three Trp residues in NSCP can be used to monitor structural and / or dynamical modifications with the state of ligation of the protein.

2g. Fluorescence quantum yield

The molecule in the fundamental state absorbs a light of intensity equal to I and reaches an excited state S_n. Then, different competitive processes, including fluorescence, will compete one with each other in order to deexcite the molecule. The rate constant (k) of the excited state is equal to the sum of the kinetic constants of the competitive processes:

$$k = k_r + k_{isc} + k_i \qquad (2.56)$$

The fluorescence quantum yield is equal to the number of photons emitted by the radiative way over that absorbed by the molecule.

Let us see how we can describe mathematically the absorption and the deexcitation processes:

Photon absorption:

$$I + {}^1M \longrightarrow {}^1M^* \tag{2.57}$$

where the fundamental singlet state and the excited singlet state of the molecules are designated by ${}^1M$ and ${}^1M^*$.

Deexcitation of ${}^1M^*$ will occur principally with the following phenomena:

$$ {}^1M^* \longrightarrow {}^1M + h\nu \quad (k_r) \tag{2.58}$$

$$ {}^1M^* \longrightarrow {}^1M + heat \quad (k_i) \tag{2.59}$$

$$ {}^1M^* \longrightarrow {}^3M^* \text{ (triplet state) } (k_{isc}) \tag{2.60}$$

$$ {}^3M^* \longrightarrow {}^1M + chaleur \quad (k_i) \tag{2.61}$$

$$ {}^3M^* \longrightarrow {}^1M + h\nu' \quad (k_p) \tag{2.62}$$

The global balance of ${}^1M^*$ can be written as

$$\frac{d[{}^1M^*]}{dt} = I - (k_r + k_i + k_{isc})\,[{}^1M^*] \tag{2.63}$$

and the global balance of ${}^3M^*$ is

$$\frac{d[{}^3M^*]}{dt} = k_{isc}\,[{}^1M^*] - (k_i + k_p)\,[{}^3M^*] \tag{2.64}$$

At equilibrium, one have

$$\frac{d[{}^1M^*]}{dt} \quad \text{and} \quad \frac{d[{}^3M^*]}{dt} = 0 \tag{2.65}$$

$$I = (k_r + k_i + k_{isc})\,[{}^1M^*] \tag{2.66}$$

and

$$k_{isc}\,[{}^1M^*] = (k_i + k_p)\,[{}^3M^*] \tag{2.67}$$

$$\phi_F = \frac{\text{emitted photons}}{\text{absorbed photons}} = \frac{k_r \, [^1M^*]}{I} = \frac{k_r}{k_r + k_i + k_{isc}} \qquad (2.68)$$

The fluorescence quantum yield of a molecule is obtained by comparing the fluorescence intensity of the molecule to that of a reference molecule with a known quantum yield

Let us consider F_2 as the fluorescence intensity of the molecule of unknown quantum yield ϕ_2 and F_1 the fluorescence intensity of the reference with quantum yield of ϕ_1. One can write:

$$\frac{F_2}{F_1} = \frac{I_o \, \varepsilon_2 \, c_2 \, 1 \, \phi_2}{I_o \, \varepsilon_1 \, c_1 \, 1 \, \phi_1} = \frac{DO_{(2)}}{DO_{(1)}} * \frac{\phi_2}{\phi_1} \qquad (2.69)$$

Therefore, in order to determine the quantum yield of a molecule, one needs to measure the optical densities of the molecule and the reference at the excitation wavelength and to calculate the sum of their intensities along their fluorescence emission spectra:

$$\phi_2 = \frac{DO_{(1)} \; \Sigma F_2}{DO_{(2)} \; \Sigma F_1} \, \phi_1 \qquad (2.70)$$

In proteins, one calculates the quantum yield from the dominant fluorescent amino acid. For example, a protein where the Trp residues are the main responsible of the protein emission, the quantum yield is determined by comparison with a solution of free L-Trp ($\Phi_F = 0.14$ at 20°C and at $\lambda_{ex} = 295$ nm). Also, quinine sulfate dissolved in 0.1 M sulfuric acid is commonly used as reference or standard molecule. Its quantum yield is 0.55 with an absorption and emission maxima at 340 and 445 nm, respectively.

Finally, one should remember that the standard and the molecule to be analyzed should be studied in the same conditions of temperature and solvent viscosity. Also, it is always better to work at low optical densities so that one avoids corrections for the inner filter effect.

In a multi-tryptophan protein, fluorescence quantum yield of the tryptophans can be additive or not. In the first case, the others do not influence each tryptophan while in the second case mainly energy transfer to or from the other tryptophan residues can influence quantum yield. In the absence of energy transfer between the tryptophan residues, mutants with one tryptophan residue are prepared, the other tryptophan residues of the protein are replaced by phenylalanine, and quantum yield of the corresponding tryptophan in each mutant is determined experimentally by comparing its fluorescence to that of free tryptophan in solution. For example, quantum yields of tryptophan residues in Barnase have been determined in the wild type and in mutants where only one tryptophan residue has been conserved.

Barnase is a ribonuclease synthesized by *Bacillus amyloliquefaciens* and then secreted into the growth medium. The same cells synthesize barstar, a potent inhibitor of barnase, and retain it inside the cell to inhibit any barnase that is not secreted. Many studies related to the interaction between barnase and barstar have been peformed in order to understand the rules, if any, of protein-protein interactions. Barnase has three tryptophan residues present in positions 35, 71 and 94 (Figure 2.28).

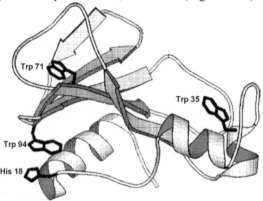

Figure 2.28. Schematic representation of the structure of barnase showing the positions of the three tryptophan residues and His 18. Source: De Beuckeleer, K., Volckaert, G. and Engelborghs, Y. 1999, Proteins : Structure, Functions and Genetics 36, 42-53. Authorization of reprint accorded by Wiley Interscience.

Quantum yields of Trp residues have been determined in the wild type and in mutants where one Trp residue has been conserved (Table 2.8).

Table 2.8. Quantum Yield Q and λ_{max} of WT and different double mutants of Barnase at two pH

Protein	Q	λ_{max}
	pH 5.8	
WT	0.078 ± 0.006	334 nm
W35FW71Y	0.017 ± 0.001	345 nm
W35FW94Y	0.131 ± 0.006	330 nm
W71YW94F	0.137 ± 0.006	333 nm
	pH 8.9	
WT	0.133 ± 0.012	340 nm
W35FW71Y	0.072 ± 0.001	350 nm
W35FW94Y	0.122 ± 0.006	330 nm
W71YW94F	0.142 ± 0.005	335 nm

Source: De Beuckeleer, K., Volckaert, G. and Engelborghs, Y. 1999, Proteins: Structure, Functions and Genetics 36, 42-53. Authorization of reprint accorded by Wiley Interscience.

Data of Table 2.8 clearly show that the sum of the individual quantum yields is not equal to the quantum yield of the wild type, indicating the possibility of energy transfer between two or three Trp residues, and / or static quenching induced by the neighboring amino acids or functional groups. The authors found that energy transfer occurs mainly between tryptophans 94 and 71.

In the additive method, the quantum yield of a protein containing more than one tryptophan residue is calculated by averaging the quantum yields of the individual isolated tryptophan residues, taking their extinction coefficient into account:

$$Q_n = \frac{\sum_{1}^{n} \varepsilon_i Q_i}{\sum_{1}^{n} \varepsilon_i} \qquad (2.71)$$

In a recent work, Verheyden et al (2003) calculated the quantum yields of tryptophan residues in different plasminogen activator inhibitor 1 (PAI-1) mutants.

PAI-1 (Fig. 2.29) is an important physiological inhibitor of tissue-type and urokinase-type plasminogen activator (Loskutoff and Schleef, 1988). It is a 50 kDa glycoprotein with 379 amino acids belonging to the serpin superfamily. The protein plays an important role in the equilibrium between blood clot formation blood clot lysis. Plasminogen activator inhibitor-1 (PAI-1), a marker of atherothrombosis, is also elevated in the metabolic syndrome and in diabetes (Devaraj et al. 2003)

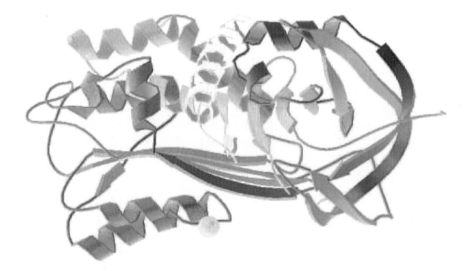

Figure 2.29. Tertiary structure of Plasminogen activator inhibitor-1.

PAI-1 possesses four tryptophans at positions 86, 139, 175 and 262. Single, double and triple mutants were constructed and the quantum yield of each mutant was measured experimentally by comparing its fluorescence with that of free tryptophan in buffer and was calculated ($Q_{i,\ calc}$) by subtraction from the quantum yield of the wild-type (Q_{wt}) and of the mutant protein where only the concerned tryptophan has been removed (-1W) ($Q_{(-i)}$).

To calculate the quantum yield of the concerned tryptophan, the calculated fluorescence intensity of the mutant protein is subtracted from the calculated intensity of the wild type and divided by its own extinction coefficient at 295 nm. Calculation of the quantum yield was performed with Equation 2.72 and the results are shown in Table 2.9.

$$Q_{i,\ calc} = \frac{Q_{wt} \times \sum_{j} \varepsilon_j - Q_{(-i)} \times \sum_{j \neq i} \varepsilon_j}{\varepsilon_i} \qquad (2.72)$$

Table 2.9. Experimental and calculated quantum Yields at 10°C of the reactive conformations of wt-PAI-1 and PAI-1 variants.

PAI-1 variants	Quantum yield	Calculated Q*
WT-PAI-1	0.085 ± 0.008	0.159 and 0.083$^{\lozenge}$
PAI-1-W86F	0.098 ± 0.007	0.185
PAI-1-W139F	0.126 ± 0.015	0.192
PAI-1-W175F	0.155 ± 0.009	0.182
PAI-1-W262F	0.060 ± 0.007	0.058
PAI-1-W86-175F	0.223 ± 0.021	0.225
PAI-1-W86-262F	0.043 ± 0.007	0.052
PAI-1-W139-175F	0.250 ± 0.021	0.233
PAI-1-W139-262F	0.092 ± 0.006	0.071
PAI-1-W86-139-175F	0.331 ± 0.033	0.127$^{\#}$
PAI-1-W86-139-262F	0.075 ± 0.009	-0.179$^{\#}$
PAI-1-W86-175-262F	0.029 ± 0.011	-0.078$^{\#}$
PAI-1-W139-175-262F	0.067 ± 0.007	-0.0383$^{\#}$
PAI-1-W86-139-175-262F	0.009 ± 0.002	-

(*) Calculated additively from data of the single tryptophans. ($\lozenge$) Calculated additively from data of single tryptophan residues W86, W139 and W175 + Q from W262 as present in wild type (=0.127). (#) Calculated substractively from data of wild type and −1W-mutants.
Source: Verheyden, S., Sillen, A., Gils, A., Declerck, P. J. and Engelborghs, Y. 2003. Biophysical Journal, 85, 501-510. Authorization of reprint accorded by the American Society of Biophysics.

Quantum yields of the tryptophan residues 86 and 262 are higher when they are single tryptophan residues than when they are in the wild-type protein (Table 2.9). This indicates that in the wild type, these tryptophans are loosing energy by energy transfer or conformational change. The quantum yields of tryptophan residues 139 and 175 are negative in the wild type. The authors explained this phenomenon by the fact that these two tryptophan residues are decreasing the quantum yield of the other two tryptophan residues.

The authors compared the calculated and the experimental quantum yields of the three tryptophan-containing proteins and they found that only the quantum yield of PAI-1-W262F was correctly calculated. In the other mutants where Trp 262 residue is present, the experimental quantum yield is smaller than the calculated one. The authors concluded that the presence of tryptophan residue 262 induces important quenching of the protein fluorescence compared to the sum of the single tryptophans. The authors attribute this quenching effect to an energy transfer phenomenon. However, when looking to the double mutant proteins, one can notice that calculated and experimental quantum yields are very close. For example, one can notice that for the PAI-1-W86-262F, PAI-1-W139-175F and PAI-1-W139-262F, the calculated quantum yields differ from the experimental quantum yields by 17.7 and 23%, respectively. The calculated and the experimental quantum yields of the PAI-1-W86-175F, where tryptophan 262 residue is present, are identical (0.225 and 0.223). These comparisons lead us to conclude that when mutants are prepared, structure of the wild type has been modified. The authors have already reported these structural modifications. Also, from a mutant to another, the structure and the protein flexibility could be altered differently, modifying by that all or some of the fluorescence parameters.

In order to measure quantum yields of an extrinsic fluorophore bound to a protein and which emits at longer wavelengths than in the U.V., standards such as 3,3'-diethylthiacarbocyanine iodide (DTC) in methanol ($\Phi_F = 0.048$) and rhodamine 101 in ethanol ($\Phi_F = 0.92$) or any other dyes can be used. For example, quantum yield of curcumin bound to HSA has been determined using the coumarin – 153 laser dye as a reference ($\Phi_F = 0.56$ in acetonitrile). Curcumin, (1,7-bis [4-hydroxy-3-methoxyphenyl]-1,6-heptadiene-3,5 dione), is a natural polyphenol found as a major pigment in the indian spice tumeric (Sharma, 1976). It shows important pharmacological activity, including anti-inflammatory, anticarcinogenic and antioxidant activity. See for example, Lin and Lin-Shiau, 2001; Metha et al. 1997). The complex (curcumin-HSA) and the reference dye were excited at 426 nm and the emission spectra were recorded from 450 to 600 nm. The quantum yield is calculated according to equation:

$$\frac{\Phi_S}{\Phi_R} = \frac{A_S}{A_R} \times \frac{(O.D.)_R}{(O.D.)_S} \times \frac{n^2_S}{n^2_R} \quad (2.73)$$

where, Φ_S and Φ_R are the fluorescence quantum yield of the sample and reference, respectively. A_S and A_R are the area under the fluorescence spectra of the sample and the reference, respectively, $(O.D.)_S$ and $(O.D.)_R$ are the respective optical densities of the sample and the reference solution at the wavelength of excitation, and n_S and n_R are the values of refractive index for the respective solvents used for the sample and the reference (Barik et al. 2003).

Dyes that intercalate DNA show a high quantum yield compared to free dyes in solution. For example, the quantum yield of fluorescence of thiazole orange dimer (TOTO) increases by a factor of 1400 when intercalated in double stranded DNA (dsDNA) (Rye et al. 1992).

2h. Fluorescence and light diffusion

There are two types of light diffusion: Rayleigh and Raman. Rayleigh diffusion occurs at wavelength equal to that of the excitation one. Thus, energy of the diffused photons is equal to the energy of the excitation photons.

Usually one should not start the emission spectrum before or at the excitation wavelength in order to avoid recording Rayleigh diffusion.

Raman diffusion is observed at a wavelength that is higher than the excitation wavelength. The diffused photons have quantum energy inferior to that of the excitation photons. Therefore, the intensity of the Raman peak is lower than that of the Rayleigh peak. Raman can perturb the emission spectrum and thus one should subtract the Raman from the emission spectrum.

In general, more the fluorescence intensity is important, less intense will be the Raman peak. In aqueous medium, the O-H bonds are responsible for the Raman spectrum. The position of the Raman peak is dependent on the excitation wavelength. It is possible to calculate the position of the Raman peak in aqueous medium with the following equation:

$$1 / \lambda_{ram} = 1 / \lambda_{ex} - 0.00034 \qquad (2.72)$$

The wavelengths are given in nanometers.

Chapter 3

FLUOROPHORES: DESCRIPTIONS AND PROPERTIES

1. Introduction

Fluorophores are small molecules that can be parts of a molecule or added to a molecule. Their fluorescence properties are interesting because they are modified with the medium. Although general laws or rules exist for the fluorescence dependence on the medium, each fluorophore has its own specific fluorescence properties. Also, a fluorophore can have one, two or more fluorescence lifetime depending on the surrounding medium. Also, in some cases, fluorescence of tryptophans in proteins is completely quenched.

Fluorophores are spread in nature and thus they can be used as natural indicators to study the structure, the dynamics and the metabolism of living cells.

2. Types of fluorophores

Fluorophores can be part of a molecule and in this case we call them intrinsic fluorophores, or they can be added to a molecule and they will be known as extrinsic fluorophores. We are going here to describe the properties of the most known fluorophores.

2a. Intrinsic fluorophores

2a1. Aromatic amino acids

In proteins, three amino acids are responsible of the absorption and fluorescence of the proteins in the U.V., tryptophan, tyrosine and phenylalanine (figures 1.7 and 3.1).

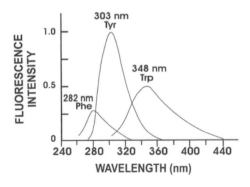

Figure 3.1. Fluorescence emission spectra of aromatic amino acids.

Tryptophan fluorescence is the most sensitive to local environment of the fluorophore. When the tryptophan is surrounded by a non-polar environment, the position of its fluorescence maximum is located at 320 nm. It shifts to 355 nm in the presence of a polar environment. This is why denaturation of a protein induces a shift of the tryptophan fluorescence to 355 nm (Figure 3.2).

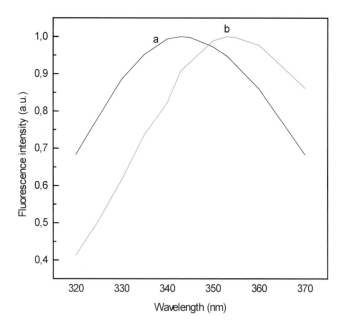

Figure 3.2. Fluorescence emission spectra flavodehydrogenase extracted from the yeast *Hansenula anomala* in phosphate buffer (λ_{em} = 443 nm) (a) and in 6 M guanidine pH 7 (λ_{em} = 453 nm) (b). λ_{ex} = 280 nm.

Also, upon binding of a ligand to a protein, Trp observables (intensity, polarization and lifetime) can be altered and one can follow this binding with Trp fluorescence.

In proteins, it is the tryptophan that is the highly fluorescent amino acid. In most of the cases, we can say that its fluorescence is dominating. The absence or the weak fluorescence of tyrosine and phenylalanine are the results of energy transfer to the tryptophan or / and to neighboring amino acids and also of proteins tertiary structure. However, if in many cases, denaturation of the proteins can induce the appearance of the tyrosine fluorescence, in revenge the fluorescence of phenylalanine is not observed. Since most of the proteins have Trp residues and since their fluorescence is very sensitive to the environment, it was normal that all the focus went to understand and to study tryptophan fluorescence.

E. A. Burstein and his collaborators (1973) classified the tryptophan in proteins into three categories, according to the position of their fluorescence maximum (λ_{max}) and the bandwidth ($\Delta\lambda$) of their spectrum:

Category 1 : λ_{max} : 330 to 332 nm and $\Delta\lambda$: 48 – 49 nm.
Category 2 : λ_{max} : 340 to 342 nm and $\Delta\lambda$: 53 – 55 nm.
Category 3 : λ_{max} : 350 to 353 nm and $\Delta\lambda$: 59 – 61 nm.

When a protein contains two classes of Trp residues, the resulting fluorescence emission spectrum will be the result of the contribution of each class. For example, the fluorescence emission spectrum of α_1-acid glycoprotein recorded with an excitation wavelength of 295 nm, shows a maximum located at 331 nm and a bandwidth of 53 nm (Figure 3.3). The spectrum is typical for protein containing Trp residues in hydrophobic and hydrophilic areas of the protein.

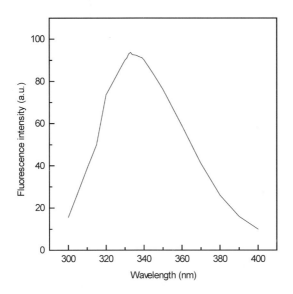

Figure 3.3. Fluorescence emission spectrum of α_1-acid glycoprotein recorded with an excitation wavelength of 295 nm. The maximum is located at 331 nm and the bandwidth of the spectrum is equal to 53 nm.

Increasing hydrostatic pressure can induce denaturation of a protein. This will lead to a red shift of the emission peak of the tryptophan fluorescence. Other spectral features such as a modification in the fluorescence intensity and / or an enlargement of the spectrum bandwidth could also be observed. Figure 3.4 displays the effect of hydrostatic pressure on the fluorescence spectrum of Trp residues of horse heart apomyoglobin. A red shift of the emission maximum and an increase in the total intensity are observed. The plot of the average emission wavelength as a function of pressure reveals a sigmoidal profile, suggesting a structural modification induced by the pressure (Fig. 3.5).

This profile is identical to that observed when pH is modified. Thus, the authors concluded that total apomyoglobin unfolding occurs via the formation of a molten globule intermediate.

Many authors consider that perturbation of the native state of a protein will induce the presence of different intermediate states between the native and the denatured forms of the protein. The most stable of these intermediate states is termed the I state and called the molten globule state.

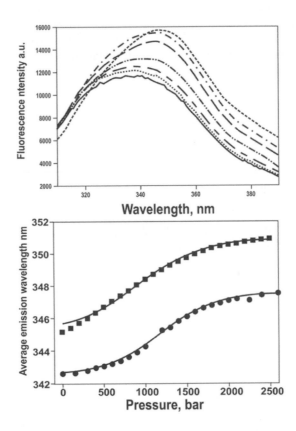

Figure 3.4 (upper panel). Dependence of the intrinsic fluorescence spectrum of native horse heart apomyoglobin at pH 6.0 in 10 mM bis-Tris buffer at 21°C on hydrostatic pressure. Hydrostatic pressure values: 0 bar (solid line); 150 bar (dotted line); 600 bar (small dashed line); 1000 Bar (dash-double dotted line); 1200 Bar (large dashed line); 1400 Bar (dash-single dotted line), and 2600 Bar (medium dashed line).

Figure 3.5 (lower panel). (◆), Average emission wavelength of native horse heart apomyoglobin at pH 6.0 in 10 mM bis-Tris buffer at 21°C as a function of pressure. (■), Average emission wavelength for the molten globule state of horse heart apomyoglobin at pH 4.2, 10 mM sodium acetate, 21°C. Solid lines represent fits to the data points. Source: Vidugiris, G. J. A. and Royer, C. A. 1998, Biophysical Journal 75, 463-470. Authorization of reprint accorded by the American Biophysical Society.

Athès et al. (1998) studied the effect of hydrostatic pressure and temperature on the fluorescence emission spectrum of *Kluyveromyces lactis* β- galactosidase (β-galactosidase galactohydrolase) which is used in food industry for the preparation of low-lactose dairy products for people suffering from lactose intolerance. The enzyme of molecular mass of 200000 contains up to 45% of its mass as carbohydrate. Figure 3.6 displays the fluorescence emission spectra of *Kluyveromyces lactis* β- galactosidase in the absence and the presence of two hydrostatic pressures. Native protein emits with a maximum equal to 332 nm. Increasing hydrostatic pressure leads to a red shift of the emission maximum accompanied with an intensity decrease. This red shift is the result of the unfolding of the protein. Intensity decrease indicates that interaction of the Trp residues with the neighboring amino acids is more important in the partially unfolded state than in the native one. In apomyoglobin (Figure 3.4), hydrostatic pressure induces an increase of the fluorescence intensity. Thus, interaction between Trp residues and the neighboring amino acids is less important in the molten state than in the native one. Figures 3.4 and 3.6 clearly indicate that unfolding and folding of proteins does not follow the same rules.

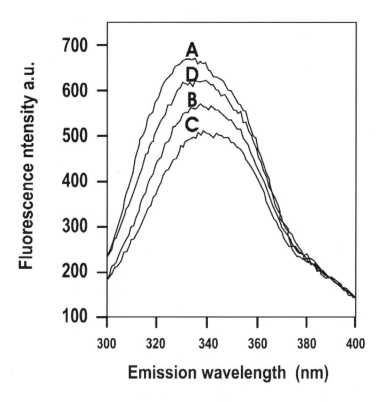

Figure 3.6. Intrinsic fluorescence emission spectra of native or pressure denatured β-galactosidase. Emission spectra of (A) 1.2 M native β galactosidase, (B) at 300 MPa, (C) at 400 MPa, (D) 1 h after release of pressure. Source: Athes, V., Lange, R. and Combes, D., 1998, Eur. J. Biochem. 255, 206-212. Authorization of reprint accorded by Blackwell Publishing.

Figure 3.7 displays effect of temperature on the fluorescence spectrum of Kluyveromyces *lactis* β- galactosidase. Up to the temperature range studied, an intensity decrease is observed. The emission peak position is not modified. This can be interpreted by the fact that the temperatures reached could not be sufficient to provoke an important molecular conformational change within the protein. The carbohydrate residues would play an essential role in protecting the protein structure from unfolding. Intensity decrease is the result of the high internal motions within the protein.

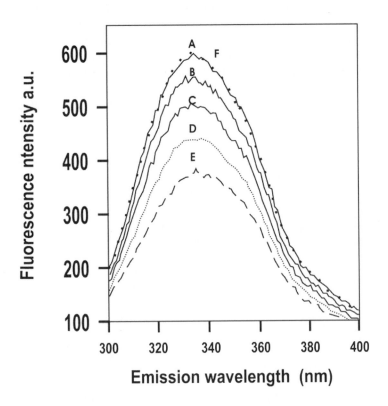

Figure 3.7. Emission spectra of 1.2 μM β-galactosidase recorded at 25°C (A, full line), 35°C (B), 45°C (C), 51°C (D), 60°C (E) and after cooling and storage overnight at 4°C (F, dotted spectrum). Source: Athes, V.,Lange, R. and Combes, D., 1998, Eur. J. Biochem. 255, 206-212. Authorization of reprint accorded by Blackwell Publishing.

Looking closely to the absorption spectra of the aromatic amino acids in figure 1.7, one can notice that the spectra overlap for many wavelengths. In proteins, when we excite at 280 nm for example, we are going to excite both Trp and tyrosine residues. If we want to observe the emission from Trp residues alone, excitation should occur at 295 nm and / or above. At these wavelengths, energy transfer from tyrosine to tryptophan does not take place and emission observed emanates from the tryptophan residues only.

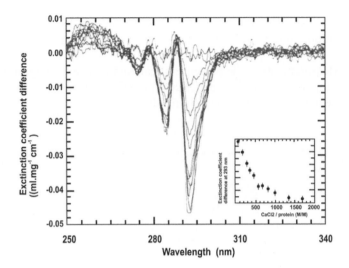

Figure 3.9. UV -difference absorption spectra of annexin V. Superposition of UV-difference absorption spectra of annexin V (1 mg/ mL) as a function of calcium concentration (from 0 to 0.046 M). Inset: titration at 293 nm resulting from the difference spectra.

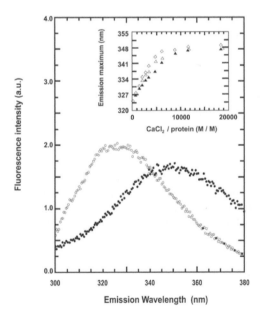

Figure 3.10. Fluorescence emission spectrum of Trp 187 in annexin V: (○) calcium-free protein and (●) calcium-bound protein. Jnset: variation of the fluorescence emission maximum as a function of the total calcium/protein mole ratio. Temperature: (Δ) 10°C, (▲) 20 °C, and (◊) 30°C. Excitation wavelength: 295 nm. Protein concentration: 0.4 mg/mL. Source : Sopkova, J., Gallay, J., Vincent, M., Pancoska, P. and Lewit-Bentley, A. 1994. Biochemistry 33, 4490-4499. Authorization of reprint accorded by the American Chemical Society.

At 20°C, the decay fluorescence parameters of the Trp-187 residue shows three lifetime populations: one major component with a center of gravity at about 800 ps (70%) and two minor components with center of gravity at 250 ± 50 ps and 1.8 ns, respectively. Measurements in the presence of calcium induce the appearance of a long lifetime (4 ns). Figure 3.11 displays the effect of binding of calcium to annexin V on the fluorescence intensity decay of the Trp187 residue of the protein.

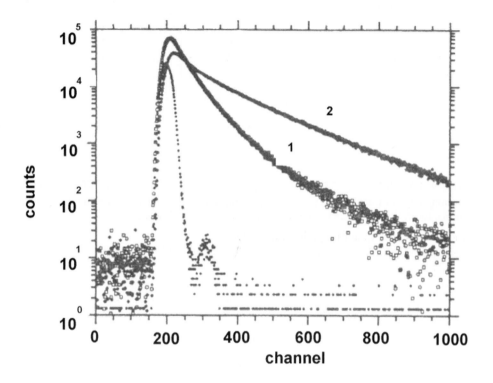

Figure 3.11. Fluorescence intensity decay of Trp 187 in annexin V: curve 1, in the absence of calcium; curve 2, in the presence of calcium ions (200 mM). Excitation wavelength: 295 nm (bandwidth, 5 nm). Emission wavelength: 320 nm (bandwidth 10 nm) in the absence of calcium and 350 nm (bandwidth, 10 nm) in its presence. Temperature: 20°C. Protein concentration: 0.4 mg/mL. Source: Sopkova, J., Gallay, J., Vincent, M., Pancoska, P. and Lewit-Bentley, A. 1994. Biochemistry 33 : 4490-4499. Authorization of reprint accorded by the American Chemical Society.

Also, the same authors found a restricted mobility for the Trp-residue in the calcium-free protein and a high mobility in the presence of calcium.

All these data show how sensitive could be the spectral parameters of Trp residue in proteins. These parameters will not depend on the internal properties of the indole itself but mainly on the ability of the protein to change its conformation whether on the local or / and global levels. Trp residue is used as a probe to follow the structural and dynamical modification occurring within the protein.

Tyrosine residues can also be used to describe the structure and the dynamics of a protein. Single stranded DNA binding protein modulates DNA replication and repair, transcription, translation, regulation and recombination. However, while most of SSB proteins activate DNA replication, *e.g. Escherichia coli* SSB and Φ29 SSB, some SSBs such as those of filamentous phages M13, are inhibitory.

In vivo, elongation process of DNA replication is dependent on the presence of a virally-encoded SSB protein, Φ29 SSB. Φ29 SSB binds to ssDNA with relatively low affinity (Ka = 10^5 M^{-1}) covering 3-4 nt per monomer, probably as a single array of protein units along the DNA.

Φ29 SSB contains 3 tyrosine residues at positions 50, 57, and 76 and lacks of tryptophan. Fluorescence spectra of Φ29 SSB free in solution and complexed to ssDNA, obtained at pH 7.5, are displayed in Fig. 3.12 ($\lambda_{ex,}$ 276 nm and $\lambda_{em,}$ 308 nm). One can notice that both excitation and emission spectra intensities decrease in presence of ssDNA. The absence of spectra overlap between the excitation and emission indicates that energy transfer between tyrosine residues in Φ29 SSB is almost non-existent.

We can notice also that there is no peak at 345 nm in the fluorescence emission spectrum, revealing that formation of tyronisate is not occurring within the protein. Therefore, interaction of tyrosine residues with the surrounding amino acids is weak. Interaction with the side-chains of the neighboring residues contributes to the formation of hydrogen bond with the tyrosine. This interaction, when it exists, induces a decrease in the fluorescence quantum yield of the tyrosines.

Quantum yield of the tyrosine in Φ29 SSB is obtained by comparing the steady-state fluorescence spectrum to that of aqueous solutions of 5-methoxy-indole (5 MeOI) of known quantum yield (Φ_{ref} = 0.28) (Hersberger and Lumry, 1976). Φ_F of Φ29 SSB = 0.065. This Φ_F value is unusually high for proteins that lack tryptophan in their primary sequence such as insulin for example. Since Φ_F of Φ29 SSB is approximately 70% of that of *N*-acetyl-L-tyrosine amide, the authors concluded that Φ29 SSB tyrosines do not appear to maintain strong interactions with other residues of the protein

The fluorescence intensity decay of Φ29 SSB can be analyzed with three exponential decays 0.4, 1.5 and 3.8 ns. We can in no way assign the fluorescence lifetimes to any particular tyrosine residues.

The fluorescence quantum yield of the protein can be estimated from the ratio τ_o / τ_R, where τ_o is the mean fluorescence lifetime equal to 1.8 ns and τ_R is the radiative lifetime of tyrosine equal to 2.7 ns (Lux et al. 1977; Wu et al. 1994). Φ_F = 0.067. The fact that steady-state and lifetime measurements yield the same value of Φ_F indicates that the three tyrosines of Φ29 SSB contribute to its intrinsic fluorescence. Therefore, ground-state deexcitation of the tyrosine fluorescence by the surrounding environment is weak.

Ionization of Φ29 SSB tyrosines, followed by the fluorescence quenching at 308 nm as the function of pH, is described by a global pattern characteristic of that of tyrosine residue (Fig. 3.13). Although the global patterns of free tyrosine and Φ29 SSB are close, we observe a difference between them. The 50% decrease of the quantum yield occurs at pH 10.2 and 10.9 for free tyrosine and Φ29 SSB, respectively. This difference results from the fact that in Φ29 SSB the tyrosine residues are in contact with other amino acids.

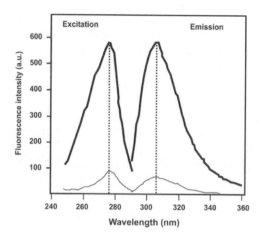

Figure 3.12. Corrected fluorescence spectra of Φ29 SSB. Data were obtained at a final concentration of 8 μM in buffer B at 25 °C in the absence (*thick curves*) and presence (*thin curves*) of saturating amounts of ssDNA (300 μM poly(dT). The excitation (λ_{em} = 308 nm) and emission spectra (λ_{exc} = 276 nm) were corrected for background emission, inner filter effect of DNA absorption and wavelength dependence of the fluorimeter response. Source: Soengas, M. S., Reyes Mateo, C., Salas, M., Ulises Acuña, A. and Gutiérrez, C. 1997, J. Biol. Chem. 272, 295-302. Authorization of reprint accorded by the American Society for Biochemistry and Molecular Biology.

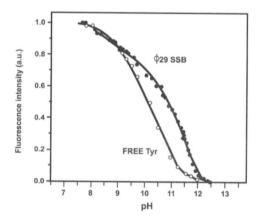

Figure 3.13. pH-dependent ionization of the phenol hydroxyl group of Φ29 SSB tyrosines. Fluorescence intensity as a function of progressive pH increase, recorded at λ_{em} = 308 nm, of a solution of Φ29 SSB (6 μM, *closed symbols*) compared with a solution of free tyrosine (5 μM, *open symbols*). A 50% reduction of the initial fluorescence was observed at pH 10.3 for free tyrosine and 10.9 for the tyrosyl residues of Φ29 SSB. Source: Soengas, M. S., Reyes Mateo, C., Salas, M., Ulises Acuña, A. and Gutiérrez, C. 1997, J. Biol. Chem. 272, 295-302.

2a2. The Co-factors

NADH is highly fluorescent with absorption and emission maxima located at 340 and 450 nm, respectively (Fig. 3.14) while NAD^+ is not fluorescent.

Fluorescence lifetime of NADH in an aqueous buffer is around 0.4 ns. Once bound to a protein, fluorescence quantum yield of NADH increases about four times.

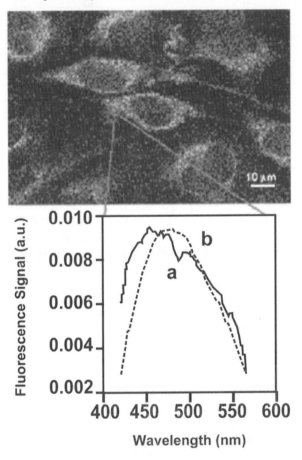

Figure 3.14. Intracellular autofluorescence is often dominated by the reduced pyridine nucleotides (NAD(P)H) and the oxidized flavins (FMN, FAD), both of which are potentially useful as cellular metabolic indicators. Mitochondrial NADH autofluorescence can be directly used as an indicator of cellular respiration. Since only the reduced form has an appreciable fluorescence yield, hypoxia, which causes an increase in the $NADH/NAD^+$ ratio, can be detected as an increase in mitochondrial autofluorescence. The figure is an autofluorescence image of HeLa cells (λ_{ex} 700 nm, average power was 10 mW at sample). The spectrum (a) was acquired by parking the excitation beam on a cell and collecting the emission through a fiberoptic coupled, LN 2 cooled CCD spectrograph. The spectrum of NADH (b) is shown for comparison. The blue shift of the autofluorescence peak is an indication of enzyme-bound NADH within the cell. Source: Piston, D.W., Masters, B. R. and Webb, W.W. 1995, Journal of Microscopy 178, 20-27. Authorization of reprint accorded by Blackwell Publishing.

FMN and FAD absorb light in the visible at 450 nm and fluoresce at around 515 nm (Fig. 3.15). Fluorescence lifetime of FMN and FAD are 4.7 and 2.3 ns, respectively.

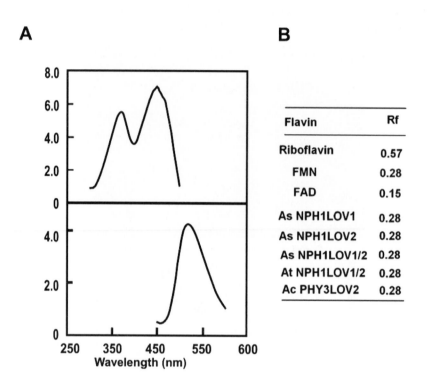

Figure 3.15. Spectrofluorometric and TLC analysis of the chromophore associated with each of the LOV (light, oxygen, or voltage) domain fusion proteins. (A) A typical fluorescence excitation spectrum (Upper) and fluorescence emission spectrum (Lower) of the chromophore released from each of the CBP fusion proteins. (B) Identification of the chromophore as FMN by TLC. The mobility of the chromophore released from each of the CBP-LOV fusions, relative to the solvent front (Rf), is indicated. Rf values for riboflavin, FAD, and FMN standards are also shown. Source: Christie, J. M., Salomon, M., Nozue, K., Wada, M. and Briggs, W. R. Proc. Natl. Acad. Scien. U.S.A. 1999, 96, 8779-8783. Authorization of reprint accorded by the National Academy of Sciences (U. S. A.)

Flavins are found in nature in many biological samples. For example, trace amounts of flavins from different types of biological samples were detected using sensitive capillary electrophoresis (CE) with laser-induced fluorescence (LIF). For example, in pooled human plasma, 8.4 ± 0.2 of FMN and 44 ± 6.8 nM of FAD were found (Britz-McKibbin et al. 2003).

2b. Extrinsinc fluorophores

2b1. Fluorescein and rhodamine

They bind covalently to the lysines and cysteines of proteins.
They absorb and fluoresce in the visible (Fig. 3.16 and 3.17).

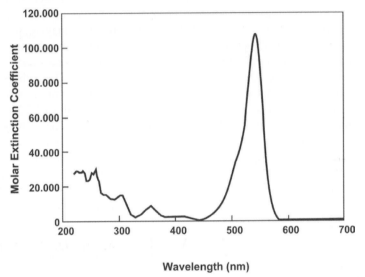

Figure 3.16. Graph of the molar extinction coefficient of Rhodamine B dissolved in ethanol. It was measured by R. A. Fuh on 6/20/95. Rhodamine B has a molar extinction coefficient of 106,000 $M^{-1}cm^{-1}$ at 542.75 nm [anonymous, "Eastman Laboratory Chemicals Catalog No," 1993].

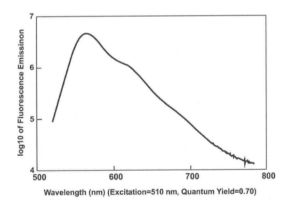

Figure 3.17. Fluorescence emission spectrum of Rhodamine B dissolved in ethanol. The spectrum was taken by R. A. Fuh on 6/20/95 using an excitation wavelength of 510 nm. See: Du, H., Fuh, R.A., Corkan, Li, A., Lindsey, J. S., 1998. Photochem. Photobiol. 68, 141-142. Curtosey from Oregon Medical Laser Center.

Fluorescence lifetime of fluorescein and rhodamine is around 4 ns and their emission spectra are not sensitive to the polarity of the medium.

Quantum yield of Rhodamine in ethanol is 0.70 (LopezÝArbeloa, F., RuizÝOjeda, P. and LopezÝArbeloa, I (1989).

Absorption and fluorescence of fluorescein are sensitive to the pH.

Fluorescein and other fluorophores such as Nile red and acridine orange can be incorporated into micellar system of poly(oxyethylene)isooctyl phenyl ether, Triton X-100 (TX-100). Experiments with these fluorophores should be performed at low concentrations (equal to or less than 10 μM), otherwise the dyes may aggregate and / or self-quenching effects would occur. Energy transfer studies have shown that energy transfer from fluorescein to Nile red is less efficient than the one observed between acridine orange and Nile red (De, S. and Girigoswami, A. 2004). The authors found that compared to the experiments carried out in neat water, the donor–acceptor distance decreases in TX-100 micelles. Thus, they concluded that micelles of TX-100 help in "anchoring" the donor and acceptor dyes, facilitating energy transfer between them. They also found that the distance separating both donor and acceptor is smaller for the acridine orange – Nile red couple (44 Å) than the fluorescein - Nile red one (50 Å). The authors explained this result by the structural differences that exist between fluorescein and acridine orange. Hydrophobic organic dyes such as Nile red (acceptor) are readily incorporated within micellar assemblies. Both acridine orange and fluorescein contain hydrophobic moieties and are also charged. However, the dianionic state of fluorescein leads it to reside closer to the bulk water phase in the micellar periphery region. In other terms, acridine orange will penetrate deeper into the micelle than fluorescein, will have less restricted motions than fluorescein and will be closer to Nile red than fluorescein. All this will induce a more efficient energy transfer between Acridine orange and Nile red.

2b2. Naphthalene sulfonate

IAEDANS and DNS bind covalently to proteins and are used to follow their conformational changes.

Other two probes, l-anilino-8-naphthalene sulfonate (ANS) and 2-p toluidinylnaphthalene-6-sulfonate (TNS):

- They bind non-covalently to proteins and membranes.

- Dissolved in a polar medium such as water, these probes display a very weak fluorescence that increases with the decrease of the polarity of the medium. When bound to proteins or membranes, fluorescence of ANS and TNS increases and their maximum shifts to the blue edge. The increase in the intensity and the shift to the shorter wavelength are more and more important with the decrease of the polarity of the binding site.

- ANS binding to proteins could occur through electrostatic interactions between the negative charge of ANS sulfonic group and the cationic groups present on the protein (Matulis and Lovrien, 1998). Upon ANS binding, proteins such as bovin serum albumin undergo external molecular compactness. The type of binding does not induce the fluorescence of the probe such as when ANS binds to a hydrophobic site of the protein (Matulis, et al. 1999). See Table 3.2 for a description of the properties of some fluorophores.

Table 3.2. Absorption and spectral properties of some used fluorophores

Fluorophore	Binding site	Absorption		Emission		
		λ_{max} (nm)	ε_{max} (x 10^{-3})	λ_{max} (nm)	Φ_f	τ_f (ns)
Dansyl chloride	Covalent bond on Cys and Lys.	330	3.4	510	0.1	13
1,5-I-AEDANS	Covalent bond on Lys and Cys.	360	6.8	480	0.5	15
Fluorescein isothiocyanate (FITC)	Covalent bond on lysines.	495	42	516	0.3	4
8-Anilino-1-naphthalene sulfonate (ANS)	non-covalent bond on proteins	374	6.8	454	0.98	16
Pyrene and derivatives	membranes	342	40	383	0.25	100
Ethidium bromide	non-covalent bond on nucleic acids.	515	3.8	600	1	26.5

Reproduced from (Cantor, R. C. and Schimmel, P. R. Biophysical Chemistry, 1980, W.H. Freeman and Company Editions, New York). Fluorescence lifetimes are mean ones.

IAEDANS can be used to study the mechanism of protein unfolding. Acyl-CoA binding protein (ACBP), an 86-residue α-helical protein is commonly used in folding studies because of its regular four-helix bundle structure (Fig. 3.18a). Although some studies have reported that folding of ACBP can be described by a two-state mechanism, other works have indicated the possibility of having a rapid formation of an ensemble of marginally stable intermediate states with optical properties close to the unfolded state. To introduce a specific fluorescence label, the I86C variant of ACBP, which contains a single cysteine residue at the C terminus, was derivatized with IAEDANS, a thiol-specific dansyl derivative. Absorption spectrum of the dansyl group overlaps the tryptophan emission spectrum from 300 to 400 nm (Fig. 3.18b), and energy transfer from the tryptophans to the dansyl group can occur. Fluorescence of IAEDANS in the unfolding state is much more important than that observed in the native one. A subsequent decrease in dansyl fluorescence in the native state is attributed to intramolecular quenching of donor fluorescence on formation of the native state.

Combination of fluorescence resonance energy transfer with stopped flow experiments helped to resolve submillisecond events in the folding of ACBP. This approach has allowed the characterization of the kinetics and stability of a transient intermediate populated on the 100-μs time scale. Thus, intermediate states between native and totally denatured do exists in ACBP as it is observed for other proteins such as apomyoglobin.

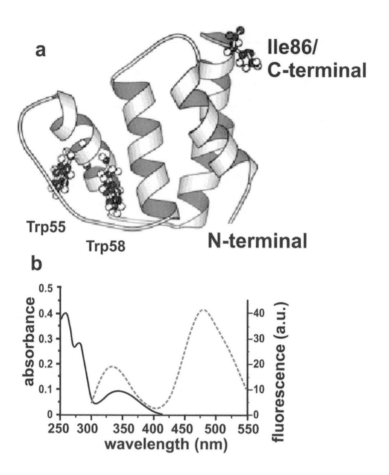

Figure 3.18. (*a*) Ribbon diagram of acyl-CoA binding protein ACBP (Kraulis, P. J. 1991, J. Appl. Cryst. 24, 946-950), based on the NMR structure coordinates of Andersen and Poulsen (Andersen, K. V. and Poulsen, F. M;, 1993, J. Biomol. NMR 3, 271-284.) The two tryptophan residues and the mutated C-terminal isoleucine are shown in ball and stick. (*b*) Absorbance spectrum (solid line) and fluorescence emission spectrum with excitation at 280 nm (dashed line) of 15 μM ACBP,I86C labeled with IAEDANS (in 20 mM Na-acetate, pH 5.3).
Source: Teilum, K., Maki, K., Kragelund, B. B., Poulsen, F. M. And Roder, H. 2002, Proc. Natl. acad. Sci. U. S. A. 99, 9807-9812. Authorization of reprint accorded by the National Academy of Sciences (U. S. A.)

The following example describes ANS binding to proteins. Apoglycoprotein A-I (apo-I), the major protein component of plasma high-density glycoprotein (HDL), activates the enzyme lecithin: cholesterol acyltransferase. It helps removing cholesterol from peripheral studies.

Figure 3.19 displays the fluorescence emission spectra of ANS bound to bovine serum albumin (BSA) (A), human apo-I (B), chicken apo-I (C) and carbonic anhydrase (D) (λ_{ex}, 295 nm). One can notice that the presence of many hydrophobic sites on BSA facilitates ANS binding to the protein. This binding is characterized by an important increase in the fluorescence intensity of the fluorophore compared to the weak fluorescence observed in water and by the shift in the fluorescence emission maximum from 515 to 480 nm.

Binding of ANS to apo A-I induces a shift of 35 nm of the fluorescence maximum accompanied with an increase in the fluorescence intensity. Since the observed intensity increase is smaller than the one observed in presence of BSA, affinity of ANS to the apo-I could be lower than that to BSA and / or there are less binding sites for ANS on the apo-I than on BSA.

In presence of carbonic anhydrase, fluorescence of ANS is not important revealing a very weak interaction between the fluorophore and the protein.

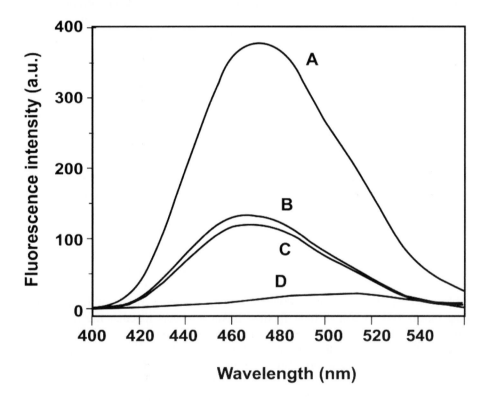

Figure 3.19. ANS fluorescence spectra were obtained in the presence of 50 *μ*g/mL bovine serum albumin (A), human apoA-I (B), chicken apoA-I (C), and carbonic anhydrase (D). Protein samples were excited at 395 nm and the emission spectra from 400 to 560 nm monitored (excitation and emission slit widths set at 5 nm). Spectra were recorded in 10 mM sodium phosphate (pH 7.5). Source: Kiss, R. S., Kay, C. M. and Ryan, R. O. 1999, Biochemistry, 38, 4327-4334. Authorization of reprint accorded by the American Chemical Society.

Let us give an example of the application of ANS fluorescence in following structural modification within a protein. We are going to describe briefly the effect of calcium binding on calmodulin on the fluorescence spectra of ANS complexed to calmodulin and of the phenylalanine residues of the protein.

Calmodulin is a Ca^{2+}-binding protein implicated in a wide array of regulatory functions in all eukaryotic systems. Because of its ubiquity and the large number of proteins with which it interacts and whose functions it modulates, CaM has fascinated scientists since its discovery in the 1970's. The protein has been the subject of countless studies, yielding a wealth of information regarding its structure and dynamics, its Ca^{2+}-binding properties, its target protein-recognition properties, and its ability to regulate many proteins *in vitro*. The fact that CaM is now included in virtually every undergraduate biochemistry textbook is a testament to its overall importance.

The sheer number of cellular processes in which CaM is implicated, and their complexity and inter-dependence make *in vivo* study of CaM challenging. The discovery of calmodulin in the budding yeast, *Saccharomyces cerevisiae,* in 1986 afforded the first opportunity to study the roles of CaM *in vivo. S. cerevisiae* contains a single gene that encodes for CaM (CMD1) whose presence is absolutely required for cell growth (Davis et al., 1986). Deletion or disruption of the CMD1 gene is lethal, implying that CaM performs at least one essential function in yeast. The lethal phenotype can be rescued by plasmid-encoded copies of yeast CaM (yCaM), allowing the use of yeast genetics to identify the essential function(s) of CaM in the organism. Over the past decade, such studies have yielded unexpected results that have dramatically altered the general view of CaM's mechanisms of action. For example, studies in yeast were the first to demonstrate that Ca^{2+}-free CaM plays essential roles in several key cellular processes. This result was initially viewed with much skepticism, as it seemed to contradict the very *definition* of CaM as the protein responsible for responding to changes in intracellular Ca^{2+} concentration. The unanticipated result in yeast does not imply that CaM has no Ca^{2+}-dependent functions in yeast (indeed, several Ca^{2+}-dependent targets have been identified), but rather that these functions are not essential for viability. Ca^{2+}-free and Ca^{2+}-bound forms of CaM are *both* active and perform distinct functions *in vivo*. Therefore, a full understanding of the mechanisms by which CaM regulates cellular functions requires detailed knowledge of all active states of the protein and their interactions with target proteins.

vCaM consists of two globular domains, each containing a pair of EF-hand Ca^{2+}-binding motifs (Fig. 3.20). A long flexible linker connects together the two globular domains. (Barbato, et al., 1992.) Each EF-hand motif consists of a helix-loop-helix structure, with the residues directly involved in binding the Ca^{2+} ion located within the loop region. The helices and loops are preformed in the "empty" state of CaM and the Ca^{2+}-induced conformational change involves a rearrangement of the helices in each globular domain relative to one another. Each domain of the apo-form resembles a four-helix bundle, with all helices roughly antiparallel ("closed" state). The interhelical angles change significantly in the Ca^{2+}-bound state, with certain helical pairs becoming almost perpendicular to each other ("open" state). This rearrangement exposes a large hydrophobic surface in each domain. The structures of target peptide/Ca^{2+}-vCaM complexes confirm that the newly exposed hydrophobic surfaces form the intermolecular contact surface.

Vertebrate calmodulin (vCaM)

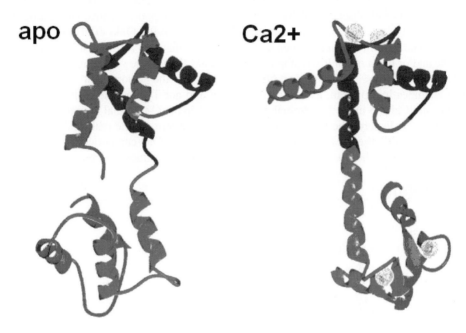

Figure 3.20. Structure of apo-vCaM, the "closed" form (PDB lCFD) and Ca^{2+}-vCaM, the "open" form. (PDB lCLL). Curtosey from Professor Rachel Klevit, University of Washington Seattle, Dept. of Biochemistry.

These structures offer many insights into the interactions responsible for CaM's Ca^{2+}-dependent functions. They are, however, only one half of the story -- similar detailed information is required for interactions responsible for CaM's Ca^{2+}-free functions.

Solution structures of Ca^{2+}-free vCaM have been determined for the completely empty state of the protein. However, the intracellular milieu contains millimolar concentrations of Mg^{2+}, and CaM binds Mg^{2+} ions in the absence of Ca^{2+}. Therefore, the form of CaM likely to be present *in vivo* under conditions of low [Ca^{2+}] is the Mg^{2+}-bound protein.

Given the 60% sequence identity between yCaM and vCaM, it is reasonable to ask why one cannot merely draw on the vast wealth of information known about vCaM to understand the structure/function relationships of yCaM. Several years ago, Klevit and collaborators started collecting NMR spectra of yCaM with the expectation of confirming its close structural similarities to vCaM. This expectation held true for the empty state of yCaM but NMR spectra of Ca^{2+}-yCaM revealed a completely unexpected result (Lee, S. Y. 2000). As reported in Lee & Klevit (2000), the two globular domains, which behave as independent entities in vCaM, interact with each other in Ca^{2+}-yCaM to form a species that the authors called the "collapsed state." The inter-domain interaction detected by NMR experiments offers structural insight for an earlier surprising result of binding cooperativity among all three Ca^{2+} ions to yCaM (the fourth EF-hand does not bind Ca^{2+}).

Although direct evidence for inter-domain interactions have not been reported for vCaM, a number of observations have provided indirect support for this, suggesting that vCaM may also assume a collapsed state (Zhang, et al. 1995, Ohki et al. 1997, Shea et al.1996). The inability to detect such a state indicates that this form must have a shorter lifetime and/or has a much lower population in vCaM than in the yeast counterpart.

yCaM offers the unusual opportunity to study the fluorescent properties of phenylalanines in a protein, as it contains neither Trp nor Tyr residues. There are 8 Phe residues in yCaM,:5 in the N-terminal domain and 3 in the C-terminal domain.

Figure 3.21 displays the fluorescence emission spectrum of Phe residues of yCaM in the absence and in the presence of 0.1, 1 and 10 mM Ca^+. One can notice that binding of calcium quenching the fluorescence intensity of the intrinsic fluorophore.

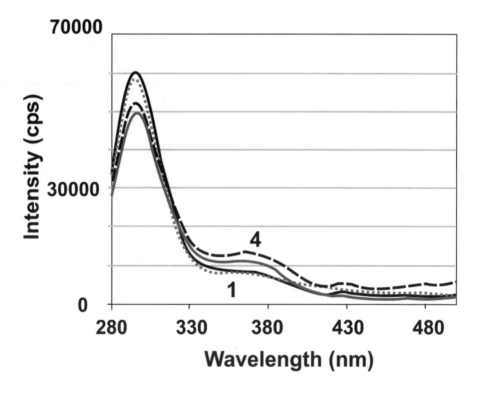

Figure 3.21. Fluorescence emission spectra of Phe from yCaM (6 µM) in absence (spectrum 1) and presence of different concentration of calcium. Spectra 2 to 4 are obtained in presence of 0.1, 1 and 10 mM Ca, respectively. λ_{ex} = 257 nm. Curtosey from Professor Rachel Klevit, University of Washington. Seattle, Dept. of Biochemistry.

Figure 3.22. shows the same experiment in presence of different concentrations of Mg^{2+}. One can notice that binding of Mg^{2+} to the protein has no effect on the fluorescence emission intensity of the Phe residues. Addition of calcium ion to the yCaM-Mg complex induces a decrease in the fluorescence intensity of the protein. Therefore, the presence of Mg on the protein does not inhibit binding of calcium.

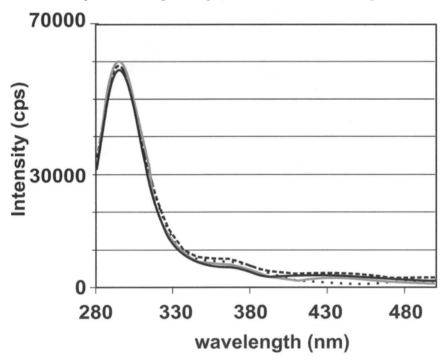

Figure 3.22. Fluorescence emission spectra of Phe from apo yCaM (6 μM) in absence and presence of different concentration of magnesium (0.1 to 10 mM) $\lambda_{ex} = 257$ nm. One can notice that the four spectra are overlapping. Curtosey from Professor Rachel Klevit, University of Washington. Seattle, Dept. of Biochemistry.

Fluorescence of ANS bound to yCaM is also sensitive to calcium binding to the protein. In fact, in presence of protein, the emission peak of ANS is shifted to the blue and its fluorescence emission increases (Fig. 3.23). Addition of calcium yields an important increase in the fluorescence emission of ANS accompanied by a blue shift. This means that binding of calcium on yCaM induces a structural modification of the protein. The environment of ANS in the yCaM-Ca$^+$complex is more shielded from the solvent than in the absence of calcium.

Binding of Mg^{2+} on yCaM does not change drastically the fluorescence emission of ANS. However, upon addition calcium ion to the yCaM-Mg complex, an important increase in the fluorescence of ANS in observed accompanied with a shift toward the shortest wavelengths (Fig. 3.24). This result is in good agreement with that observed with the Phe residues of the protein.

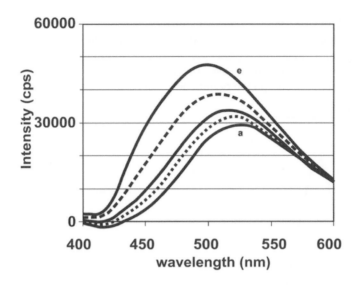

Figure 3.23. Fluorescence emission spectra of 4 μM ANS free in solution (a), bound to 6 μM yCaM in absence of calcium (b) and bound to 6 μM yCaM in presence of 0.1 (c), 1 (d) and 10 mM calcium (e). λ_{ex} = 360 nm. Curtosey from Professor Rachel Klevit, Dept. of Biochemistry. University of Washington. Seattle,

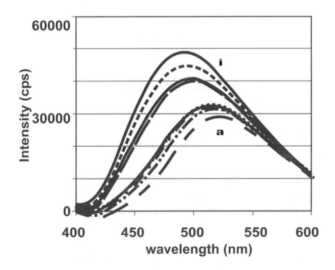

Figure 3.24. Fluorescence emission spectra of 4 μM ANS free in solution (a), bound to 6 μM yCaM in absence of calcium or magnesium (b), bound to 6 μM yCaM in presence of 0.1 (c), 1 (d) and 10 mM magnesium (e) and bound to 6 μM yCaM in presence of 0.1 mM Ca-10 mM Mg (f), 1 mM Ca-10 mM Mg (g), 10 mM Ca-10 mM Mg (h) and 20 mM Ca-10 mM Mg (i), λ_{ex} = 360 nm. Curtosey from Professor Rachel Klevit, University of Washington, Seattle.

2b3. Nucleic bases

Puric and pyrymidic bases are fluorescent.

Their fluorescence is weak in aqueous medium but increases by a factor of 10 in a medium of pH 2 and by a factor of 100 at 77 K.

The quantum yield of the nucleic acids depends largely on the temperature.

Their fluorescence lifetime is weak.

The fluorescence polarization is high even at ambient temperature.

At low pH, the emission maximum depends strongly on the temperature.

Puric bases fluoresce at least three times more than pyrimidic bases. The guanidine is the most fluorescent basis.

DNA and RNA display in the native state a very weak fluorescence, to the difference of the transfer RNA of yeast (t-RNA) that contains a highly fluorescent basis called Y. basis.

At 77 K, fluorescence of DNA and RNA increases. This fluorescence is owed to the formation of dimers of adenine and thymine, while the fluorescence of the guanine and the cytosine is inhibited.

Ethidium bromide, acridine and Hoechst 33258 intercalate between DNA and RNA Bases inducing by that a fluorescence increase.

Ethidium bromide emits red fluorescence when bound to DNA. The fluorescent probe is trapped in the base pair of DNA (Fig.3.25) inducing an increase in its fluorescence intensity. Also, ethidium bromide can bind to double-stranded and single-stranded DNA and by RNA.

Hoechst 33258 binds to the minor groove of double stranded DNA with a preference for A-T sequence (Pjura, 1987).

Single brand DNA does not induce an important increase of the fluorescence of the fluorophore.

RNA interferes very little with the DNA-Hoechst complex.

It is possible to study migration of DNA on gel by staining the gel with ethidium bromide solution.

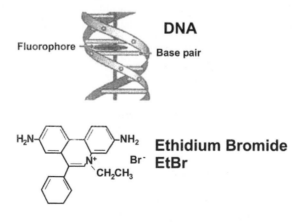

Figure 3.25. Binding of ethidium bromide to DNA.

Interaction between DNA and proteins induces very often, structural modifications in both interacting molecules. Characterization of such modifications in DNA can be obtained with 2-aminopurine (2 AP) that is a highly fluorescent isomer of adenine. 2AP does not alter DNA structure, forms a base pair with thymine, can be selectively excited since its absorption is red-shifted compared to that of nucleic acids and aromatic amino acids and its fluorescence is sensitive to the conformational change that occurs within the DNA (Rachofsky et al. 2001).

2b4. Ions detectors

Some fluorophores are used to detect the presence of the calcium in cells. For example, the maximum of the fluorescence excitation spectrum of Fura-2 shifts toward short wevelengths in presence of calcium accompanied with an increase of the intensity (Fig. 3.26).

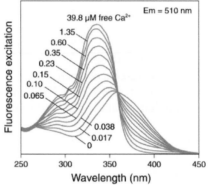

Figure 3.26. Fluorescence excitation spectra of Fura-2 at different concentrations of calcium. Source: Molecular Probes.

On the other hand, the maximum of the fluorescence emission spectrum of indo-1 shifts from 490 to 405 nm in presence of calcium. This shift is accompanied by an increase of the fluorescence intensity with the increase of the calcium concentration (Figure 3.27).

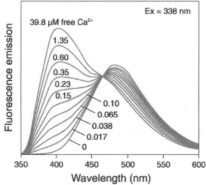

Figure 3.27. Fluorescence emission spectra of indo-1 at different concentrations of calcium. Source: Molecular Probes.

Another fluorophore whose fluorescence is sensitive to calcium binding is 2-[(2- bis [carboxymethyl] amino - 5 - methylphenoxy) - 6 - methoxy-8-bis [carboxymethyl] aminoquinole, known also as Quin2, is a calcium sensitive probe. Its fluorescence is modified upon complexation with calcium or / and upon binding to proteins such as human serum albumin.

Figure 3.28 displays fluorescence emission spectra of Quin-2 in presence of increased concentrations of human serum albumin. We can notice that the fluorescence intensity of the probe increases accompanied with a shift of the emission maximum to short wavelengths. The fluorescence intensity increase indicates that the probe binds to human serum albumin and the blue shift reveals the hydrophobic nature of the binding site compared to free fluorophore in solution.

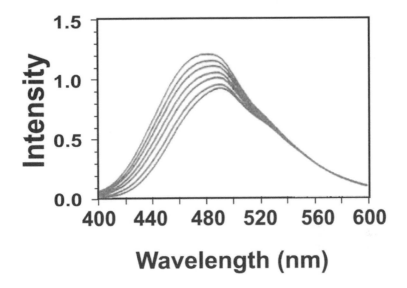

Figure 3.28. Effect of human serum albumin on the steady-state emission spectra of Quin2. Emission spectra were taken for Quin2 in the presence of 17 mM EGT A at protein concentrations ranging from 0-4.4 µM. The concentration of Quin2 was 34 µM. All data were obtained at 20°C. The buffer contained 10 mM TRIS, pH7.2, 0.1 M KC1, and 5 mM EGT A. The excitation wavelength was 325 nm. Source: Hirshfield, K. M, Toptygin, D., Grandhige, G., Kim, H., Packard., B. Z. and Brand, L. Biophys. Chem. 1996, 62, 25-38.

Figure 3.29 shows the fluorescence emission spectra of 10 µM Quin-2 at different concentrations of calcium (0 to 40 µM) in the absence (A) and in the presence (B) of 15.3 µM human serum albumin. From the data of figure 3.29, the authors generate steady-state emission spectra of the protein (Fig. 3.30a), Quin-2 (Fig. 3.30b), the Quin-2-calcium complex (Fig. 3.30c) and the Quin-2 – protein complex (Fig. 3.30d). One can notice that Quin-2 fluorescence emission spectrum in presence of human serum albumin is blue shifted compared to Quin-2 free in solution. The Quin-2-calcium complex is however red-shifted.

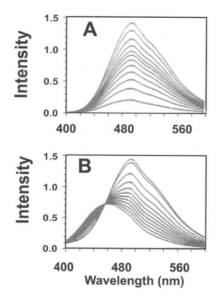

Figure 3.29. Effect of protein-probe interactions on the calcium-binding capacity of Quin2. Panel (A) shows the steady-state emission spectra obtained on a titration of Quin2 with calcium in the absence of human serum albumin. Panel (B) was generated from a titration of Quin2 with calcium in the presence of 15.3 µM HSA. The concentration of calcium ranged from 0-40 µM. The concentration of Quin2 was 10 µM. The titration was performed at 20°C. Source: Hirshfield, K. M, Toptygin, D., Grandhige, G., Kim, H., Packard., B. Z. and Brand, L. Biophys. Chem. 1996, 62, 25-38.

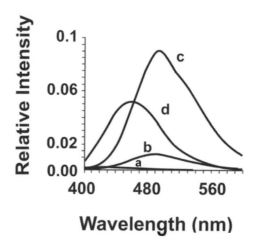

Figure 3.30. Theoretical steady-state emission spectra of HSA(a). Quin2(b). the Quin2-calcium complexee), and the Quin2-HSA complex(d). These spectra were generated from figure 3.25. The intensities represent intensities per unit concentrations of the species given. Source: Hirshfield, K. M, Toptygin, D., Grandhige, G., Kim, H., Packard., B. Z. and Brand, L. Biophys. Chem. 1996, 62, 25-38.

Titration of constant concentration of human serum albumin with Quin-2 induces a concomitant decrease of both the fluorescence intensity and lifetime of the Trp residues of the protein (Table 3.3).

Quin-2 binds to human serum albumin and forms a complex with the protein. Thus, we are in presence of a static quenching. In this case, one should not observe any variation in the fluorescence lifetime of the Trp residues of the protein. However, this is not the case (Table 3.3) and the decrease in the fluorescence lifetime is in the same order of that of the intensity. This is explained by the presence of an energy transfer from the Trp residues to Quin-2.

Table 3.3. Effect of Quin 2 on Trp emission and lifetimes in human serum albumin.

[Quin 2]. (μM)	α_1	τ_1	α_2	τ_2	α_3	τ_3	τ_o	I_{ss}
0	0.08	0.4	0.13	2.4	0.15	7.0	3.9	1.37
0.8	0.09	0.4	0.13	2.4	0.15	6.9	3.9	1.33
2.5	0.08	0.4	0.13	2.5	0.13	6.9	3.7	1.23
5.8	0.10	0.4	0.13	2.3	0.11	6.6	3.2	1.08
13.8	0.11	0.3	0.13	2.1	0.09	6.1	2.6	0.85
21.5	0.12	0.3	0.12	2.0	0.07	5.9	2.2	0.70

$\lambda_{ex} = 295$ nm and $\lambda_{em} = 350$ nm. Pre-exponential terms are given as absolute alphas. I_{ss} refers to the steady – state fluorescence intensity. Source: Hirshfield, K. M, Toptygin, D., Grandhige, G., Kim, H., Packard., B. Z. and Brand, L. Biophys. Chem. 1996, 62, 25-38.

However, the decrease in both the fluorescence intensity and lifetime upon energy transfer is not a general case. In fact, binding of TNS to α_1-acid glycoprotein induces a decrease in the fluorescence intensity of the Trp residues of the protein without any change in the fluorescence lifetime although energy transfer from Trp residues to TNS is occurring (Albani et al. 1996).

2b5. Carbohydrates fluorescent probes

Carbohydrates are not fluorescent. Therefore, one usually binds fluorophores covalently to small carbohydrates so that to carry out interaction or conformational studies between these carbohydrates and proteins. Also, fluorophores can be bound covalently to the proteins, and thus interaction between small carbohydrates and the protein is carried out by following variation of the fluorophores observables. See for examples (Monsigny et al. 1980, Khan et al. 1988, Yu and Pettigrew, 2003).

Calcofluor White (Fig. 3.31) is a fluorophore that binds to carbohydrate residues. Type of interaction depends on the secondary structure of the carbohydrate residues and fluorescence parameters of calcofluor are sensitive to this spatial secondary structure.

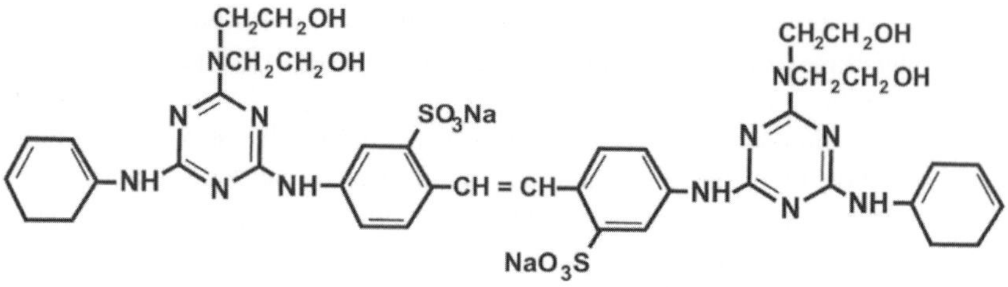

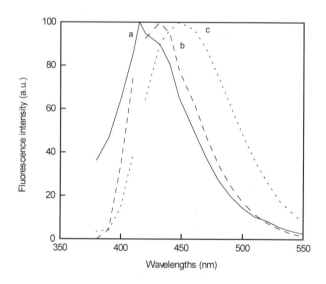

Figure 3.31. Chemical structure of calcofluor white.

Dissolved in water, fluorescence maximum of Calcofluor White is located between 435 and 438 nm (Fig. 3.32.b), while in an alcohol such as isobutanol or when bound to human serum albumin it fluoresces at 415 nm (Fig. 3.32a). In presence of α_1- acid glycoprotein, fluorescence maximum of calcofluor shifts toward 439 nm when the fluorophore is at low concentrations and toward 448 nm when it is present at high concentrations (Fig. 3.32c). The shift compared to water observed in presence of α_1-acid glycoprotein is the result of calcofluor binding on the carbohydrates (40% in weight) of the protein.

Figure 3.32. Normalized emission spectra of 0.9 µM of calcofluor free in 10 mM phosphate, 0.143 M NaCI buffer, pH 7 (b) (λ_{max} 432 nm), in the presence of 50 µM of HSA (λ_{max} 415 nm) (a) and of 250 µM of calcofluor in the presence of 8.5 µM of α_1-acid glycoprotein (λ_{max} 447 nm) (c). λ_{ex} 300 nm and temperature = 20°C. Source: Albani, J. R. and Plancke, Y. D., 1998 and 1999, Carbohydrate Research . 314, 169-175 and 318, 194-200

2b6. Oxidation of tryptophan residues with N-bromosuccinimide

Oxidation of indole by N-bromosuccinimide (NBS) (Fig.3.33) induces a decrease in the absorption of the indole at 280 nm and the appearance of a major peak at 261 nm (ε_M = 10300) and a minor peak at 309 nm (ε_M = 1630) (Ramachandran and Witkop, 1967).

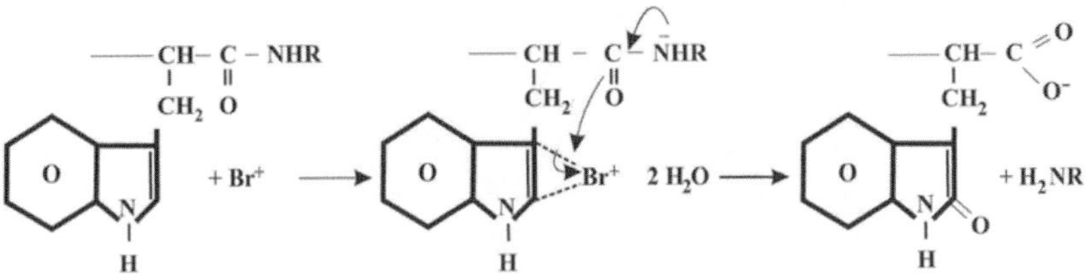

Figure 3.33. Oxidation of indol group with N-bromosuccinimide at pH 5.

Addition of NBS to proteins allows titration of tryptophan residues, mainly those present at the surface of the protein, and helps to study energy transfer within the protein and to follow local and/or global conformational change. One should be aware to the fact that addition of NBS to a protein may modify its conformation and structure and thus all fluorescence parameters. Therefore, in general, before comparing the data obtained in the absence and in the presence of NBS, one should be sure that the global and the local structure of the protein is the same in both cases.

UDP – glucose 4 – epimerase catalyses a reversible transformation between UDP-Glc and UDP-Gal. A tightly bound pyridine nucleotide NAD is involved in the mechanism of epimerization. Epimerase from the yeast *Kluyveromyces fragilis* is a homodimer containing 1 mol of NAD noncovalently bound per mol of enzyme. The NAD has fluorescence properties identical to those of NADH, due to the presence of the pyridine, and thus fluoresces with an emission maximum located at 435 nm. The NAD fluorescence is used to monitor events occurring at the active site of the enzyme.

The role of the Trp residues in the catalytic site of epimerase has been investigated by studying the energy transfer inhibition from Trp residues to NAD.

Excitation of Trp residues of epimerase at 295 nm induces an emission spectrum with two maxima at 330 and 435 nm. These peaks correspond to the emission of the Trp residues and of NAD respectively. The peak at 435 nm is result of energy transfer from Trp residues to NAD.

Addition of increased concentrations of NBS to a solution of epimerase induces a decrease in the fluorescence intensity of Trp residues and NAD. Thus, Trp residues act as an energy donor to NAD. Epimerase activity was monitored in presence of NBS parallel to the fluorescence intensity of the Trp residues of the enzyme. The results show that although the loss of activity is complete, the fluorescence intensity decrease does not reach zero. This means that not all of the Trp residues are involved in the catalytic site (Figure 3.34).

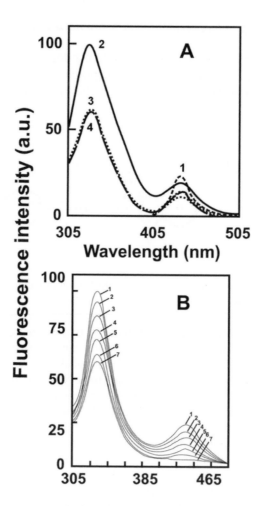

Figure 3.34. **A**, fluorescence properties of native epimerase from *K. fragilis.* 1 μM solution of epimerase in 20 mM phosphate buffer, pH 6.7, at 25°C was used for recording the following spectra: (- - -) coenzyme fluorescence emission spectrum of native epimerase on excitation at 353 nm, (▲ ▲) fluorescence emission spectrum of native epimerase on excitation at 280 nm, (—) tryptophan fluorescence emission spectrum of native epimerase on excitation at 295 nm, (...) tryptophan fluorescence emission spectrum of epimerase in the presence ofUMP (1 mM) on excitation at 295 nm. **B**, effect of NBS on tryptophan fluorescence and tryptophan to coenzyme fluorescence energy transfer. Tryptophan fluorescence and tryptophan to coenzyme fluorescence energy transfer in epimerase was monitored as a function of incremental additions of an aqueous solution of NBS. 2- μl aliquots of an aqueous solution of NBS (4.5 mM) were added to 900 μl of epimerase solution (1 μM, in 20 mM phosphate buffer, pH 6.7). After 2-3 min of incubation following each addition, the fluorescence emission spectrum was recorded by exciting the protein at 295 nm. The molar ratios of NBS to enzyme for the curves 2-7 were 10 (2), 20 (3), 30 (4), 40 (5), 50 (6), and 60 (7), respectively. Curve 1 represents the native enzyme. Source: Ray, S., Mukherji, S. and Bhaduri, A. 1995, J. Biol. Chem. 270, 11383-11390. Authorization of reprint accorded by the American Society for Biochemistry and Molecular Biology.

2b7. Nitration of tyrosine residues with tetranitromethane (TNM)

Tetranitromethane (TNM) is a highly potent pulmonary carcinogen. The toxic and carcinogenic mechanism(s) of TNM and related nitro compounds are unknown. Because TNM specifically nitrates tyrosine residues on proteins, the effects of TNM on phosphorylation and dephosphorylation of tyrosine, and subsequent effects on cell proliferation can be investigated.

Nitration of phenolic compounds leads to the formation of 3-nitro tyrosine (molar extinction coefficient = 14400 at 428 nm) (Fig. 3.35). The reaction is very specific for phenolic compounds.

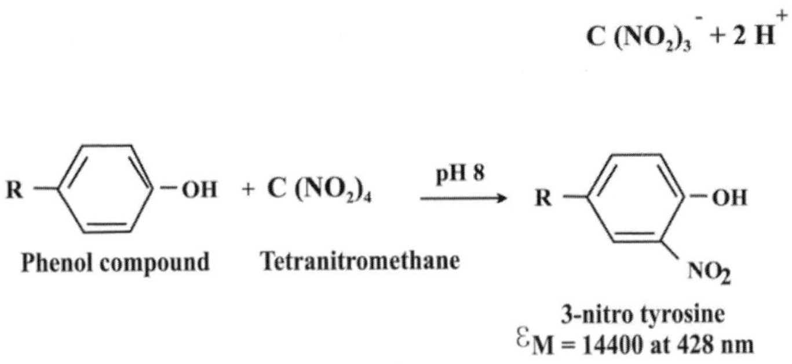

$$C (NO_2)_3^{-} + 2 H^{+}$$

Phenol compound **Tetranitromethane** →(pH 8) **3-nitro tyrosine** $\varepsilon_M = 14400$ at 428 nm

Figure 3.35. Nitration of Tyrosine with TNM. Sources: Riordan, J.F., Wacker, W.E.C. & Vallee, B.L. 1966, J Am. Chem. Soc. 88, 4104-4105 and Froschle, M., Ulmer, W. and Jany, K. D. 1984, Eur. J. Biochem. 142, 533-540. Authorization of reprint accorded by Blackwell Publishing.

Chemical nitration of functionally important tyrosine residues by tetranitromethane was often found to inactivate or alter the properties of the enzyme under investigation. It was only after the detection of *in vivo* nitrotyrosine formation in inflammatory conditions that the physiological aspects of nitrotyrosine metabolism came to attention. Abundant production (1–120 μM) of nitrotyrosine was recorded under a number of pathological conditions such as rheumatoid arthritis, liver transplantation, septic shock, and amyotrophic lateral sclerosis (Balabanli et al. 1999).

CGTases (EC 2.4.1.19) are bacterial enzymes that facilitate the biosynthesis of cyclodextrins from starch through intramolecular transglucosylation. The primary structures of most of these enzymes have been published, and the three-dimensional structure of B. *circulans* CGTase has been established. Studies of the molecular mechanism of transglucosylation have indicated that amino acids such as histidine and tryptophan are implicated in such mechanisms. Nitration of CGTase with TNM induces a loss of the enzyme activity, a decrease of the enzyme affinity towards the β-CD co-polymer accompanied with a loss of the tryptophan fluorescence (Villette et al. 1993). We are going to show now the effect of tyrosine nitration on the CGTase conformation.

Fluorescence emission maximum of CGTase is located at 338 nm, the bandwidth of the spectrum is equal to 55 nm (Figure 3.36a). Thus, both embedded and surface tryptophan residues contribute to the protein fluorescence.

Although CGTase contains many tyrosine residues, the absence of a shoulder or a peak at 303 nm (Figure 3.36b) when excitation is performed at 273 nm suggests that tryptophans are responsible for the protein emission.

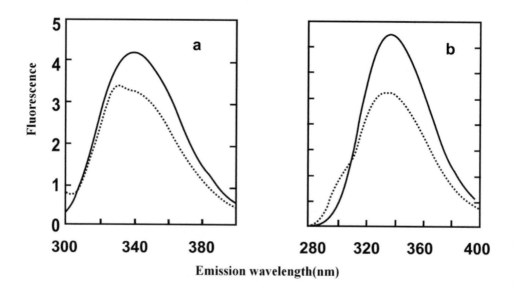

Figure 3.36a and b. Fluorescence of CGTase obtained at λ_{ex} = 295 (a) and 273 nm (b). (—) : native enzyme, (----) : 8 mM TNM modified protein. Source: Villette J.R, Helbecque, N. , Albani, J. R. , Sicard, P. J. and Bouquelet, S. J. 1993, Biotechnol. Appl. Biochem. 1993 17, 205-216. Authorization of reprint accorded by Portland Press.

The tryptophan fluorescence intensity of the native CGTase is 8-fold higher than that of a L-tryptophan solution with respect to the total amount of tryptophan. This suggests that CGTase supports an energy transfer from tyrosyl to tryptophan residue(s).

The fluorescence intensity of CGTase tryptophan residues decreases (Figure 3.37) after treatment of the enzyme with increasing concentrations of TNM and purification of the modified enzyme on co-polymer. The tryptophan fluorescence intensity of the 8 mM-TNM-modified CGTase (0.03 μM^{-1} tryptophan) is close to that of free L-tryptophan (0.034 μM^{-1}). The loss of tryptophan fluorescence observed during the nitration of the enzyme may then be related to the elimination of this energy transfer.

CGTase nitration induces an 11.5 nm shift of the fluorescence maximum to the shorter wavelengths (λ_{max} = 326.5 nm instead of 338 nm for 8 mM TNM), suggesting a relative increase of the buried tryptophan fluorescence. The (Tyr→Trp) energy transfer may then involve solvent-exposed residue(s).

After excitation at 273 nm of the nitrated CGTase (Figure 3.36b), tyrosine fluorescence appears as a typical shoulder of the peak at 302 nm. This phenomenon, not shown in the native enzyme, is consistent with elimination of a (Tyr → Trp) energy transfer during the nitration of the enzyme.

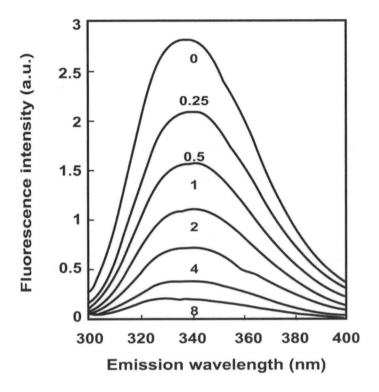

Figure 3.37. Tryptophan fluorescence spectra of CGTase after 18 hours of nitration with 0, 0.25, 0.5, 1, 2, 4 and 8 mM TNM. λ_{ex} = 295 nm. Source: Villette J.R, Helbecque, N. , Albani, J. R. , Sicard, P. J. and Bouquelet, S. J. 1993, Biotechnol. Appl. Biochem. 1993 17, 205-216. Authorization of reprint accorded by Portland Press.

Circular dichroism spectra were recorded on native and 8 mM TNM-treated CGTases in the 210-250 nm and 250-320 nm regions (data not shown) in order to investigate the effects of the nitration both on some secondary structural elements of the protein and on the conformational environment of the aromatic residues.

The spectrum of the native enzyme in the first region shows two typical peaks at 212 and 220 nm. After nitration of the enzyme with 8 mM TNM the ellipticity decreases and the 212 nm peak seems to disappear. This result suggests a slight impairment of the CGTase conformation during its nitration by TNM.

The 250-320 nm spectrum shows an important change around 280 nm for nitrated CGTase in comparison with that of the native enzyme. This is in part due to tyrosine nitration. In particular, the dichroism observed around 300 nm in the differential spectrum (results not shown) strongly suggests that the conformational environment of some tryptophan residues has been changed during nitration.

These results, supporting a conformational impairment of the enzyme during nitration would account for the loss of the tryptophan fluorescence by the elimination of a (Tyr →Trp) energy transfer (removing of the chromophores), and hence enzyme inactivation.

3. Effect of the environment on the fluorescence observables

3a. Polarity effect on the quantum yield and the position of the emission maximum

Quantum yield and fluorescence emission maximum are sensitive to the surrounding environment. This can be explained as follows:

Fluorophore molecules and the amino acids of the binding sites (case of an extrinsic fluorophore such as TNS, fluorescein, etc…) or the amino acids of their microenvironment (case of Trp residues) are associated by their dipoles. Upon excitation, only the fluorophore absorbs the energy. Thus, dipole of the excited fluorophore has an orientation different from that of the fluorophore in the ground state. Therefore, fluorophore dipole-solvent dipole interaction in the ground state is different from that in the excited state (Figure 3.38).

The new interaction is unstable. To reach stability, fluorophore molecules will use some of their energy to reorient the dipole of the amino acids of the binding site and of their microenvironment. The dipole reorientation is called the relaxation phenomenon Emission occurs after relaxation.

More the dipole of the environment is important, i.e., more the polarity of the environment is important, more energy should be released by the fluorophore to reorient the dipole of the microenvironment. Thus, more the environment polarity increases, less energy will be left for the photon emission and more the emission maximum of the fluorophore will be shifted to the red.

In an apolar medium, the fluorophore at the excited state will induce the formation of a dipole within the environment. Formation of a new dipole needs less energy than reorientation of an already existing dipole. Thus emission from an apolar environment yields an emission spectrum with a maximum located in the blue compared to the emission occurring from a polar environment.

Also, since the intensity and thus the quantum yield are dependent on the number of emitted photons, they will be lower when emission occurs from a polar environment. One should mention that fluorescence parameters are more sensitive to the environment than absorption spectra and / or extinction coefficients.

Figure 3.39 displays the fluorescence emission spectra of ANS dissolved in solvent of different polarity. We notice that ANS fluorescence intensity, quantum yield and emission maximum is dependent on the polarity of the solvent.

The same phenomenon is observed for other fluorophores such as TNS or PRODAN. Figure 3.40 displays the position of the emission maximum of TNS as a function of solvent polarity (Z).

PRODAN exhibits a fluorescence spectrum with a maximum located at 392 nm in cyclohexane and at 523 nm in water. This important sensitivity of PRODAN to the solvent seems to be owed to an important change in the dipolar moment of the molecule after excitation. This dipolar moment modification would be the result of the delocalization of the positive and negative charges within the molecule.

When the fluorophore is bound to a protein, its fluorescence will be dependent on the polarity of the surrounding amino acids.

Relaxation studies on Prodan and on Patman in presence of membrane vesicles have shown that the fluorophore Patman is more embedded in the polar region of the membrane than Prodan. This is probably because of the long palmitoyl chain in Patman (Hutterer et al. 1996).

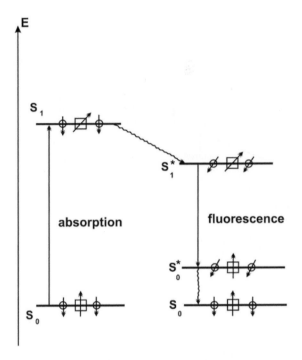

Figure 3.38. Relation between fluorescence emission and dipole orientation. The square characterizes the fluorophore dipole and the circle the solvent dipole.

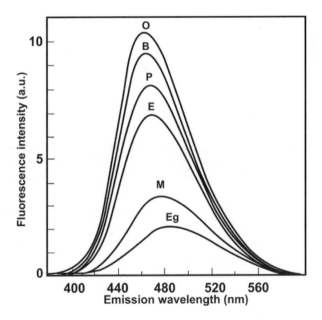

Figure 3.39. Fluorescence emission spectra of ANS in solvents of different polarities. O:Octanol, B:Butanol, P:Propanediol, E:Ethanol, M: Methanol and Eg: Ethylene-glycol. Source: Stryer, L. S. 1965, J. Mol. Biol. 13, 482-495.

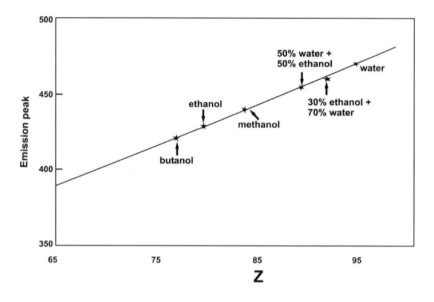

Figure 3.40. Position of the emission maximum of TNS as a function of solvent polarity (Z).

3b. Effect of the viscosity on the fluorescence emission spectrum

Fluorescence spectra are also dependent on the rigidity of the medium. Relaxation phenomenon (reorientation of the dipole environment) occurs much easily when the medium is fluid. In such case, emission will occur after relaxation. This is the case when relaxation is faster than fluorescence, i.e. the relaxation lifetime τ_r is shorter than the fluorescence lifetime τ_0. This occurs when the binding site is flexible and the fluorophore can move easily. Displacements of fluorophore molecules will be representative of the time scale and amplitude of the motions of the surrounding protein matrix. Emission from a relaxed state does not change with the excitation wavelength. This can be explained by the fact that whatever the value of the excitation wavelength, emission will occur always from the same energy level.

When the binding site is rigid, fluorescence emission occurs before relaxation. In this case, excitation at the longer wavelength edge of the absorption band photoselects population of fluorophores energetically different from that photoselected when the excitation wavelength is shorter.

When the red edge excitation is performed, the energy $h\nu^{edge}$ of the electronic transition is equal to $E_e^{edge} - E_g^{edge}$, where $E_e - E_g$ refers to the difference of energy between excited and ground states; $h\nu^{edge}$ is lower than $h\nu^{sw}$, the energy of the electronic transition that occurs at short wavelengths. Thus excitation at the red edge gives a fluorescence spectrum with a maximum located at higher wavelength than that obtained when excitation is performed at short wavelengths.

Thus, the observation of red shift of $\lambda_{em,max}$, upon red shift in λ_{ex}, indicates that the system meets with the $\tau_0 < \tau_r$ condition. This means a decreased mobility of the fluorophore on its binding site with respect to the dipolar matrix of the site.

In other words, when the surrounding environment of the fluorophore or when its binding site is rigid, the emission maximum of the fluorophore will shift to higher wavelengths with the excitation wavelength. In general, emission maxima are compared if the spectra are symmetric. Otherwise, the centers of gravity should be compared. The red-edge excitation spectra method can be applied to Trp residues (Fig. 3.41) and to extrinsic fluorophores such as TNS, calcofluor or any other fluorophores.

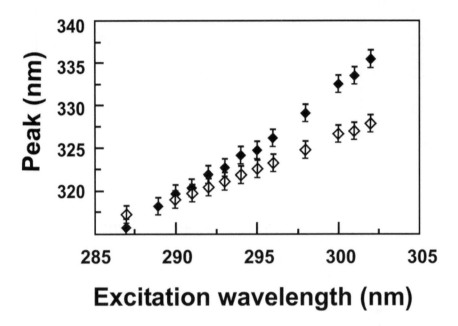

Figure 3.41. Red-edge excitation of holo-Phe110Ser azurin. Dependence of the fluorescence spectrum peak of holo-Phe110Ser as a function of the excitation wavelength, at 1 bar (filled symbols) or at 2000 bar (open symbols). At 1 bar, the shift of the emission peak is larger than that at 2000 bar. The decrease in the spectral shift in presence of hydrostatic pressure suggests a higher relaxation of the Trp48 microenvironment, due to the destabilization of the protein structure. Source: Mei, G., Di Venere, A., Campeggi, F. M., Gilardi, G., Rosato, N., De Matteis, F. and Finazzi-Agrò, A. 1999, European Journal of Biochemistry 265, 619-626. Authorization of reprint accorded from Blackwell Publishing.

Finally, we should indicate that emission from a low polar or apolar environment yields a fluorescence emission spectrum located at low wavelength such as when the environment is rigid. Also, a highly dynamic environment yields a fluorescence emission spectrum with a maximum located at long wavelengths such as when the polarity of the environment is high. Therefore, one should be careful in interpreting the results when work is performed on macromolecules.

An example to illustrate this. Ascorbate oxidase is a copper-containing enzyme which catalyzes a redox between vitamin C and molecular oxygen. The protein is a homodimer of monomers, each containing three domains and 14 tryptophan residues.

X-ray diffraction studies (Messerschmidt et al. 1989; 1992) have shown that six tryptophan residues / monomer are completely shielded from the solvent, six others are buried but not completely and two are exposed to the solvent. Figure 3.42 displays the position of the emission peak of protein emission with excitation wavelength, in absence and presence of 1.3 M CsCl.

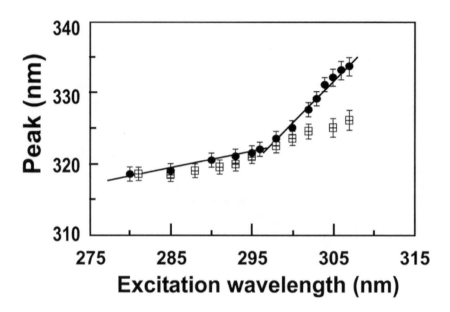

Figure 3.42. Red-edge fluorescence emission of ascorbate oxidase. Maximum fluorescence emission versus excitation wavelength for Ascorbate oxidase (circles) and for ascorbate oxidase in presence of 1.3 M CsCl (squares). Solid lines represent the best linear fits of circles. Source: Di Venere, A., Mei, G., Gilardi, G., Rosato, N., De Matteis, F., McKay, R., Gratton, E. and Finazzi Agro, A. 1998. Eur. J. Biochem. 257, 337-343. Authorization of reprint accorded from Blackwell Publishing.

We notice that up to 295 nm, the peak moves slowly to longer emission wavelength, above 297 nm this dependence becomes steeper and within 10 nm (297 –307 nm) the spectrum shifts to the red by about 11 nm. The authors correlated this effect to a significant different exposure of the classes of tryptophan residues to the solvent. In fact, when the experiment was carried out in the presence of CsCl, two slopes were still visible but the difference was much less pronounced. This experiment by itself shows that the protein contains tryptophan residues that are exposed to the solvent.

Addition of cesium quenches the fluorescence of these tryptophans and the main fluorescence will occur mainly from buried Trp residues. In the present data, we do not have any direct proof of the dynamic aspect. Although one should expect that the tryptophan residues that are exposed to the solvent would rotate much faster than the other tryptophan residues. The authors showed this result by measuring the anisotropy as a function of wavelength in the absence and in the presence of 1.3 M CsCl (Di Venere et al. 1998).

3c. Effect of the environment on the fluorescence lifetime

A fluorophore can have one or several fluorescence lifetimes, depending on several factors:
- A ground state heterogeneity resulting from the presence of equilibrium between different conformers. Each conformer presents a specific fluorescence lifetime.
- Different non-relaxing states of the Trp emission. These relaxed states may arise from one single Trp residue.
- Internal protein motion.
- Presence in the studied macromolecule, for example a protein, of two or several tryptophans having each different microenvironments and thus different emissions.

The origin of the heterogeneity of the fluorescence lifetime will be developed in a chapter 7.

In conclusion, the nature of the solvent can affect all the fluorescence parameters. Some examples, Table 3.3 shows the fluorescence lifetime measured with the frequency method of the nitrobenzoxadiazole dissolved in different solvents. One can notice that the fluorescence lifetime is solvent dependent.

Table 3.3. Fluorescent lifetime of Nitrobenzoxadiazole in various solvents

Solvent	τ_o
Water	0.92 ns
Methanol	5.35 ns
DMSO	7.15 ns
Ethylacetate	10.93 ns

Curtosey from Amersham Biosciences.

Table 3.4 displays fluorescence lifetime, quantum yield and position of the emission maximum of tryptophan in basic, neutral and acid solvents. One can notice that the three parameters are not the same in the three mediums. The different types of protonation explain this variation.

Table 3.4. Fluorescence observables of tryprophan in solvents of different pH.

	Φ_F	λ_{max}	τ_o
Basic	0.41	365 nm	9 ns
Neutral	0.14	351 nm	3.1 ns
Acid	0.04	344 nm	0.7 ns

Figure 3.43 displays the fluorescence emission spectra of protoporphyrin IX dissolved in HCl (a) and in water in presence of small quantity of NaOH (b). The fluorescence lifetime of the porphyrin is equal to 3.7 and 0 ns in HCl and in water-NaOH mixture, respectively. In water, the fluorophore is completely aggregated inducing by that a fluorescence lifetime equal to 0 ns.

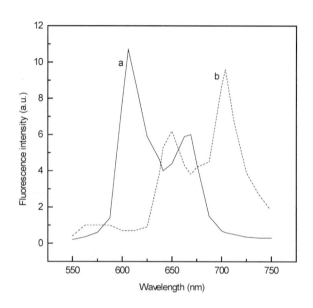

Figure 3.43. Fluorescence emission spectra of protoporphyrin IX dissolved in HCl (a) and in water mixed with small amount of NaOH (b).

Also, it is important to mention that in many cases, fluorescence lifetimes are dependent on the structure of the fluorophore itself (see chapter 8, paragraphs 3b and 4e).

3d. Relation between fluorescence and a specific sequence in a protein

Green fluorescence protein isolated from coelentrates displays a principal function that is to transduce the blue chemiluminescence of the protein aequorin into green fluorescent light by energy transfer. Wild type GFP from jellyfish displays an absorption spectrum with two peaks, a major one at 395 and a minor one at 475 nm, with extinction coefficient of 30000 and 7000 M^{-1} cm^{-1}, respectively. The fluorescence maximum is located at 509 nm. GFP fluorescence originates from a Ser-Tyr-Gly-sequence that forms a p-hydroxybenzylidene-imidazolidone. Although this peptide sequence exists in many proteins, none of these peptides display structural and fluorescence characteristics identical to those observed for the peptide sequence in GFP. Number of mutants has been constructed; one of them is 5 to 6 times more fluorescent than the wild type. This was possible by mutating a serine to threonine at position 65 in the protein sequence (Heim et al. 1995) (see also chapter six).

Chapter 4

FLUORESCENCE QUENCHING

1. Introduction

Interaction between a fluorophore and a molecule or between two molecules induces usually perturbation or modification in the fluorescence parameters such as intensity, quantum yield or/ and lifetime. These fluorescence perturbations will allow to explain and to understand the nature and the origin of interactions between the different molecules. Since we are applying fluorescence measurements to follow these interactions, then one should expect to record modifications occurring within the range of the fluorescence lifetime.

In general, one can observe different types of fluorescence quenching, collisional, static, thermal and energy transfer at distance known also as Förster energy transfer. In the present chapter we are going to deal with all types of energy quenching but energy transfer at distance. This will be detailed in chapter six.

Collisional quenching occurs when the fluorophore and another molecule diffuse in the solution and collide one with each other. In this case, the two molecules do not form a complex. Fluorescence parameters allow to determine quenching constants that are dependent on the dynamics of the studied system..

Static quenching is observed when two molecules bind one to the other forming a complex. Intrinsic fluorophore such as the aromatic amino acids or extrinsic fluorophores such as ANS or fluorescein can be used to follow the interaction between the two interacting molecules. Fluorescence studies will allow deriving the binding parameters of the complex such as the stoichiometry and the binding constant. In these types of studies, the fluorophore is bound to one of the two interacting molecules. Each time the two molecules form a complex, fluorescence parameters such as intensity, anisotropy or / and lifetime of the fluorophore are monitored.

One should indicate here that direct physical interaction between the fluorophore and the ligand is not necessary. We can observe a long or short-distance effect depending on whether the fluorophore is within the interaction area or not. All of some of the fluorescence observables will be modified upon ligand binding on the macromolecule.

2. Collisional quenching : the Stern-Volmer relation

Macromolecules display continuous motions. These motions can be of two main natures: the molecule can rotate on itself following precise axis of rotation and it can have a local flexibility. The latter can differ from an area of the macromolecule to another. This local flexibility, called also internal motions, allows to different small molecules such as the solvent molecules to diffuse along the macromolecule. This diffusion is in general dependent on the importance of the local internal dynamics. Also, the fact that the solvent molecules can reach the interior hydrophobic core of macromolecules such as proteins means clearly that the term hydrophobicity should be considered as relative and not as absolute. Although the core of a protein is composed mainly by hydrophobic amino acids, it is constantly in contact with the solvent. The internal dynamics of the proteins allow and facilitate this permanent contact. Also, this internal motion will permit to small molecules such as oxygen to diffuse within the core of the protein. Since oxygen is a collisional quencher, analyzing the fluorescence data in presence of different oxygen concentrations yields information on the internal dynamics of macromolecules.

We have defined the fluorescence lifetime as the time spent by the fluorophore at the excited state. Collisional quenching is a process that is going to depopulate the excited state in parallel to the other processes we have already described in the Jablonski diagram. Therefore, the fluorescence lifetime of the excited state will be lower in presence of collisional quencher than in its absence.

The velocity of fluorophore deexcitation can be written as:

$$v = k_q \, [F] \, [Q] \tag{4.1}$$

where k_q is the bimolecular quenching constant and it is expressed in $M^{-1} \, s^{-1}$, [F] is the fluorophore concentration, it is held constant during the experiment and [Q] is the quencher concentration. If [Q] is higher than [F], the system can be considered as a first-pseudo order with a constant equal to $k_q \times [Q]$.

$$v = k' \, [F] = k_q \, [Q] \, [F] \tag{4.2}$$

The quantum yield in the absence of the quencher is equal to:

$$\phi_F = \frac{k_r}{k_r + k_i + k_{isc}} \tag{4.3}$$

In the presence of a quencher, the quantum yield will be equal to :

$$\phi_{F(Q)} = \frac{k_r}{k_r + k_i + k_{isc} + k_q \, [Q]} \tag{4.4}$$

$$\frac{\phi_F}{\phi_{F(Q)}} = \frac{k_r + k_i + k_{isc} + k_q \, [Q]}{k_r + k_i + k_{isc}} = \frac{k_r + k_i + k_{isc}}{k_r + k_i + k_{isc}} + \frac{k_q \, [Q]}{k_r + k_i + k_{isc}} =$$

$$1 + k_q \, [Q] \, \frac{1}{k_r + k_i + k_{isc}} = 1 + k_q \, \tau_o \, [Q] = 1 + K_{SV} \, [Q] \tag{4.5}$$

where K_{SV} is the Stern-Volmer constant and τ_o is the mean fluorescence lifetime of the fluorophore in absence of quencher. Equation 4.5 is called the Stern-Volmer equation.

Since the fluorescence intensity is proportional to the quantum yield, the Stern-Volmer equation can be written as:

$$I_o / I = 1 + K_{SV} [Q] = 1 + k_q \tau_o [Q] \qquad (4.6)$$

where I_o and I are respectively the fluorescence intensities in the absence and presence of quencher (Stern and Volmer, 1919). Plotting I_o / I as a function of [Q] yields a linear plot with a slope equal to K_{SV}.

Figure 4.1 displays the Stern-Volmer plot for the fluorescence intensity quenching of zinc protoporphyrin IX by oxygen. The slope of the plot yields values of K_{SV} and of k_q equal to 22.6 M^{-1} and 1.19 x 10^{10} M^{-1} s^{-1}, respectively. This value of k_q is equal to that found for oxygen when fluorescence quenching of free tryptophan in solution with oxygen is performed. This value of k_q is the higest value one can measure in a homogeneous medium. Within macromolecule such as proteins, important oxygen diffusion leads to a high value of k_q. More the internal fluctuations of the protein are important, highest will be the value of k_q .

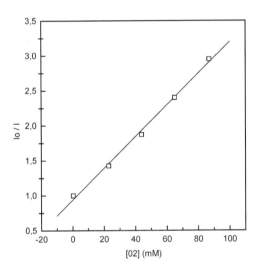

Figure 4.1. Oxygen quenching of ZnPPIX in water at 35°C. λ_{ex} = 514.5 nm and λ_{em} = 590 nm. Slope = Ksv = 22.6 M^{-1}. k_q = 22.6 / 1.9 ns = 1.19 x 10^{10} M^{-1} s^{-1} .

The Stern-Volmer plot can be obtained also from quenching of fluorescence lifetime. In fact, fluorescence lifetime in the absence of quencher is equal to:

$$\tau_o = \frac{1}{k_r + k_i + k_{isc}} \qquad (4.7)$$

and in presence of quencher, the fluorescence lifetime is equal to :

$$\tau_{(Q)} \quad = \quad \frac{1}{k_r + k_i + k_{isc} + k_q\,[Q]} \tag{4.8}$$

$$\tau_o / \tau_{(Q)} = 1 + k_q\,\tau_o\,[Q] = 1 + K_{SV}\,[Q] \tag{4.9}$$

Equation 4.9 is identical to Eq.4.6. Therefore, in presence of collisional quenching, we have the following equation (Figures 4.2 and 4.3):

$$I_o / I = \tau_o / \tau_{(Q)} \tag{4.10}$$

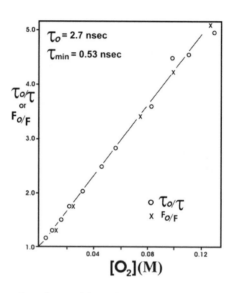

Figure 4.2.   Stern-Volmer plot of quenching of oxygen intensity and lifetime of tryptophan in solution.   Source: Lakowicz, J. R. and Weber, G. 1973.   Biochemistry, 12, 4161-4170. Authorization of reprint accorded by the American Chemical Society.

The most common quenchers are oxygen, acrylamide, iodide and cesium ions. More the probability of collisions between fluorophore and quencher is important higher will be the value of k_q.   Oxygen is a small and uncharged molecule, thus it can diffuse easily. Therefore, the bimolecular diffusion constant k_q observed for oxygen in solution is the most important between the all cited quenchers.

Acrylamide is an uncharged polar molecule, thus it can diffuse within a protein and quenches fluorescence emission of Trp residues. The quencher should be able to collide with tryptophan whether it is on the surface or in the interior of a protein. Nevertheless, Trp residues, mainly those buried within the core of the protein, are not all reached by acrylamide. For a fully exposed tryptophan residue or for a Trp free in solution, the upper value of k_q found with acrylamide is $6.4 \times 10^9 \ M^{-1} \ s^{-1}$.

Cesium and iodide ions will quench Trp residues that are present at or near the surface of the protein. Iodide ion is more efficient than cesium ion, i.e., each collision with the fluorophore induces a decrease in the fluorescence intensity and lifetime, which is not the case when cesium is used. Also, since cesium and iodide ions are charged, their quenching efficiency will depend on the charge of the protein surface. For a free Trp and Tyr in solution, the highest value of K_{SV} we found with iodide are 16 and 19 M^{-1}, respectively, and thus, the corresponding k_q values are 5.8 x 10^9 and 5.3 x 10^9 M^{-1} s^{-1}, respectively.

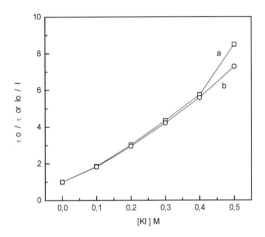

Figure 4.3. Stern-Volmer plot of quenching of fluorescein intensity and lifetime with iodide. Data are from Table 2.5.

Fluorescence intensity quenching with iodide can be used to find out whether ethidium bromide added to DNA is intercalated into the double helical DNA or present at the surface of the DNA. In fact, since DNA is negatively charged, addition of iodide ion to a DNA-EB complex will not decrease the fluorescence intensity of the ethidium bromide if the fluorophore is intercalated into the DNA. However, if the fluorophore is bound to the DNA surface, its fluorescence will be quenched by iodide ion.

3. The different types of dynamic quenching

When a protein possesses two or several Trp residues and when quenchers such as iodide, cesium or acrylamide are used and if all the Trp residues are not accessible to the quencher, the Stern-Volmer equation yields a downward curvature. In this case, we are in presence of a selective quenching (Fig. 4.4B). The linear part of the plot allows calculating the value of the Stern-Volmer constant corresponding to the interaction between the quencher and the accessible Trp residues. The fluorescence of these residues has been quenched.

Upon complete denaturation and loss of the tertiary structure of a protein, all Trp residues will be accessible to the quencher. In this case, the Stern-Volmer plot will display an upward curvature.

In summary, Inhibition of the protein fluorescence having two or several Trp residues can yield three different representations for the Stern-Volmer equations, depending on the accessibility of the fluorophore to the quencher. (a) An exponential plot, i.e., all residues are accessible to the quencher or fluorescence is dominated by one single residue, (b) a downward curvature, i.e. fluorescence is heterogeneous, residues do not have an identical accessibility to the quencher, (c) linear plot, i.e., fluorescence is heterogeneous and the accessibility of residues to the inhibitor slightly differs (Fig. 4.4A and 4.5) (Eftink and Ghiron, 1976; Di Venere et al. 1998).

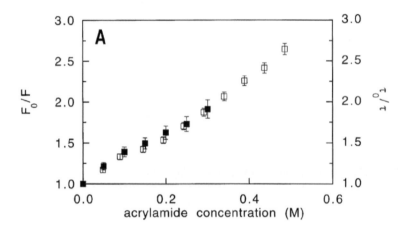

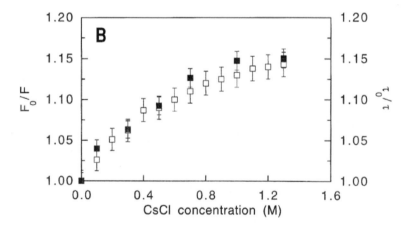

Fig. 4.4. Acrylamide and CsCI quenching effects on the ascorbate oxidase emission properties. Steady-state (open squares) and dynamic (filled squares) quenching of ascorbate oxidase by acrylamide (A) and CsCI (B) upon excitation at 293 nm. Source: Di Venere, A., Mei, G., Gilardi, G., Rosato, N., De Matteis, F., McKay, R., Gratton, E. and Finazzi Agro, A. 1998. Eur. J. Biochem. 257, 337-343. Authorization of reprint accorded by Blackwell Publishing.

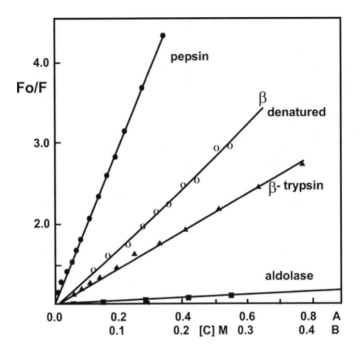

Figure 4.5. Acrylamide quenching of the milti-tryplophan conlaining proteins: pepsin, trypsin and aldolase. Darkened symbols (scale A) are for native proteins. Open symbols (scale B) are for denatured proteins in 6.7 M Gdn.HCI. Source: Eftink, M. R. and Ghiron, C. A. 1986, Biochemistry 15, 672-680. Authorization of reprint accorded by the American Chemical Society.

The Myb oncoprotein binds specifically DNA by a domain that consists of three imperfect repeats, R_1, R_2 and R_3, each containing 3 tryptophans. The tryptophan fluorescence of the $R_2 R_3$ domain is used to monitor structural flexibility changes occurring upon DNA binding to $R_2 R_3$.

Upon complex formation with a double stranded DNA, the intensity of Trp residues of $R_2 R_3$ decreases by about 50%. This intensity decrease is accompanied by a shift of the fluorescence emission maximum from 336 to 332 nm (Fig. 4.6) and by a decrease in the mean fluorescence lifetime from 1.46 to 0.71 ns. These results indicate that Trp residues environment is not the same in presence of DNA. However, this does not mean necessarily that the environment of all the six tryptophans has been modified.

Figure 4.7 displays the Stern-Volmer plot of the fluorescence lifetime quenching by acrylamide of free $R_2 R_3$ and $R_2 R_3$ – DNA complex. The slope of the Stern-Volmer plot yields the Stern-Volmer constant that is an indication of the accessibility of the Trp residues to the quencher. From the slopes of the plots, it is evident that Trp residues in $R_2 R_3$ are less accessible to acrylamide in the $R_2 R_3$ – DNA complex. This result is consistent with that observed in figure 4.5, i.e., DNA binding induces the shielding of some or all Trp residues from the solvent.

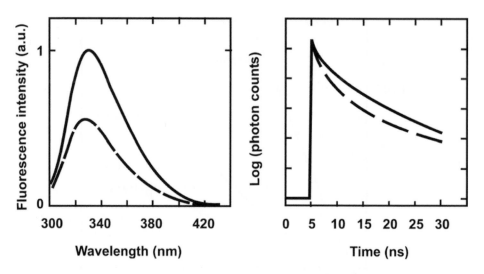

Figure 4.6. Left panel: Emission spectra of free R_2R_3 (-) and R_2R_3 bound to $DNA_{mim\ 16}$ (- - -). Right panel: Experimental Trp fluorescence intensity decay of free R_2R_3 (-) and of the R_2R_3-DNA complex (- - -). [R_2R_3] free and bound to DNA = 2.5×10^{-7} M, λ_{ex} = 290 nm. Source: Zargarian, L., Le Tilly, V., Jamin, N., Chaffotte, A., Gabrielsen, O. S., Toma, F. and Alpert, B. 1999, Biochemistry, 38, 1921-1929. Authorization of reprint accorded by the American Chemical Society.

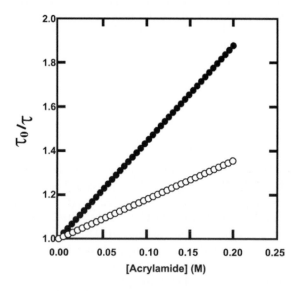

Figure 4.7. Stern-Volmer plots of the Trp fluorescence quenching of R_2R_3 by acrylamide: free R_2R_3 (·) and R_2R_3-DNA complex (○). τ_0 and τ are the averaged Trp fluorescence decay times, without and with acrylamide, respectively. The different slopes ($k_q \tau_0$) of the two species show the change of their k_q values: λ_{ex} = 290 nm; [R_2R_3] free and bound to DNA = 2.5×10^{-7} M. Source: Zargarian, L., Le Tilly, V., Jamin, N., Chaffotte, A., Gabrielsen, O. S., Toma, F. and Alpert, B. 1999, Biochemistry, 38, 1921-1929. Authorization of reprint accorded by the American Chemical Society.

When a selective quencher occurs, it is possible to determine the accessible fluorophores fraction to the quencher. Let us consider two populations of fluorophore, one accessible to the quencher and the second not. At high quencher concentrations, the fluorescence of the accessible fluorophores will be completely quenched. Thus, residual fluorescence originates from inaccessible fluorophores, i.e., fluorophores buried in the hydrophobic core of the protein.

The fluorescence intensity Fo recorded in absence of quencher is equal to the sum of the fluorescence intensities of the accessible (F_a) and not accessible (F_b) populations to the quencher:

$$F_o = F_a + F_b \qquad (4.11)$$

In presence of quencher, only the fluorescence of the accessible population decreases according to the Stern-Volmer equation. The fluorescence recorded at a defined quencher concentration [Q] is :

$$F = \frac{F_{o(a)}}{1 + Ksv\,[Q]} + F_{o(b)} \qquad (4.12)$$

where Ksv is the Stern-Volmer constant of the accessible fraction.

$$F_o - F = \Delta F = \frac{F_{o(a)} \times K_{SV}\,[Q]}{1 + K_{SV}\,[Q]} \qquad (4.13)$$

$$\frac{F_o}{\Delta F} = \frac{F_o + F_o \times K_{SV}\,[Q]}{F_{o(a)} \times K_{SV}\,[Q]} = \frac{F_o}{F_{o(a)} \times K_{SV}\,[Q]} + \frac{F_o}{F_{o(a)}} \qquad (4.14)$$

$$\frac{F_o}{\Delta F} = \frac{1}{f_a \times K_{SV}\,[Q]} + \frac{1}{f_a} \qquad (4.15)$$

where f_a is the fraction of the accessible fluorophore population to the quencher. Therefore, if one plots $F_o\,/\,\Delta F$ as a function of $1\,/\,[Q]$, a linear plot is obtained whose slope is equal to $1\,/\,(f_a\,K_{SV})$ and an intercept equal to $1\,/\,f_a$. Equation 4.15 is called the Lehrer equation (Lehrer, 1971). Figure 4.8 displays the fluorescence spectra and the plot of the modified Stern-Volmer equation of fluorescence intensity quenching of Trp residues of $R_2\,R_3$ in presence of iodide.

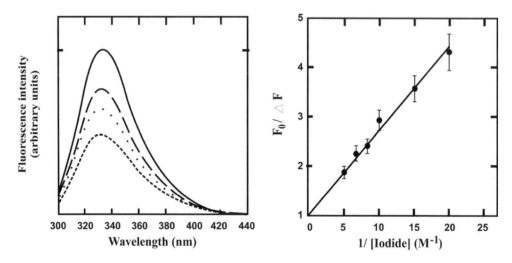

Figure 4.8. Left panel: Emission spectra of R_2R_3 (2.5×10^{-7} M) as a function of iodide concentrations: (-) R_2R_3 without iodide; (- - -) R_2R_3 with 0.05 M; (···) 0.1 M; and (---) 0.2 M iodide concentrations. Right panel: Lehrer plot of the Trp fluorescence quenching of R_2R_3 by iodide: F_0 = Trp fluorescence intensity of free R_2R_3; F = Trp fluorescence intensity of R_2R_3 in the presence of iodide; $\Delta F = F_0 - F$, decrease of Trp fluorescence intensity of R_2R_3 induced by Trp-iodide collisions. Corrective amounts of NaCl were added in order to have the same ionic strength in each sample. Source: Zargarian, L., Le Tilly, V., Jamin, N., Chaffotte, A., Gabrielsen, O. S., Toma, F. and Alpert, B. 1999, Biochemistry, 38, 1921-1929. Authorization of reprint accorded by the American Chemical Society.

The accessibility of a fluorophore to the solvent and thus to the quencher depends on its position within the protein. For example, buried Trp residues should have a lower accessibility to the solvent than those present at the surface. This means that the value of f_a will be more important for the accessible fluorophores than for the non accessible ones. Figure 4.9 displays the Stern-Volmer plots at three emission wavelengths of fluorescence intensity quenching of Trp residues with KI of the protein *Vicia fava* agglutinin. One can notice that fluorescence quenching at 310 nm is not observed indicating that buried Trp residues are not easily accessible to the solvent.

Figure 4.10 displays the fluorescence emission spectra of LCA (a) (λ_{max} = 330 nm), of the inaccessible Trp residues (b) (λ_{max} = 324 nm) obtained by extrapolating to $[I^-]$ = ∞, and of the quenched Trp residues (c) obtained by subtracting spectrum (b) from spectrum (a). The emission maximum of the accessible Trp residues is located at 345 nm, a characteristic of an emission from Trp residues near the surface of the protein. Figure 4.10 indicates that both classes of Trp residues contribute to the fluorescence spectrum of LCA (Albani, 1996).

However, it is important to mention here that the presence of 5 Trp residues makes the analysis by the modified Stern-Volmer equation very approximate. Selective quenching cannot in no way resolve the fluorescence emission spectrum of each Trp residue. However, it allows quantifying the percentage of accessible fluorophores to the quencher.

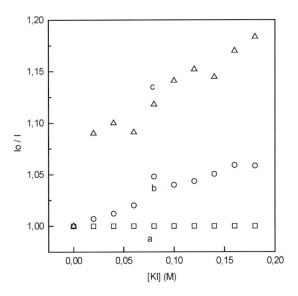

Figure 4.9. Stern - Volmer plot for the fluorescence intensity quenching of Trp residues of *Vicia fava* agglutinin. $\lambda_{ex} = 295$ nm and $\lambda_{em} = 310$ (a), 330 (b) and 350 nm (c).

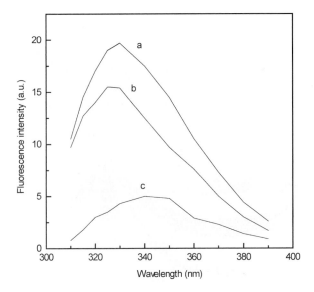

Figure 4.10. Fluorescence emission spectra of LCA (a), of Trp residues exposed to KI (c) and of Trp residues buried in the protein matrix (b). $\lambda_{ex} = 295$ nm. Source: Albani, J. R. 1996, J. Fluoresc. 6, 199-208. Authorization of reprint accorded by Kluwer Academic Publishing.

When a protein contains two Trp residues one located at the surface and the second buried in the protein core and when it is possible to assign for each Trp residue a specific lifetime, quenching experiments with a selective quencher yield Stern-Volmer plots that differ from a Trp residue to another. For example, immunophilin FKBP59 of apparent molecular mass equal to 59 kDa is a protein present in the hetero-oligomeric complexes containing nontransformed, non-DNA-binding and steroid receptors. It can be found associated with the heat shock protein hsp90 included in nontransformed steroid receptor complexes. Also, FKBP59 binds the immunosuppressant FK506 and thus it belongs to the FK506-binding protein class. The protein has peptidyl-prolyl cis-trans isomerase (PPIase) activity. Its association with steroid receptors and heat-shock proteins suggests that it may play a role in the modulation of the transcriptional control of glucocorticoid steroid and progesterone receptors (Tai et al., 1994; Renoir et al., 1995). By its isomerase activity, it may contribute to the proper folding of proteins in the cell (Schmid, 1991). FKBP59 is organized in three globular domains plus a four C-terminal domain. The N-terminal domain is responsible for the immunophilin character of the whole protein. FKBP59 contains two Trp residues located in two distinct regions of the protein (Fig. 4.11).

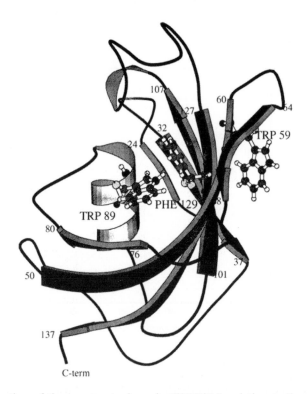

Figure 4.11. Location of the two tryptophans in FKBP50-I, relative to the protein backbone three-dimensional structure. Source: Rouvière, N., Vincent, M., Craescu, C. T. and Gallay, J. 1997, Biochemistry, 36, 7339-7352. Authorization of reprint accorded by the American Chemical Society.

Three fluorescence lifetimes 0.44, 2.82 and 6.2 ns are attributed to the Trp residues. Quenching of the fluorescence lifetimes with acrylamide yields the Stern-Volmer plots of Fig. 4.12. One can notice that acrylamide has an important effect on the value of the longest fluorescence lifetime, while it has no effect on the shortest one. Although the intermediate fluorescence lifetime is not possible to assign to any of the two Trp residues, Stern-Volmer plots observed for the longest and shortest lifetimes clearly indicate that the longest lifetime can be assigned to the surface Trp residue while the shortest lifetime can be attributed to the buried Trp residue. The Trp residue exposed to the solvent has weak interactions with the amino acids of the surrounding environment and thus has a long fluorescence lifetime. In the contrary, the buried Trp residue is surrounded by amino acids and its fluorescence is quenched by different mechanisms such as collisional quenching with the neighboring amino acids or by electron or / and proton transfer (Rouviere et al. 1997).

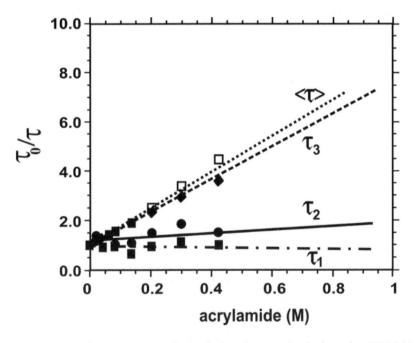

Figure 4.12. Stern-Volmer representation of the time-resolved data for FKBP50-I and complexed with FK506. Source: Rouvière, N., Vincent, M., Craescu, C. T. and Gallay, J. 1997, Biochemistry, 36, 7339-7352. Authorization of reprint accorded by the American Chemical Society.

Increasing the temperature from 5 to 40°C induces a significant decrease in both shorter and longer fluorescence lifetimes. The fluorescence lifetime decreases from 6.96 to 4.82 ns (a 30% decrease) while the shorter lifetime decreases from 0.45 to 0.35 ns (a 22% decrease). Therefore, interaction between the exposed tryptophan residue and the solvent is more important than that occurring between the buried tryptophan residue and the neighboring environment.

Selective quenching allows also to put into evidence and to study a conformational change within a protein. ADAR2 (adenosine deaminase that acts on RNA) is a ~80-kDa protein that efficiently deaminates the R/G site of GluR-B pre-mRNA sequences in vitro (O'Connell et al. 1997). This enzyme has an RNA binding domain (RBD)1 located in the C-terminal catalytic domain. Deamination of adenosine (A) in the mRNA results in inosine (I) at that position. Because inosine is translated as guanosine (G), the editing reaction causes a functional A to G replacement.

RNA editing is a term used to describe the structural alteration, insertion, or deletion of nucleotides in RNA that changes its coding properties. When the modification occurs in messenger RNA (mRNA), it can result in the translation of a protein sequence different from that predicted by the DNA sequence of the gene. Thus, this process plays an important role in creating functional diversity in the protein products of gene expression.

Reaction mechanism of ADARs is the requirement for duplex RNA structure in the ADAR substrate. The necessary trajectory of an attacking water molecule for hydrolytic deamination of adenosine makes it likely that the reactive nucleotide is flipped out of the duplex during reaction. This has been demonstrated by following fluorescence of 2-aminopurine (2 AP) -modified substrate. 2-AP is a fluorescent adenosine analog whose quantum yield and emission maximum are sensitive to the environment (Ward et al. 1969). Single-stranded oligonucleotides containing 2-AP are highly fluorescent. When present in a base-paired duplex, the 2-AP fluorescence is quenched. These characteristics make 2 AP an interesting probe to follow base-flipping in nucleotides.

ADAR2 possesses 5 Trp residues all located in the C-terminal domain (Yang et al. 1997). Binding of the protein to double-stranded RNA was first examined by measuring fluorescence modification of tryptophans. At saturating amount of dsRNA (200 nM), a 20% enhancement in the fluorescence intensity is observed without any modification in the position peak of the emission (λ_{em} = 345 nm) (Fig. 4.13). Single-stranded RNA has no effect on the ADAR2 emission as expected for a double-stranded RNA binding protein.

Fluorescence intensity quenching of Trp residues of ADAR2 and of ADAR2-RNA complex was performed with acrylamide. A Stern-Volmer plot shows that tryptophan quenching by acrylamide is homogeneous for ADAR2 alone (Fig. 4.14A). However, the Stern-Volmer plot of the tryptophan fluorescence quenching of ADAR2·RNA complex shows a downward curvature. This suggests that within the complex the tryptophans are not equally accessible to the quencher. The modified Stern-Volmer plot yields a y intercept equal to 4.4, i.e., an accessible fraction of 22.8 % (Fig. 4.14B). The main significance of Figure 4.14 is that a conformational change occurred at the C-terminal of the protein upon RNA binding. The value of the accessible fraction could be an indication of the number of Trp residues accessible to acrylamide or the percentage of the fluorescence accessibility occurring from more than one Trp residue.

In conclusion, the results indicate that RNA binding domain (RBD) of an ADAR alters the conformational dynamics of the RNA, indicating that this protein domain plays more than just a recognition role in the ADAR2-editing reaction. Finally, analysis of the ADAR2 tryptophan fluorescence in the presence and absence of RNA indicates the protein undergoes conformational changes upon substrate binding. Acrylamide quenching suggests that one of the five tryptophans in the catalytic domain is more exposed in the protein·RNA complex (Yi-Brunozzi et al. 2001).

A.

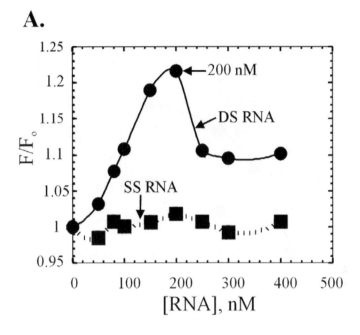

B.

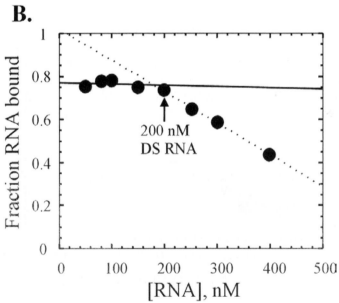

Figure 4.13. A: plot of the change in tryptophan fluorescence of 400 nM ADAR2 with increasing amounts of double-stranded RNA (DS R/G) (●) and single-stranded RNA (SS R/G) (■). B: fraction of DS R/G duplex bound to 400 nM ADAR2 with increasing concentrations of added RNA under the conditions of tryptophan fluorescence measurements obtained using a gel mobility shift assay. Source: Yi-Brunozzi, H.Y., Stephens, O. M. and Beal, P. A. 2001, J. Biol. Chem., 276, 37827-37833. Authorization of reprint accorded from The American Society for Biochemistry and Molecular Biology.

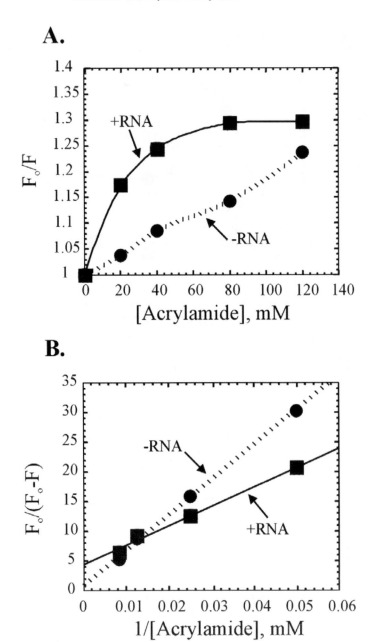

Figure 4.14. Acrylamide quenching study of tryptophan fluorescence using 400 nM ADAR2 and 200 nM DS R/G. A: Stern-Volmer plot of acrylamide quenching in the presence (■) and (●) absence of RNA. The downward curvature is indicative of a heterogeneous population of tryptophans. *B*, modified Stern-Volmer plot in the presence (■) and (●) absence of RNA. Source: Yi-Brunozzi, H.Y., Stephens, O. M. and Beal, P. A. 2001, J. Biol. Chem., 276, 37827-37833. Authorization of reprint accorded from The American Society for Biochemistry and Molecular Biology.

α-Toxin is a soluble hemolytic protein exotoxin secreted by *Staphylococcus aureus* considered as the major factor contributing to the pathogenicity of *S. aureus*. The toxin is composed of 293 amino acids and has a molecular weight of 33,400 (Kehoe et al., 1983; Gray and Kehoe, 1984). α-Toxin damages membranes by the formation of nonspecific oligomeric pores in the target membranes, which lead to cell lysis. These pores are large in size, diameters of 1 to 2 nm have been reported (Fussle et al., 1981; Menestrina, 1986). The protein contains 8 Trp residues and thus their fluorescence can be used to study the interaction between the protein and a membrane. The emission maxima of the native and denatured toxin in 8 M urea were observed at 333 and 352 nm, respectively. This does not mean that the tryptophans in the native state are all buried in the protein core. Quenching studies with iodide are necessary to reveal the fraction of tryptophan present at the protein surface.

In the native state, only 8.6% of the fluorescence was quenched with an effective K_{sv} equal to $0.38\,M^{-1}$. The accessible fraction f_a is found equal to 0.14, indicating that most of the tryptophan residues of the toxin are inaccessible to iodide. In the denatured state, 60% of the initial fluorescence is quenched by iodide, and the Stern-Volmer plot shows a negative deviation from linearity. The value of f_a is 0.63. These results indicate that some of the tryptophan residues are still buried in the protein core even on denaturation of the toxin in 8 M urea. All these results clearly point out to the fact that most of the tryptophan residues in the native toxin must be deeply buried in the toxin interior. Also, one should indicate here that since in the denatured state the tryptophan accessibility to the solvent is not 100%, this means that: 1) the protein is not completely unfolded, 2) there is a negative charge near some of the Trp residues.

Quenching studies performed with acrylamide that can penetrate the protein interior and quench buried tryptophans show that only 16% of the tryptophans in α-toxin were quenched. The value found for K_{SV} is $0.58\,M^{-1}$, a value much lower than that obtained to the Stern-Volmer quenching constant for a fully exposed tryptophan as in the case of *N*-acetyl-L-tryptophanamide in water ($K_{SV} = 17.5\,M^{-1}$) (Eftink and Ghiron, 1976), indicating the buried nature of most of the tryptophans in α-toxin.

Iodide quenching experiment performed on membrane-bound-α-toxin induces a Stern-Volmer plot with a negative deviation from linearity and gives for f_a and K_{SV} values equal to 0.78 and $2.96\,M^{-1}$, respectively. These results suggest that most of the buried tryptophan residues of α-toxin in the soluble form are exposed to the iodide and thus to the solvent in the membrane-bound form. Therefore, upon binding to the membrane, α-toxin would display a modification in its tertiary structure.

The authors exposed briefly the several structural changes that α-toxin must go through before reaching its functional pore-forming state. The three following steps are mentioned: 1) binding of monomer to the membrane, probably accompanied by a mild denaturation of the monomer at the interface, resulting in a conformational change and formation of a molten globule state of the monomer; 2) formation of heptamer by lateral diffusion in the plane of the membrane (interfacial region) and stabilization of the heptamer by formation of intermonomeric contacts; and 3) a second conformational change, resulting in the spontaneous insertion of the loop into the membrane and formation of the membrane active heptameric pore (Raja et al. 1999). The presence of the tryptophan residues in α-toxin could be an important key to the interaction of the protein with the membrane as the result of their hydrophobic and hydrophilic properties. This property helps the tryptophans to float in the interfacial region of the membrane.

4. Static quenching

4a. Theory

The dynamic quenching we described in the previous three sections occurs within the fluorescence lifetime of the fluorophore, i.e. during the lifetime of the excited state. This process is time dependent. Fluorescence quenching can also take place by the formation at the ground state of a non-fluorescent complex. When this complex absorbs light, it immediately returns to the fundamental state without emitting any photons. This type of complex is called static quenching and it can be described with the following equations:

$$F + Q \longrightarrow FQ \qquad (4.16)$$

The association constant K_a is equal to

$$Ka = \frac{[FQ]}{[F]_l \ [Q]_l} \qquad (4.17)$$

where $[FQ]$ is the complex concentration, and $[F]_l$ and $[Q]_l$ the concentrations of free fluorophore and quencher.

Going from the fact that the total fluorophore concentrations is

$$[F]_o \ = \ [F]_l + [FQ] \qquad (4.18)$$

and that bound fluorophore does not fluoresce, replacing Equation 4.18 in 4.17 yields

$$Ka = \frac{[F]_o - [F]_l}{[F]_l \ [Q]_l} \quad ; \quad Ka \, [Q]_l = \frac{[F]_o - [F]_l}{[F]_l} = \frac{[F]_o}{[F]_l} - 1$$

$$[F]_o \ / \ [F]_l \ = 1 + \ K_a \, [Q]_l \qquad (4.19)$$

Since the concentrations are proportional to the fluorescence intensities, Eq.4.19 can be written as

$$I_o / I \ = 1 + K_a \, [Q]_l \qquad (4.20)$$

If we consider the concentration of bound quencher very small compared to the added quencher concentration, then free quencher concentration is almost equal to the total added concentration. Thus, Eq.4.20 can be written as

$$I_o \ / \ I \ = 1 + K_a \ [Q] \qquad (4.21)$$

Therefore, plotting $I_o \ / \ I \ = f \, ([Q])$ yields a linear plot whose slope is equal to the association constant of the complex (Fig. 4.15).

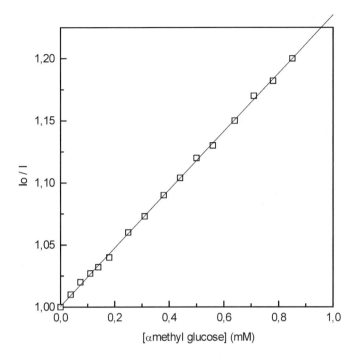

Figure 4.15. Determination of the binding constant of α – methylglucose on *Lens culinaris* agglutinin. The dissociation constant obtained from the slope of the plot is equal to 4.25 ± 1.5 mM, an associaion constant equal to $2.35 \times 10^2 \, M^{-1}$. The plot shown is from 9 experiments. λ_{ex} = 280 nm and λ_{em} = 330 nm. The same data were obtained when the emission wavelength was 310 or 350 nm. This value of the association constant for the glucose-LCA complex is close to that found for the LCA-mannose complexe ($K_a = 5.6 \times 10^2 \, M^{-1}$).

In the static quenching, one observes only the fluorophore that is still emitting, i.e., free fluorophore in solution. Complexed fluorophore does not fluoresce and is not observed in the derived equations. Fluorescence lifetime of free fluorophore is the same whether all fluorophore molecules are free or some are complexed. Thus, fluorescence lifetime of fluorophore does not change upon increasing quencher concentration. This implies the following equation:

$$\tau_o / \tau_{(Q)} = 1 \qquad (4.22)$$

Therefore, in the static quenching, to the difference of the dynamic quenching, one observe an intensity decrease only. Binding of TNS to α_1-acid glycoprotein induces a decrease in the fluorescence intensity if the Trp residues of the protein. Fluorescence lifetime of the intrinsic fluorophore is not modified. The variation of the Trp residues intensity and lifetime are analyzed by plotting the intensities and the lifetimes in the absence and presence of TNS as a function of TNS concentration (Fig. 4.16). It is clear from the figure that TNS - α_1-acid glycoprotein interaction is of a static nature.

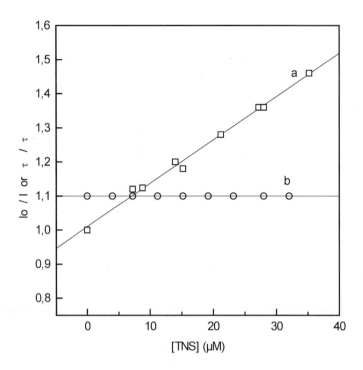

Figure 4.16. Quenching of the tryptophan fluorescence of α_1-acid glycoprotein (4 µM) by TNS at 20°C. Intensity (a) and lifetime (b) variation. Adapted from Figure 3 of : Albani, J., Ros, V., Willaert, K. and Engelborghs, Y. 1995. Photochem. Photobiol. 62, 30-34. Authorization of reprint accorded by the Americal Society for Photochemistry.

Dale Spencer wrote in his thesis that "static quenching is due to the reduction of the number of free fluorophore molecules available for excitation after complex formation in the ground state and not to depopulation of the excited state throughout the lifetime" (Spencer, 1970).

In the static quenching, one uses intrinsic or extrinsic fluorophore to probe the interaction between two macromolecules. The fluorophore should be bound to one of the two proteins only (case of the extrinsic fluorophores) or should be part of it (case of the intrinsic fluorophores) if we want to perform such experiments. Also, binding parameters of the fluorophore-macromolecule complex can be determined by following fluorescence modification of the fluorophore observables.

Usually, when studying protein-protein interaction, fluorescence variation of the probe can indicate the type of the interaction. For example, one can find out whether the interaction between the two proteins is cooperative or not, etc... Two examples to illustrate this, we used 2,p-toluidinylnaphthalene-6-sulfonate (TNS) as a fluorescent probe to study the interaction between cytochrome c and cytochrome b_2 core and to follow the binding of cytochrome b_2 core on flavodehydrogenase. The three proteins are extracted from the yeast *Hansenula anomala* (see chapter 1, paragraph 6f).

4b. Cytochrome c – cytochrome b₂ core interaction

Interaction between cytochrome c and cytochrome b_2 core was investigated with stopped-flow spectrophotometry. Electron – exchange rate was found modulated by ionic strength, following the Debye-Hückel relationship with a charge factor $Z_1 Z_2 = -1.9$ (Capeillère-Blandin and Albani, 1987). In order to understand the meaning and the importance of the charge factor in the stability of the proteins complex, one should compare the value found with those of the flavocytochrome b_2 – cytochrome c and the flavodehydrogenase domain – cytochrome c complexes that occur with charge factor equal to -5.7 (Capeillère-Blandin, 1982) and -4.4 (Capeillère-Blandin et al. 1980), respectively. Therefore, the presence of the flavodehydrogenase domain increases the contribution of the electrostatic interactions in the control of cytochrome b_2 reactivity. This simply means that interaction between cytochrome c on one side and flavocytochrome b_2 and its two functional domains on the second side is electrostatic.

Also, this means that the cytochrome c – cytochrome b_2 core interaction is weaker than the interaction observed between cytochrome c with the two other proteins. Electrostatic interaction would pre-orient the proteins before any physical contact, facilitating the formation of a complex. Thus, when the interactions are weak, the pre – orientation is not adequate at every collision and the probability of formation of a complex is weak

In order to study cytochrome c – cytochrome b_2 core interaction by fluorescence spectroscopy, we first prepared apocytochrome b_2 core and then we investigated the binding affinity between cytochrome c and TNS-apocytochrome b_2 core complex. Binding of cytochrome c leads to a decrease in the fluorescence intensity of the TNS bound to apocytochrome b_2 core (Fig. 2.7). This decrease follows a hyperbolic curve (Fig. 4.17) (Albani, 1985) suggesting that interaction between cytochrome c and apocytochrome b_2 core is not cooperative. One can notice also that interaction between the two proteins is ionic strength dependent. At 20 mM ionic strength, the stoichiometry of the cytochrome c-apocytochrome b_2 core complex is equal to 1:1 and the dissociation constant K_d is equal to 6.3 μM. At 100 mM, the stoichiometry of the complex is not reached and the dissociation constant of the complex is found equal to 31 μM.

The dependence of the Kd on the ionic strength can be fitted as a Debye-Huckel plot, $- \log K_d = f(\sqrt{I})$ based on the relation

$$-\log K_d = -\log K_{d(O)} + + 2A Z_1 Z_2 \sqrt{I} \qquad (4.23)$$

where K_d is the dissociation constant at ionic strength I, $K_{d(O)}$ is the dissociation constant at zero ionic strength, $2A = 1$, and $Z_1 Z_2$ is the product of the charges intervening in the two proteins interaction. The slope $Z_1 Z_2$ obtained is equal to -4 instead of the value, -2.6, found from the electron transfer studies between the cytochrome c and the cytochrome b_2 core (Capeillère-Blandin and Albani, 1987). This result suggests that one or two addition al charges are present in the cytochrome c-apocytochrome b_2 core interaction compared to the cytochrome c-cytochrome b_2 core situation. We cannot explain from our results the origin of these charges which are very important at high ionic strengths and which decrease the affinity between the interacting proteins.

Thus the K_d found at 100 mM ionic strength for the cytochrome c-apocytochrome b_2 core case is higher than the K_d for the cytochrome c-cytochrome b_2 core case in this ionic strength range. However, at 20 mM ionic strength, the effects of electrostatic charges are less then at 100 mM and the K_d value obtained, 6 ± 2 µM, is the same for both these protein complexes. This value is in the same range of that (1 µM) estimated at 10°C by other researchers for the dimer Zn cytochrome c-cytochrome b_2 core complex (Thomas et al. 1983). Hence, using the Debye-Huckel equation and considering $Z_1 Z_2 = -2.6$, we can estimate the value of the K_d of the cytochrome c-cytochrome b_2 core complex, at 100 mM ionic strength, to be equal to 18 µM, which is about a factor of 2 less than the value found from our original estimate.

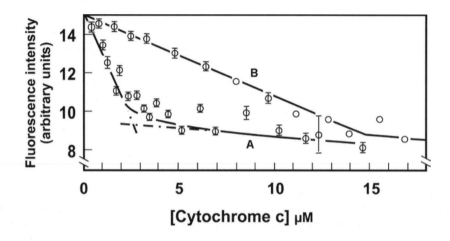

Figure 4.17. Fluorescence quenching of the TNS bound to the apocytochrome b_2 core as a result of cytochrome c-apoprotein interaction. Aliquots of cytochrome c (0.4 µM each) are added to a solution of 2.2 µM apocytochrome b_2 core and 7 µM TNS. The fluorescence intensities are normalized to a constant final volume and then corrected for the absorbance at 320 and 433 nm. At 20 mM ionic strength, one cytochrome *c* is bound per molecule of apocytochrome b_2 core. The plot of the TNS fluorescence intensity in the absence of cytochrome *c* over that in the presence of cytochrome *c* vs [cytochrome c] gives a line with a slope equal to the association constant of the complex, cytochrome c-apocytochrome b_2 core. At 20 mM ionic strength, the mean value of the dissociation constant obtained from four experiments is equal to 6.3 ± 2 µM. Source: Albani, J. 1995, Archives of Biochemistry and Biophysics. 243, 292-297.

4c. Cytochrome b_2 core – flavodehydrogenase interaction

Although it has been established that both flavodehydrogenase and cytochrome b_2 core interacts with cytochrome c, we do not know how the cytochrome b_2 core and flavodehydrogenase react with each other. X-ray diffraction studies performed on flavocytochrome b_2 from *Saccharomyces cerevisiae* indicated that two cytochrome b_2 cores are well ordered in the crystal lattice and the two others are disordered, completely absent from the electron-density map (Zia et al. 1990).

Interaction between cytochrome b_2 core and flavodehydrogenase was performed by recording fluorescence intensity change of TNS bound to the flavoprotein at different concentrations of cytochrome b_2 core.

Addition of cytochrome b_2 core to a solution of TNS does not modify the fluorescence of the probe. However, upon addition of the cytochrome to a solution of FDH-TNS complex, an increase in TNS fluorescence was observed, as the result of an increase in the affinity between the probe and the flavoprotein (λ_{ex}, 320 nm) (Albani, 1993). X-ray diffraction studies have indicated that FMN is buried in the flavin-binding domain of flavocytochrome b_2 (Zia et al. 1990), and fluorescence studies have shown that binding of TNS to the FDH does not induce any release of the flavin from its binding site (Albani, 1993). Titration of a constant concentration of FDH-TNS complex with cytochrome b_2 core yields a sigmoidal curve for the TNS intensity increase (Fig. 4.18) (Albani, 1997). Thus, interaction between cytochrome b_2 core and FDH is cooperative.

In this case, the plot of 1 / I vs 1 / [cytochrome b_2 core] is curved and approaches a horizontal line that intersects the 1 / I axis at 1 / I_{max} , the maximum intensity reached when the binding site of the cytochrome b_2 core on the FDH is saturated.

Analysis of the data of Fig. 4.18 with the Hill equation (Fig. 4.19) (Eq. 4.24)

$$Log (F / F_{max} - 1) = n \log [\text{cytochrome } b_2 \text{ core}] - \log K' \qquad (4.24)$$

yields n, the number of binding sites, and K', the apparent dissociation constant.. The maximum slope (n = 3.34) indicates that there are at least four cytochrome b_2 core binding sites on FDH because n must be higher than or equal to n. Binding of the cytochrome b_2 core to FDH follows a cooperative manner. One interpretation of the data is that the four cytochromes do not have the same affinity for the flavoprotein. These results are consistent with those found by X-ray diffraction studies (Zia et al. 1990), *i.e.* there are two classes of cytochromes with different affinities for the FDH protomers.

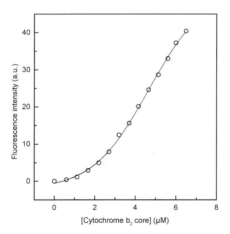

Figure 4.18. Titration of $[FDH]_4$ - TNS complex with cytochrome b_2 core. Binding of cytochrome b_2 core to the complex was followed by the fluorescence change of TNS. [TNS] = $[FDH]_4$ = 5 µM. λ_{ex} = 320 nm and λ_{em} = 440 nm. Source: Albani, J. R., 1997, Photochem . Photobiol. 66, 72 – 75. Authorization of reprint accorded by the American Society for Photobiology.

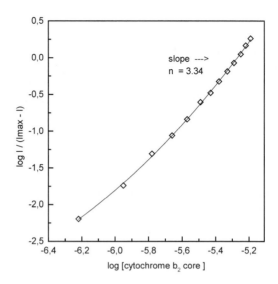

Figure 4.19. The Hill representation of the TNS - [FDH]₄ - cytochrome b₂ core interaction. Source: Albani, J. R., 1997, Photochem . Photobiol. 66, 72 – 75.

4d. Determination of drug binding to α_1 – acid glycoprotein

In terms of function, α_1 – acid glycoprotein is probably best known as a plasma drug binder of basic, as well as some acidic and neutral drugs. The outcome of the binding of drugs to α_1 – acid glycoprotein is a decrease in the free, and therefore, active concentration of the drug in the plasma that subsequently reduces the level of therapeutic effect. Recently it has been discovered that the interaction of drugs with α_1 – acid glycoprotein can be studied using fluorescence spectroscopy, a technique that is based on the intrinsic fluorescence of tryptophan residues present in the polypeptide backbone of α_1–acid glycoprotein. Upon addition of a drug to α_1 – acid glycoprotein solution, the fluorescence intensity of the protein solution will be quenched, should there be binding between the drug and protein, due to a masking of the fluorescent contributions of tryptophan residues. After excitation at 280 nm, the emission spectrum of α_1–acid glycoprotein is recorded between 300 and 400 nm. The fluorescence peak is located at 335 nm.

The measurement of α_1–acid glycoprotein-drug binding by spectrofluorimetry is best illustrated by an example. Following the addition of 500 µM theophylline, a drug commonly used to treat chronic asthma, the peak fluorescence of α_1–acid glycoprotein (0.5mg/mL; 12.2 µM) was found to decrease by 95%, indicating a high degree of binding between the protein and theophylline. The addition of 250 µM theophylline resulted in a decrease in α_1–acid glycoprotein fluorescence by 83%, while 125 µM drug resulted in 60% quenching of α_1–acid glycoprotein fluorescence (Figure 4.20).

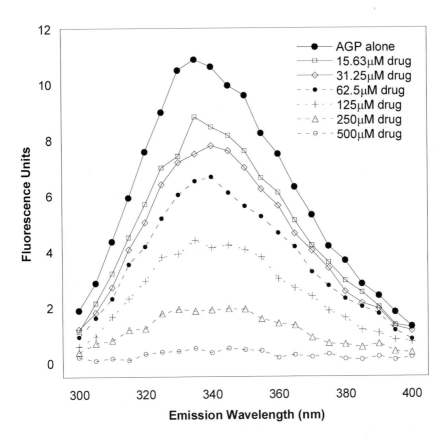

Figure 4.20. Binding of α_1–acid glycoprotein to theophylline. Increasing concentrations of theophylline results in increased levels of quenching of the fluorescence of the protein Trp residues. λ_{ex} = 280 nm and λ_{em} = 335 nm. Unpublished results by Sarah Paterson BSc(Hons) and Kevin Smith PhD, Department of Bioscience, University of Strathclyde, Royal College Building, 204 George Street, Glasgow, G1 1XW, Scotland. A work supported by the Caledonian Research Foundation.

The quenching, or percentage decrease in α_1–acid glycoprotein fluorescence following addition of drug, can be plotted as a function of drug concentration, and the data fitted using a quadratic binding equation (Parikh et al, 2000) (Fig.4.21) :

$$\% \text{ Quenching} = C_1 \frac{S - \sqrt{S^2 - 4\,D_T\,P_T}}{2} - C_2\,D_T \qquad (4.25)$$

where D_T = total drug concentration; P_T = total protein concentration; $S = D_T + P_T + K_d$; C_1, C_2 and K_d are unknown constants, determined by least squares curve fitting

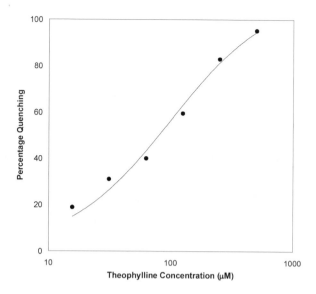

Figure 4.21. Quenching curve showing the degree of binding between AGP and theophylline at varying concentrations. These experiments show that there is concentration dependent binding of theophylline to α_1–acid glycoprotein. Unpublished results by Sarah Paterson BSc(Hons) and Kevin Smith PhD, Department of Bioscience, University of Strathclyde, Royal College Building, 204 George Street, Glasgow, G1 1XW, Scotland. A work supported by the Caledonian Research Foundation.

5. Comparison between dynamics and static quenching

We saw in the previous sections that a collisional quenching induces a decrease in both fluorescence intensity and lifetime while in presence of a static quenching only the intensity decreases, the fluorescence lifetime remains unchanged.

It is clear from the examples we have given and from the definitions of the two types of quenching, that static quenching occurs when a complex is formed between two molecules. Collisional quenching does not need the formation of a stable complex with a long lifetime. The fluorophore and the quencher enter into collision inducing the decrease of the fluorescence lifetime and intensity of the fluorophore.

Collisional quenching needs a direct physical contact between the fluorophore and the quencher, which is not the case in the static quenching. In the latter case, the fluorophore should be on one of the two interacting molecules and its fluorescence should be sensitive to the interaction between the two molecules. A direct physical contact between the fluorophore and the quencher is not necessary.

Thus, taking into consideration the above definitions, it will be possible to distinguish between the two types of quenching without measuring the fluorescence lifetime at different quencher concentrations.

l) Dynamic quenching is diffusion dependent. The coefficient of diffusion D of a molecule is given by the Stokes-Einstein equation:

$$D = k_q \, T \, / \, 6 \, \pi \, \eta \, R \tag{4.26}$$

where k_q is the Boltzman constant, equal to the bimolecular diffusion constant and η the solvent viscosity. Since D is proportional to T/η and k_q to D, the value of k_q is proportional to T/η. Thus, an increase of the temperature will induce an increase of the diffusion coefficient and of k_q (Fig. 4.22). The diffusion dependent of the dynamic quenching means that in highly viscous solutions, dynamic quenching is difficult to occur.

In presence of static quenching, increasing the temperature destabilizes the formed complex and thus decreases its association constant. Figure 4.23 shows the variation of the dissociation constant of protein L and a human κ light chain with the temperature (Beckingham et al. 1999). Protein L is a multidomain cell-wall protein isolated from Peptostreptococcus magnus. It contains repeated domains that are able to bind to Igs without stimulating an immune response. It binds exclusively to the kappa –chain on two binding sites. The function of protein L in vivo is not clear but it is thought that it enables the bacteria to evade the host's immune system. Two binding sites for kappa - chain on a single Ig-binding

2) The formation of a complex at the fundamental state can be characterized by the disruption of the absorption spectrum of one of the two molecules that forms the complex, as it is the case for binding of bichromic cyanine dyes (Fig. 4.24) to DNA (Fig. 4.25) or in the interaction between ANS and apomyoglobin (Fig. 1.15).

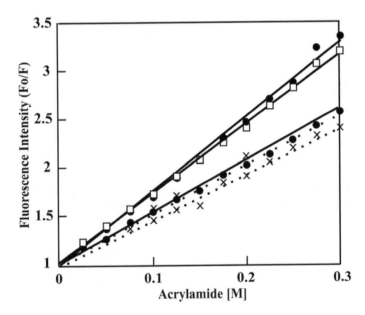

Figure 4.22. Acrylamide quenching of cyt b_5 at 35°C (dashed lines) and 50°C (solid lines) for wild type (·), mutant S18C:R47C$_{ox}$ (×), and mutant S18C:R47C$_{red}$ (□). The susceptibility of S18C:R47C$_{ox}$ to acrylamide quenching does not increase with temperature. In contrast, K_{SV} increases significantly for wild type, S18D, and S18C:R47C$_{red}$. Source: Storch, E. M., Grinstead, J. S., Campbell, A. P., Daggett, V. and Atkins, W. M. 1999, Biochemistry, 38, 5065- 5075. Authorization of reprint accorded by the American Chemical Society.

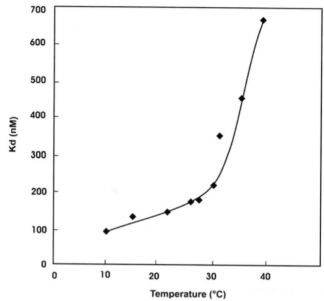

Figure 4.23. Variation of K_d for the binding reaction between F39W and κ-chain at various temperatures. Source: Beckingham, J. A., Bottomley, S. P., Hinton, R., Sutton, B. J. and Gore, M. G. 1999. Biochem. J. 340, 193-199. Authorization of reprint accorded by Portland Press.

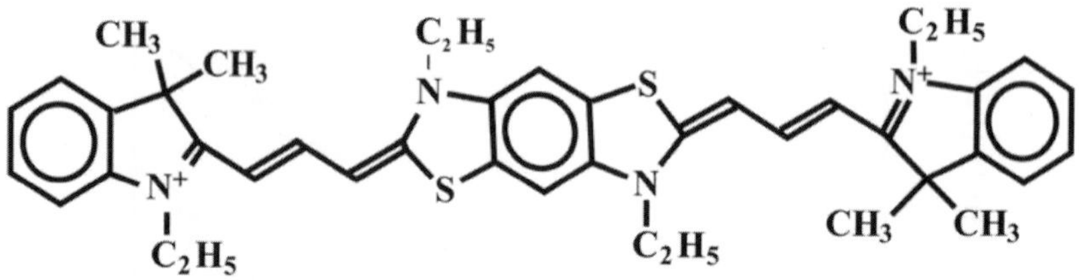

Figure 4.24. Structure of a bichromic cyanine dye synthesized by Prof. Felix Mikhailenko in the Institute of Organic Chemistry of the Academy of Science of Ukraine.

Cyanine dyes possess high affinity for DNA, their quantum yields increase considerably upon binding. Also, due to their important absorption in the red and near infrared spectral region, they are photostable and monotoxic and thus they are considered potential photoactive compounds for photodynamic therapy such as detection of human breast cancer. Three-dimensional reconstruction of optical signatures based on diffuse optical tomography can be performed. This technique employs diffuse light that propagates through tissue, at multiple projections yielding three-dimensional quantified tomographic images of the internal optical properties of organs (Ntziachristos and Chance, 2001).

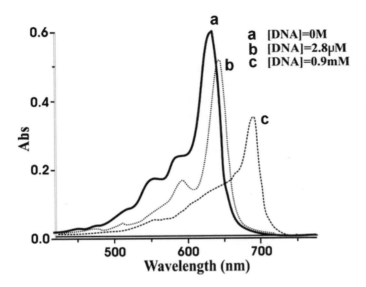

Figure 4.25. Absorption spectra of 5 μM bichromic cyanine dye in absence and presence of two DNA concentrations. Source: Schaberle, F. A., Kuz'min, V. A. and Borissevitch, I. E. 2003, Biochimica Biophysica Acta. 1621, 183-191.

6. Combination of dynamic and static quenching

In several cases, dynamic and static quenching can occur together. In this case, the Stern-Volmer plot of the fluorescence intensity quenching yield a linear plot accompanied by an upward curvature. The Stern-Volmer equation can be written as:

$$F_0 / F = (1 + K_{SV} [Q]) (1 + K_a [Q]) \qquad (4.27)$$

where K_{SV} and K_a are respectively the dynamic and static constants. The latter is simply the association constant of the complex. Equation 4.27 can be written also as:

$$F_0 / F = 1 + (K_{SV} + K_a) [Q] + K_{SV} K_a [Q]^2 \qquad (4.28)$$

or

$$\frac{F_0 / F - 1}{[Q]} = (K_{SV} + K_a) + K_{SV} K_a [Q] \qquad (4.29)$$

Plotting $(F_0 / F - 1) / [Q]$ as a function of [Q], one obtains a linear graph with a slope equal to $(K_{SV} + K_a)$ and a y-intercept equal to $(K_{SV} * K_a)$.

An example, fluorescence quenching of 10-methyl acridinium chloride (MAC) by the nucléotide guanosine 5 ' – monophosphate (GMP) is a dynamic and static quenching combination. The Stern-Volmer plot (Fig. 4.26), obtained from the intensity quenching, indicates that at high quencher concentration a static quenching is observed. The dynamic constant can be obtained from the Stern-Volmer plot of the lifetime quenching.

Plotting the data of figure 4.26 with equation 4.29 allows separating the dynamic and the static constants (Fig. 4.26) (Kubota et al. 1979). One can notice that the dynamic constant can be obtained from the lifetime data or / and by plotting the asymptote to the intensities plot at low quencher concentrations.

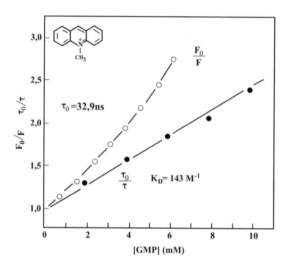

Figure 4.26. Stern-Volmer plots of the fluorescence (intensity and lifetime) quenching of 10-methyl acridinium chloride (MAC) by the nucléotide guanosine 5 ' - monophosphate (GMP).

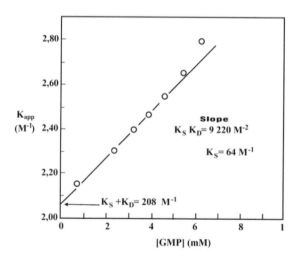

Figure 4.27. Determination of the static (K_S) and dynamic (K_D) constants for MAC and GMP interaction.   Source: Kubota, Y., Matoda, Y., Shi-Genuare, Y. and Fujisaki, Y. 1979, Photochem. Photobiol. 29, 1099-1106.

7. Thermal intensity quenching

Deexcitation of a fluorophore occurs via the different competitive mechanisms we have already described in the Jablonski diagram. The global rate constant with is the inverse of the fluorescence lifetime can be considered as equal to the sum of the different rates of the competitive mechanisms. Thus one can write :

$$1 / \tau_o = k_r + k_{isc} + k_i + \Sigma\, k_{qi} \qquad (4.30)$$

where k_r is the radiative constant, k_{isc} the intersystem crossing constant, k_i the constant corresponding to deexcitation due to the effect of the temperature and $\Sigma\, k_{qi}$ the different quenching mechanisms such as proton transfer or electron transfer. In presence of energy transfer at distance (Förster energy transfer), another rate constant should be added with is k_T.

Working at constant temperature allows controlling the different parameters of the system and one can detail and quantify the mechanisms intervening in the depopulation of the excited state of the fluorophore.

However, what will happen when we measure fluorescence lifetime or intensity at different temperatures? Free in solution, a fluorophore can be temperature dependent or temperature independent. In the first case, variation of temperature will affect both lifetimes and intensity. In the second case, none of these parameters will be affected. Let us see now how are affected the other constant rates with temperature? Modifying the temperature does not affect k_{isc}, since the intersystem-crossing is temperature independent.

The radiative constant k_r is equal to

$$k_r = Q_F / \tau_o \qquad (4.31)$$

where Q_F is the fluorescence quantum yield and τ_o is the mean fluorescence lifetime. Increasing the temperature will increase Brownian motions and thus facilitating loss of energy via dynamic quenching by the solvent. This will induce a decrease of the fluorescence quantum yield and in parallel of the fluorescence intensity. Fluorescence lifetime is less sensitive to temperature than quantum yield and thus observed fluorescence lifetime decrease with temperature is not proportional to the decrease observed for the quantum yield or the intensity. Therefore, upon increasing the temperature, k_r decreases. If k_r is temperature-independent, this means that the fluorescence quantum yield and lifetime decrease proportionally in the same range with temperature. This is not true for many fluorophores.

The constant k_i or thermal constant increases with temperature. This constant is called the solvent constant since the latter is considered as the responsible for the temperature dependence of the fluorescence lifetime. This constant allows determining the activation energy of the fluorophore with the classical Arrhenius theory:

$$k_i = A \exp (- E / RT) \qquad (4.32)$$

where A is the temperature independent factor equal to $10^{15} - 10^{17}\, s^{-1}$.

E is the Arrhenius activation energy expressed in kcal / mole, R is the molar gas constant and T the temperature in Kelvin degrees. Plotting Ln k_i as a function of 1 / T yields the value of E.

However, it is difficult to measure k_i . Taking into consideration the fact that not all constant rates are temperature dependent and neglecting the presence of an energy transfer, Eq.4.30 can be simplified to:

$$1 / \tau_o = k_r + k_i \qquad\qquad (4.33)$$

Thus, Eq.4.32 can be written as:

$$k_i = 1 / \tau - k_r = A \exp (- E / RT) \qquad\qquad (4.34)$$

Since the fluorescence lifetime can be obtained experimentally and k_r calculated, it will possible to obtain the value of E by plotting Ln $(1 / \tau - k_r)$ as a function of 1 / T. One should indicate that since k_r is at least 10 times less than 1 / τ , it is no more taken into account in Eq.4.34.

Figures 4.28 and 4.29 display respectively the fluorescence lifetime of FMN with temperature and the Arrhenius plot obtained with Eq.4.34.

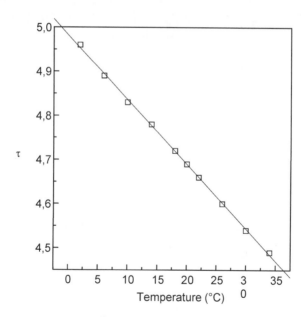

Figure 4.28. Fluorescence lifetime variation of FMN with temperature. Source: Spencer, R.D. 1970. PhD Thesis.

We can notice that the fluorescence lifetime decreases when the temperature increases. This lifetime decrease is linear, and at 34°C fluorescence lifetime value is

10% lower than the one recorded at 2.5°C. This decrease is significant and reflects the high sensitivity of FMN fluorescence to temperature. Surrounding environment can play an important role in the dependence of the lifetime or / of any other fluorescence parameter on the temperature.

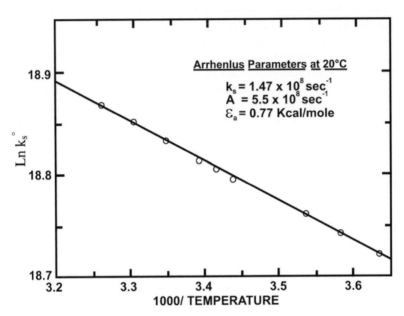

Figure 4.29. Arrhenius plot of the rate of radiationless transitions of FMN in water. Source: Spencer, R.D. 1970. PhD Thesis.

The value of the activation energy of FMN is 0.77 kcal / mole. This value is smaller than that (4.2 kcal / mole) found for NADH in the same conditions. This means clearly that the difference in the radiotionless transitions is characteristic of the structures of the studied compounds.

Activation energy of the FMN-AMP complex was also determined and was found equal to 5.6 kcal / mole, revealing that the structure around the fluorophore and the nature of the interaction between the two molecules will also play an important role in the definition of the activation energy. Therefore, one cannot deal with a free fluorophore in solution as when the same fluorophore is a part of a protein or bound to it.

In macromolecules such as proteins, temperature will have a non negligible effect on the Förster energy transfer. This phenomenon depends on the distance between the donor and acceptor and on their relative orientation. Since the proteins exert dynamical and structural modifications with temperature, distance between donor and acceptor and their relative orientation can be modified inducing by that a variation in the energy transfer mechanism. Thus, energy transfer is temperature dependent especially if the temperature induces important structural and dynamical fluctuations within the protein. See for example Table 4.1 where some of the energy transfer data between two IAEDANS-FITC pair in Ca- and Mg- G-actin are shown at different temperatures.

During muscle contraction, flexibility of actin filaments would play an important role in the actin-myosin interaction. Actin monomer is divided into two domains (the small and the large domains), and both of these domains consist of two subdomains (numbered 1 and 2 in the small domain and 3 and 4 in the large one). The small domain plays an important role in the formation of intermolecular contacts between neighboring protomers in the actin filament. The fluorescent probes IAEDANS attached to cysteine 374 residue in the monomer, and FITC attached to Lys-61 (Fig. 4.30) are suitable fluorophores to study energy transfer at distance. In fact, emission spectrum of IAEDANS overlaps absorption spectrum of FITC. Thus, excitation of IAEDANS will generate a fluorescence emission characteristic of fluorescein (see Table 3.2). It is important to remind that measuring energy transfer parameters is possible only if binding of the external probes to actin does not induce by itself a structural modification in the protein.

Table 4.1. Measured and calculated FRET parameters for the IAEDANS-FITC pair in Ca- and Mg-G-actin.

Temperature (°C)	R_0 (nm)		E(%)		R(nm)	
	Ca^{2+}	Mg^{2+}	Ca^{2+}	Mg^{2+}	Ca^{2+}	Mg^{2+}
6.6	5.19	5.28	69.39 (0.42)	70.53 (0.50)	4.53 (0.08)	4.56 (0.08)
10.2	5.13	5.23	68.85 (0.45)	69.90 (0.39)	4.50 (0.08)	4.54 (0.08)
14.0	5.06	5.17	67.78 (0.40)	68.96 (0.38)	4.49 (0.08)	4.53 (0.08)
17.8	5.02	5.11	66.54 (0.38)	68.21 (0.36)	4.48 (0.09)	4.50 (0.09)
21.5	4.96	5.04	64.94 (0.23)	67.14 (0.28)	4.48 (0.09)	4.47 (0.09)
25.8	4.90	4.98	63.44 (0.32)	65.69 (0.35)	4.47 (0.09)	4.46 (0.09)
29.5	4.84	4.90	61.08 (0.42)	64.31 (0.41)	4.49 (0.09)	4.45 (0.09)
33.7	4.78	4.84	58.71 (0.79)	62.56 (0.86)	4.51 (0.09)	4.44 (0.09)

Source: Nyitral, M., Hild, G., Lakos, Zs and Somogyi, B. 1998. Biophys. J. 74, 2474 - 2481. Authorization of reprint accorded by the American Biophysical Society.

The data show that while the distance R that separates the two fluorescent probes does not change significantly with temperature, the distance R_o at which energy transfer occurs at 50% and the energy transfer E decrease with increasing temperatures. Local fluctuations exist within the two subdomains 1 and 2. This flexibility is not identical within the two subdomains and is dependent on the nature of the metal ion bound to the subdomains.

Efficiency of energy transfer at high temperatures is dependent on the nature of the ion complexed to the protein. This could be the result of the difference in the local dynamics that seems to be modified with the nature of the ligand.

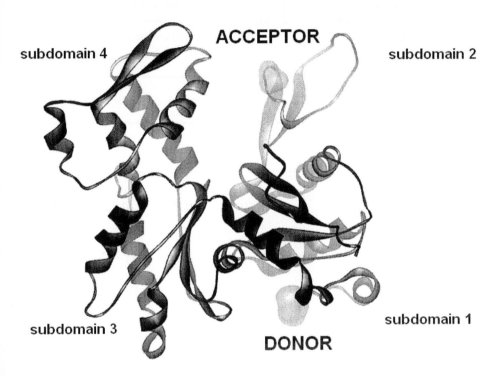

Figure 4.30. Schematic representation of the three-dimensional atomic structure of rabbit skeletal actin (Kabsch, W., Mannherrz, H. G., Shuck, D., Pai, E. F. and Holmes, K. C., 1990. Nature. 347, 37-44) showing the position of the two residues, Lys-61 (acceptor) and Cys-374 (donor), that were modified. Because the three COOH-terminal residues (involving Cys-374) are not resolved in the crystal structure (Kabsch et al., 1990), the labeled position of the donor molecule should be taken as an approximation. The four subdomains are also labeled (the coordinates were obtained from the Brookhaven Protein Data Bank, file 1ATN). Source: Nyitral, M., Hild, G., Lakos, Zs and Somogyi, B. 1998. Biophys. J. 74, 2474 - 2481. Authorization of reprint accorded by the American Biophysical Society.

When dealing with macromolecules such as proteins for example, k_i will characterize the thermal dependence of the protein matrix. Temperature modifies proteins motions, and thus k_i will characterize proteins dynamics.

Since this dynamics is characterized by observing the fluorescence variation of a specific fluorophore within the protein, k_i will characterize the motion of the environment of the fluorophore.

Table 4.2 shows the fluorescence lifetimes of the two Trp residues of immunophilin FKBP59 as a function of the temperature. One can notice that the long fluorescence lifetime is highly affected by the temperature. The activation energy was calculated from the Arrhenius representation using either the mean excited-state lifetime or the longest lifetime. Activation energy was found equal to 14.5 and 24 kcal.mol^{-1}, respectively. These high values indicate that Trp59, although located at the protein surface, needs important thermal energy to provide efficient collisions with either the solvent molecules or amino acid side chains. Thus, the microenvironment of the fluorophore is more rigid than expected for an exposed Trp residue to the solvent.

Table 4.2. Excited-state lifetime (ns) distribution parameters of FKBP59-I fluorescence emission as a function of temperature

Temperature (°C)	τ_1	τ_2	τ_3
	f_2	f_2	f_3
5	0.45 (0.15)	2.41 (0.07)	6.96 (0.78)
10	0.46 (0.16)	2.87 (0.07)	6.77 (0.77)
15	0.42 (0.18)	2.63 (0.07)	6.51 (0.75)
20	0.44 (0.18)	2.82 (0.07)	6.20 (0.75)
30	0.39 (0.19)	1.91 (0.03)	5.52 (0.78)
40	0.35 (0.19)	1.99 (0.05)	4.82 (0.76)

Source: Rouvière, N., Vincent, M., Craescu, C. and Gallay, J. 1997, Biochemistry 36, 7339-7352. Authorization of reprint accorded by the American Chemical Society.

In general, fluorescence intensity is more sensitive to temperature than lifetime, thus fluorescence thermal quenching followed by fluorescence intensity will allow probing the local dynamics of the fluorophore. Also, it will be possible to understand the nature of the fluorophore interaction with its binding site.

Figure 4.31 displays the effect of temperature on fluorescence intensity of native and guanidine unfolded AEDANS-Rnase. Increasing the temperature from 10 to 30°C induces a decrease in the fluorescence intensity of both states of proteins. One can notice that the intensity decrease of the native protein is more affected by the temperature than the guanidine unfolded protein. This thermal quenching is the consequence of the rapid movements of the protein structure around the fluorescent probe.

These movements occur during the lifetime of the excited state and their rate is temperature dependent.

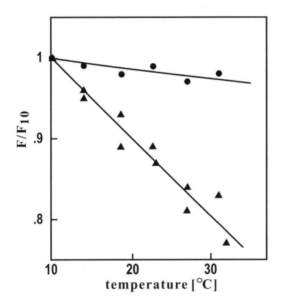

Figure 4.31. Effect of temperature on fluorescence intensity of native (▲) and guanidine unfolded (●) AEDANS-RNase. Intensities are expressed relative to that measured at 10°C for the same sample (native or unfolded). Buffer is 50 mM cacodylate, pH 6.5, in the absence or presence of 6 M guanidine hydrochloride. Protein concentration is 10^{-5} M. Excitation wavelength: 350 nm, emission wavelength: 476 nm. Source: Jullien, M., Garel, J.R., Merola, F. and Brochon, J-C. 1986, Eur. Biophys. J. 13, 131-137. Authorization of reprint accorded by Springer-Verlag.

Also, one can determine the activation energy from the fluorescence emission spectra, by plotting the inverse of the gravity centers of the spectra as a function of 1/T. In this case, the following equation is applied to calculate the activation energy E :

$$1/c = k_1 + k_o \exp(-E/RT) \tag{4.35}$$

where c is the center of the distribution component, k_1 the temperature-independent radiative rate, k_o the non-radiative thermal frequency factor, and E the activation energy for thermal quenching (figure 4.32).

Structure and Dynamics of Macromolecules

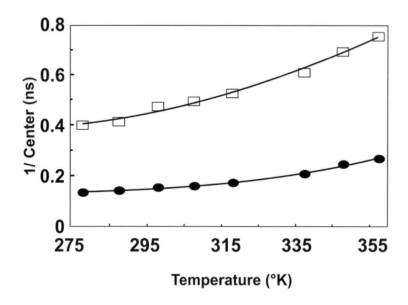

Figure 4.32. Temperature dependence of the reciprocal of S. *solfataricus* β-glycosidase distribution centers. Long-lived component (●); shortlived component (□). The continous lines represent the best fit of the data to Eqn .4.35. The activation energy was found equal to 19.26 kJ / mol and 24.7 kJ / mol for the short and the long components, respectively. Source: Bismuto, E., Irace, G., D'Auria S., Rossi, M. and Nucci, R. 1997. Eur. J. Biochem. 244, 53-58. Authorization of reprint accorded by Blackwell Publishing.

Before going further, we have to be specific in confirming that we are not dealing here with protein denaturation, since the range of temperature where the experiments are carried out are far from denaturating temperatures.

Let us see the different types of interactions that can exist within a protein. This can be done by looking to the global profile of the emission intensity with temperature (-40 to 30°C) of a protein.

Aspartate transcarbamylase from *Escherichia coli* displays a fluorescence intensity decay with two major components with fractional intensities around 50 ± 5% depending on the temperature. Figure 4.33 displays the fluorescence intensities of each of the two components as a function of the temperature (Royer et al. 1987).

The profile of component 1 corresponding to the longest lifetime is not at all monotonous with temperature. In fact, between –40 to –20°C, fluorescence is quenched by increasing temperature. This is due to the increase in motion that brings the tryptophan residue in contact with a nearby quencher. The authors assumed ground state complex formation because the fluorescence lifetime profile does not follow that of the fluorescence intensity. Between –20 and 0°C, the intensity of first component increases with temperature. This was explained as the result tryptophan larger motions decreasing by that the probability of ground – state complex formation.

Finally, above 0°C increasing temperature results in quenching of first component. Above 0°C, collisional quenching becomes more and more important decreasing by that the fluorescence intensity in parallel to the fluorescence lifetime.

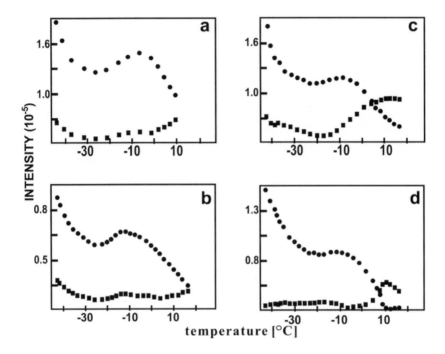

Figure 4.33. Fluorescence intensities vs. temperature (°C). (●) Cornponent 1 and (■) component 2 for (a) ATCase, (b) ATCase in presence of N-(phosphonoacetyl)-L-aspartate (PALA), (c) ATCase in presence of ATP, and (d) ATCase in presence of CTP. Source: Royer, C.A., Tauc, P., Hervé, G. and Brochon, J-C. 1987. Biochemistry 26, 6472- 6478. Authorization of reprint accorded by the American Chemical Society.

Therefore, if a fluorophore free in solution displays mainly collisional quenching with the solvent characterized by the constant k_i, in proteins, the same fluorophore can be within a ground-state complex or can have important dynamical motions.

The second case is observed at the highest temperatures as we can see from the experiment performed with aspartate transcarbamylase. Since the constant k_i characterizes for a free fluorophore in solution the solvent collisional quenching, one should determine the activation energy within a protein or a macromolecule at temperatures where dynamic quenching is dominant over ground-state complex formation.

Figure 4.33 shows that the second component profile does not follow at all the first one. This comes from the fact that a same environment does not surround the two Trp residues. Therefore, each Trp residue microenvironment will be characterized by a specific intensity profile. This leads us to conclude that thermal fluorescence intensity quenching will in fact describe the importance of the interaction between the fluorophore and its surrounding environment.

Finally, it is important to indicate that the global profiles are slightly affected in presence of ligands.

Thermal Fluorescence intensity quenching yields information on the local environment surrounding the fluorophore. Plotting the fluorescence intensity at the maximum as a function of temperature yields in general a linear plot with a slope equal to the variation of the fluorescence intensity per °C and thus we are going to call this variation, thermal structural fluctuations. Figure 4.34 displays the fluorescence intensity decrease of the Trp residues of α_1-acid glycoprotein as a function of temperature. The slope equal to -0.935%, characterizes the temperature perturbation of the environment surrounding the three Trp residues present in the protein. The thermal intensity variation is in this case a mean value between the different environments, polar and hydrophobic, of the protein. Between 6 and 40°C, we have mainly dynamic quenching.

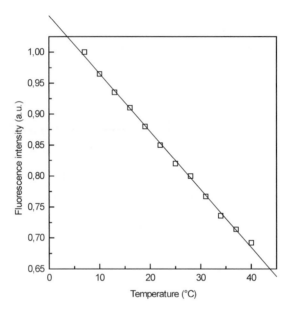

Figure 4.34. Fluorescence intensity decrease of Trp residues of α_1-acid glycoprotein with temperature. $\lambda_{ex} = 295$ nm ; $\lambda_{em} = 330$ nm.   Intensity variation : - 0.935 %

Figure 4.35 displays intensity variation of TNS bound to α_1-acid glycoprotein with temperature. Two slopes are observed, one equal to -6% and the second -1% per °C. The first slope characterizes the thermal structural fluctuation of the TNS binding site environment, while the second slope characterizes the global structural fluctuations of the whole protein. TNS is bound tightly to α_1-acid glycoprotein on the hydrophobic site of its pocket and thus its motion will follow that of the protein. The pocket is exposed to the solvent and thus TNS is in permanent contact with the solvent. The high value of the first slope clearly means that TNS binding site is surrounded by a highly dynamics protein matrix. This result is in good agreement with anisotropy measurements performed on Trp residues showing that the pocket of α_1-acid glycoprotein is highly dynamic (Albani, 1996). Also, since TNS is exposed to the solvent, thermal quenching with solvent can occur very easily.

At the breaking temperature, the decrease in the fluorescence intensity is much slower, i.e., we are here studying the thermal intensity quenching of the more compact protein matrix. One can notice that the second slope observed with TNS (- 1% per °C) is almost equal to the slope obtained when experiments were performed with Trp residues (-0.935% per °C).

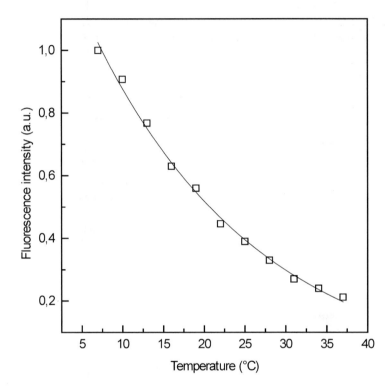

Figure 4.35. Fluorescence intensity decrease of the fluorescence intensity of TNS bound to α_1-acid glycoprotein with temperature. λ_{ex} = 320 nm ; λ_{em} = 425 nm. We observe two slopes for the intensity variation: - 6 % per °C and -1% per °C. The breaking point is at 26.5°C.

The same studies performed on the 5 Trp residues of the protein *Lens culinaris* agglutinin yield a thermal structural fluctuation value equal to –1.45% per °C (Fig. 4.36). Thermal Fluorescence intensity quenching of TNS bound tightly to LCA shows two slopes equal to –2.5% per °C and –1.6% per °C. The breaking temperature is equal to 20°C. One can notice that there is no big difference between the two slopes (Fig. 4.37). This means that TNS bound on LCA is surrounded by protein matrix and is not fully exposed to the solvent.

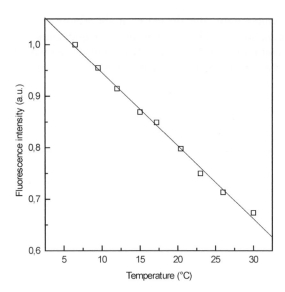

Figure 4.36. Fluorescence intensity decrease of the fluorescence intensity of Trp residues of LCA with temperature. λ_{ex}= 295 nm and λ_{em} = 330 nm. The slope characteristic of the variation with temperature is - 1.45% per °C.

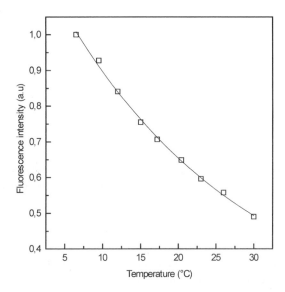

Figure 4.37. Fluorescence intensity decrease of the fluorescence intensity of TNS bound to LCA with temperature. λ_{ex} = 320 nm and λ_{em} = 425 nm. Two slopes are observed for the intensity variation if we consider this variation as linear with temperature: - 2.5% per °C and -1.6% per °C.

When the fluorescence spectrum is recorded as a function of temperature, one can notice that emission bandwidth increase with temperature is dependent upon the fluorophore environment and the fluorophore itself. Figure 4.38 displays the fluorescence emission spectrum of Trp residues of the protein *Lens culinaris* agglutinin. One can notice that in the range of studied temperatures, a shift to the red was not observed and thus we are far from denaturing temperatures. Also, the emission bandwidth (54 nm) does not change with the temperature.

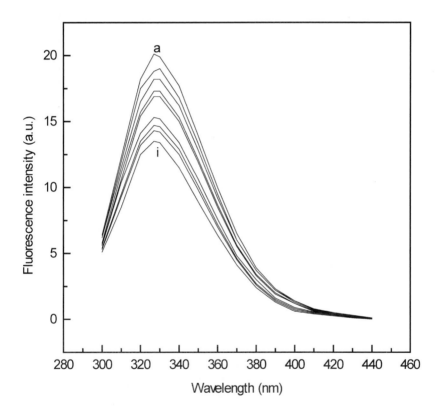

Figure 4.38. Fluorescence spectra of Trp residues of *Lens culinaris* agglutinin at different temperatures. λ_{ex} = 295 nm. Spectrum a : 6.5°C. Spectrum b : 9.5°C. Spectrum c : 12°C. Spectrum d : 15°C. Spectrum e : 17.8°C. Spectrum f : 20.4°C. Spectrum g : 23°C. Spectrum h : 26°C. Spectrum i : 30°C

Figure 4.39 displays the fluorescence emission spectrum of TNS bound to the protein *Lens culinaris* agglutinin recorded at different temperatures. The bandwidth of the spectra increases from 78 to 98 nm upon temperature increase.

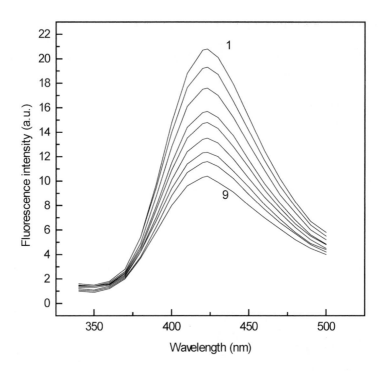

Figure 4.39. Fluorescence spectra of TNS bound to *Lens culinaris* agglutinin as function of temperature. λ_{ex} = 320 nm. Spectrum 1 : 6.5°C. Spectrum 2 : 9.5°C. Spectrum 3 : 12°C. Spectrum 4 : 15°C. Spectrum 5 : 17.8°C.   Spectrum 6 : 20.4°C.   Spectrum 7 : 23°C. Spectrum 8 : 26°C. Spectrum 9 : 30°C

Fluorescence intensity quenching of fluorescein bound to LCA has also been performed in absence and presence of two glycoproteins, lactotransferrin and serotransferrin. LCA is primarily specific for α-mannopyranosyl residues. The lectin recognizes α-mannopyranosyl end-groups or those substituted at the 0-2 position. Additional requirements for strong binding involve the presence of an L-fucose residue α-1.6-linked to an N-acetylglucosamine (GlcNac-l) which is linked to the protein via a N-glycosamine bond.

Serotransferrins from blood plasma and lactotransferrins from mammalian milk are glycoproteins that bind to the LCA. Besides their role in the iron transport and in the inhibition of the growth of microorganisms, the human sero- and lactotransferrin share the following common properties: (i) their molecular mass is around 76 kDa; (ii) they are constituted of a single polypeptide chain of 679 and 691 amino acid residues for the sero- and the lactotransferrin, respectively, organized in two lobes originating from a gene duplication; (iii) each lobe binds reversibly one $Fe3+$ ion; (iv) the protein moiety presents a high degree of homology (about 62%) (v) they are glycosylated (6.4% by weight).

The two lobes correspond to the N-terminal and C-terminal halves of the molecules, and are tightly associated by non-covalent interactions. Also, both are joined by a connecting short peptide of 12 and 11 amino acids in sero- and lactotransferrin, respectively. The three-dimensional pictures of the two proteins are perfectly superimposable with very few differences.

The human serotransferrin contains two carbohydrates of the N-acetyllactosaminic type, located in the C-terminal lobe of the polypeptide chain. The two glycosylation sites (Asn-413 and 611) may be occupied by bi-, tri- and tetra-antennary carbohydrates.

Carbohydrates of human lactotransferrins are located in both N- and C-domains, at three glycosylation sites (Asn-137, 478 and 624).

Carbohydrates of human serotransferrin are not fucosylated, while those of human lactotransferrin have an α-l,6-fucose bound to the N-acetylglucosamine residue linked to the peptide chain, and an α-l,3-fucose bound to the N-acetyllactosamine residues.

Studies have shown that α-l,3-fucose residue does not play any role in the interaction between LCA and glycopeptides isolated from human lactotransferrin or chemically synthesized (Kornfeld et al. 1981). However, the same work showed that α-l,6-fucose is important for a tight binding. Also, X-ray diffraction studies have proved that the α-1,6 fucose is essential for attaining the proper binding conformation of the carbohydrates (Bourne et al. 1993).

Fluorescence studies have shown that the association constants (Ka) are equal to 9.66 μM⁻¹ and 0.188 μM⁻¹ for LCA-LTF complex and LCA-STF complex, respectively. Therefore, the affinity between LCA and lactotransferrin is 50 times higher than that found between LCA and serotransferrin, the α-l,6-fucose playing an important role in this difference (Fig. 4.40) (Albani et al. 1997).

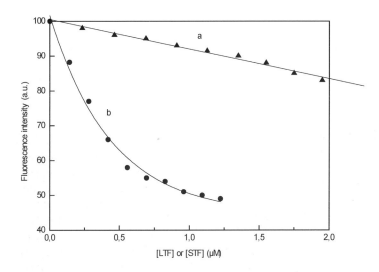

Figure 4.40. Fluorescence intensity quenching of fluorescein bound to *Lens culinaris* agglutinin, as a result of serotransferrin-LCA (a) and of lactotransferrin-LCA (b) interactions. [LCA] = 0.7 μM. Source: Albani, J. R. , Debray, H., Vincent, M. and Gallay, J. 1997, Journal of Fluorescence. 7, 293-298. Authorization of reprint accorded by Kluwer Academic Publishing.

Interaction between LCA and STF is identical to that observed between LCA and the glycoprotein *Vicia fava* agglutinin (VFA), a protein that does not contain an α-1,6-fucose. In fact, the association constant of the LCA-VFA complex, measured with fluorescence studies, is equal to $0.159\ \mu M^{-1}$ (Fig. 4.41).

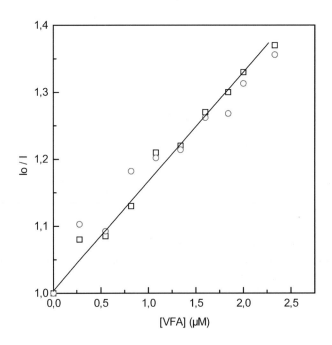

Figure 4.41. Determination of the dissociation constant of the VFA-LCA complex. From the slope of the plot, we obtain a dissociation constant equal to 6.3 μM, an association constant equal to $0.159\ \mu M^{-1}$. The fluorescence of fluorescein covalently bound to LCA (0.28 μM) was followed in presence of increasing concentrations of VFA. λ_{ex} = 495 nm and λ_{em} = 515 nm.

Collisional and thermal intensity quenching of FITC bound to LCA in absence and presence of LTF or STF allowed us to find out that the α-1,6-fucose plays a fundamental role in the dynamics of the complexed proteins. Dynamic fluorescence quenching of FITC with iodide analyzed by the Stern-Volmer equation yields K_{SV}, equal to 9.608 ± 0.273 and $3.795 \pm 0.295\ M^{-1}$, for fluorescein free in buffer (Fig. 4.42a) and for that bound to LCA (Fig. 4.42b), respectively (Albani, 1998). Thus, iodide is accessible to the bound FITC on LCA. However, this accessibility is lower than that observed for free fluorescein in solution, because amino acid residues that decrease the frequency of the collisions with the quencher surround the probe. The bimolecular quenching constant of iodide is equal to 2.402 ± 0.068 and 1.160 ± 0.090 x $10^{9}\ M^{-1}\ s^{-1}$ when the interaction occurs with free fluorescein and fluorescein bound to LCA, respectively. Thus, diffusion of iodide in the vicinity of the fluorescein when bound to LCA is hindered by the surrounding amino acid residues.

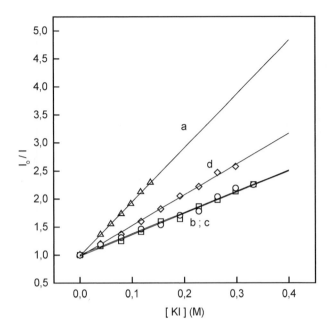

Figure 4.42. Fluorescence quenching with iodide of FITC free in solution (a), bound to LCA (b), bound to LCA-LTF complex (c) and to LCA-STF complex (d). Source: Albani, J. R. 1998. Biochim. Biophys. Acta. 1425, 405-410.

K_{SV} (a) = 9.608 ; kq = 2.402 x 10^9 $M^{-1}s^{-1}$. K_{SV} (b) = 3.795 ; kq = 1.160 x 10^9 $M^{-1}s^{-1}$

K_{SV} (c) = 3.776 ; kq = 1.155 x 10^9 $M^{-1}s^{-1}$. K_{SV} (d) = 5.476 ; kq = 1.675 x 10^9 $M^{-1}s^{-1}$

In presence of LTF, the Stern-Volmer and the bimolecular quenching constants are equal to 3.776 ± 0.276 M^{-1} and 1.155 ± 0.087 x 10^9 M^{-1} s^{-1}, respectively. Thus, binding of LTF does not affect the diffusion of Kl in proximity of the fluorescein, i.e. dynamics of the microenvironment of the probe did not change with the presence of LTF.

Binding of STF to the LCA-FITC complex yields a Stern-Volmer constant equal to 5.476 ± 0.188 M^{-1} and a bimolecular diffusion constant of 1.675 ± 0.06 M^{-1} s^{-1}. Thus, binding of the glycoprotein to the LCA-FITC complex increases the diffusion of iodide around the fluorescein, i.e. the fluctuations around the probe are more important when STF is bound to the LCA-FITC complex.

Since the fluorescein is randomly located on the protein surface, the calculated bimolecular diffusion constant is a mean one characterizing the mean dynamics of the amino acids of LCA.

Increasing the temperature from 5 to 35°C induces a decrease in the fluorescence intensity emitted by FITC bound to LCA, whether the glycoproteins are bound to the lectin or not. This decrease is not accompanied by a shift in the emission maximum wavelength and is linear with temperature with a relative change of 0.656% (Fig. 4.43b), 0.889% (Fig. 4.43c) and 0.488% (Fig. 4.43d) for FITC-LCA, FITC-LCA-LTF and FITC-LCA-STF complexes, respectively (Albani, 1998).

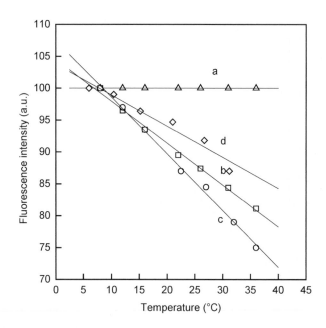

Figure 4.43. Thermal fluorescence intensity quenching of free FITC (a), of FITC bound to LCA (b), in presence of LCA-LTF complex (c) and in presence of LCA-STF complex (d). Source: Albani, J. R. 1998. Biochim. Biophys. Acta. 1425, 405-410.

The fluorescent probe by itself shows no thermal quenching (Fig. 4.43a). Therefore thermal quenching is not related to environment polarity alone. The presence of a structured matrix around the fluorophore induces the sensitivity of the fluorophore to temperature. The origin of this thermal quenching is the fast motions of the microenvironment surrounding the fluorophore.

Fluorescein molecules are covalently bound to lysine residues and are randomly distributed on LCA molecules, thus our results are explained in terms of mean global dynamics of LCA.

Fluorescence intensity quenching with iodide (Fig. 4.42) indicates that diffusion of KI is inhibited by the amino acids of LCA. Binding of LTF to the LCA-FITC complex does not affect the diffusion (k_q) and the dynamic constants (K_{SV}). Thus, the dynamics of the amino acids of LCA do not change in presence of LTF.

Binding of STF to the LCA-FITC complex induces an increase of the Stern-Volmer and of the bimolecular diffusion constants (k_q increases from 3.795 M^{-1} for the LCA-FITC complex, to 5.476 M^{-1} in presence of STF, and k_q increases from 1.16 x 10^9 to 1.675 x 10^9 M^{-1} s^{-1}). Thus, the accessibility of iodide to the fluorescein increases and the dynamics of LCA are more important in presence of STF. The more the fluctuations of the protein matrix are important, the more the diffusion of the quencher is facilitated and the more k_q is higher.

In the fluorescence intensity quenching (thermal and with iodide), it is the fluorescein environment consisting of amino acids (thermal quenching) and of amino acids and solvent dipoles that is relaxing around the excited fluorescein. In the fluorescence anisotropy experiments, on the other hand, the displacement of the emission dipole moment of the fluorescein is monitored. In the first approach, it is the environment that is either fluid or rigid. In the second approach, the restricted reorientational motion of the fluorophore is followed.

X-ray diffraction studies on the complex *Lathyrus ochrus* isolectin ll-glycosyl fragment of human lactotransferrin, have indicated that the oligosaccharide adopted an extended conformation, suggesting that the peptide part of the glycopeptide had no influence on the binding (Bourne et al. 1994).

X-ray diffraction and fluorescence studies performed on LCA have indicated that the carbohydrate binding site is flexible (Loris et al. 1993). The carbohydrates in LTF and STF are highly flexible (Dauchez et al. 1992). This flexibility is maintained when the glycoproteins are bound to the LCA (Albani et al. 1997). Otherwise, the carbohydrates would not conserve their extended conformation and we should observe a more compact structure between LCA and the glycoproteins, i.e. we should observe a decrease in the bimolecular diffusion constant of KI.

In presence of LTF, diffusion of Kl and thermal quenching are less important than when STF is bound to LCA. Thus, if the flexibility of the carbohydrates is maintained when the glycoproteins are bound to LCA, the α, l,6 fucose decreases the global flexibility of the complex compared to that observed in the absence of the sugar.

This observation is in good agreement with what we know from the literature on the motions of carbohydrates. All N-linked sugars such as those of lacto- and serotransferrin possess a high freedom of rotation around the glycosidic bonds. This rotation will induce a high flexibility of the molecules. Also, glycans are often solvent exposed on the protein surface and in most glycoprotein structures known to date, only the 1-4 sugar residues most proximal to the glycosylation site are immobilized (Wyss and Wagner, 1996).

In presence of STF, the high mobility of the carbohydrates increases that of the amino acids of LCA (K_{SV} and k_q increase and the thermal quenching effect decreases). This mobility decreased in presence of the strong interaction with the α-1,6- fucose. In fact, since the affinity between LCA and lactotransferrin is 50 times higher than that between LCA and serotransferrin, the α-1,6-fucose playing a major role in its high affinity (Albani et al. 1997), the carbohydrate – LCA bond is very weak. Only interaction between the α-1,6-fucose and LCA is strong. This will induce global and local dynamics almost identical to those observed for LCA-FITC alone. Therefore, the bond between the fucose and LCA inhibits the effect of the high mobility of the other carbohydrates of the glycoproteins.

In conclusion, these results show that: (1) dynamics of LCA are increased as the result of the carbohydrate residues flexibility. (2) presence of α,1,6 fucose inhibits the effect of the other carbohydrate residues as the result of the tight bond between the sugar and the lectin.

Thermal intensity quenching can be performed also to study the dynamics of the carbohydrate residues of α_1-acid glycoprotein. This protein contains 40% carbohydrate by weight and has up to 16 sialic acid residues (10-14% by weight). The fluorescent probe calcofluor white binds to the carbohydrate residues of α_1-acid glycoprotein and its fluorescence parameters are sensitive to the secondary structure of the carbohydrates of the protein.

Increasing the temperature from 6 to 35°C induces a decrease in the fluorescence intensity emitted by calcofluor bound to the sialylated and asialylated forms of α_1-acid glycoprotein (Fig. 4.44) (Albani, 2003). The intensity decrease is more important in the sialylated form of the protein. This decrease is not accompanied by a shift in the emission maximum wavelength (data not shown). The fluorescence of calcofluor itself shows no thermal quenching (Fig. 4.44a). Therefore, thermal quenching is not related to environment polarity alone. The presence of a structured conformation (carbohydrate residues) around the fluorophore induces the sensitivity of the fluorophore to temperature. The origin of this thermal quenching is the carbohydrate residues flexibility. Our results indicate that this flexibility is more important for the sialylated α_1-acid glycoprotein than for the asialylated one. This result is in good agreement with the fact that anisotropy studies and red-edge excitation spectra have shown that sialic acid displays important local motions while the carbohydrate residues proximal to the glycosylation site have restricted motions.

Since calcofluor binds to the pocket of α_1-acid glycoprotein, the observed thermal quenching could also be the result of a local flexibility that originates from both the carbohydrate residues surrounding the fluorophore and the amino acids of the glycosylation site within the pocket. In fact, flexibility and local motions were observed for Trp residues in both sialylated and asialylated α_1-acid glycoprotein.

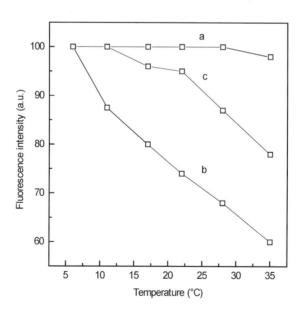

Figure 4.44. Fluorescence thermal quenching of calcofluor free in solution (a), bound to sialylated α_1 - acid glycoprotein (b) and bound to asialylated α_1 - acid glycoprotein (c). The concentration of the fluorophore (5 µM) is equal to that of the protein. $\lambda_{ex} = 330$ nm and $\lambda_{em} = 435$ nm. Source: Albani, J. R. 2003, Carbohydrate Research 338, 1097-1101.

8. Photoquenching

This is known usually under the name of photobleaching. It occurs when a very high light intensity excites the fluorophore, the dye will "photobleach" meaning that the high intensity light has rendered the dye unable to fluoresce. However, in many cases we do not need to excite the fluorophore with an intense light. In fact, most of the fluorophores are unstable even when excitation occurs with a low intensity. The result of this light sensitivity is the decrease of the fluorescence intensity of the recorded spectrum. Photochemical destruction is observed in proteins when they are irradiated for a long time at U.V. wavelengths. Also, excitation at the absorption maximum facilitates the phenomenon. An example, α - crystallin, a major protein of the lens, shows thermal aggregation of crystallins and other proteins. Fluorescence of α - crystallin is dominated by Tryptophans 9 and 60.

Exposure of α - crystallin solutions to UV light, induces an important decrease of Trp residues fluorescence intensity (Fig. 4.45). The absence of a shift in the maximum of the spectrum is evidence that UV irradiation does not modify the position of the Trp residues inside the protein. Thus, U.V. irradiation induces a photochemical destruction of the Trp residues.

Finally, figure 4.46 describes the role of α - crystallin in preventing aggregation of β - crystallin.

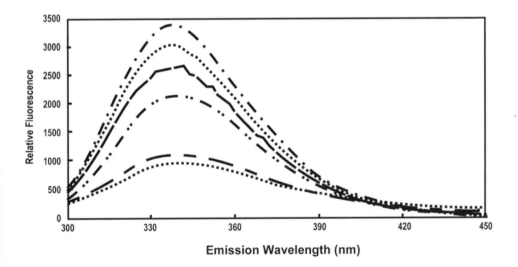

Emission Wavelength (nm)

Figure 4.45. Change of tryptophan fluorescence spectra of α-crystallin upon exposure to UV irradiation at 280 nm. The fluorescence spectra were recorded by setting the excitation wavelength at 290 nm. The concentration of α-crystallin used in the experiments was 1 mg/ml. The emission spectra from top to bottom correspond to those crystal in solutions irradiated at UV 280 nm for 0, .5, 10, 15, 30 and 60 min, respectively, with the solution of the longest exposure showing the lowest fluorescence. Source: Lee, J-S., Liao, J-H., Wu, S-H., Chiou, S.H. 1997, J. Prot. Chem. 16, 283-289. Authorization of reprint accorded by Kluwer Academic Publishers.

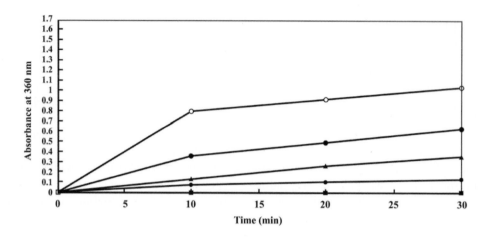

Figure 4.46. Thermal aggregation of β-crystallin at 60°C in the presence of α-crystallin irradiated with UV at 280 nm. Rapid aggregation of β - crystallin during incubation at 60°C is inhibited in presence of α - crystallin. Inhibition is 100% efficient in presence of native α - crystallin. Also, inhibition is less important when the time of U.V. irradiation of α - crystallin is longer. The concentration of β-crystallin used in these experiments was 0.37 mg/ml, whereas that of α-crystallin was 0.1 mg/ml. All concentrations of cryslallin solutions used in the experiments were estimated from their respective extinction coefficients at 280 nm (0.83, 2.23. and 2.26 for α, β and γ crystallins, respectively, at 1 mg/ml). β-Crystallin with native (α-crystallin (■); β-cryslallin with α-cryslallin irradiated for 5 min (♦); β -cryslallin with α-crystallin irradiated for 15 min (▲); β-crystallin with α-crystallin irradiated for 60 min (●); and β-crystallin only (□). . Source: Lee, J-S., Liao, J-H., Wu, S-H., Chiou, S.H. 1997, J. Prot. Chem. 16, 283-289. Authorization of reprint accorded by Kluwer Academic Publishers

Chapter 5

FLUORESCENCE POLARIZATION

1. Aims and Definition

Biological systems (membranes, DNA, proteins, etc…) display specific motions, global rotation and local dynamics, that are dependent on the structure, the environment and the function of the system. These motions differ from a system to another and, within one system local motions are not the same. The most known example is that of membrane phospholipids where the hydrophilic phosphates are rigid and the hydrophobic lipid is highly mobile. Polarized light is a good tool to put into evidence and to study the different types of rotations a molecule can undergo.

Natural light is unpolarized, it has no preferential direction. Thus, in order to study the dynamics of molecules during excited state lifetime, it is important to photoselect some of the molecules. This can be performed by exciting the fluorophores with a polarized light and by recording the emitted light in a polarized system. When excitation is performed with a polarized light (very definite orientation), absorption of the fluorophore will depend on the orientation of its dipole in the ground state compared to the polarized excitation light. Only fluorophores with dipoles perpendicular to the excitation light will not absorb. Fluorophores with dipoles parallel to the excitation light will absorb the most. Thus, a polarized excitation will induce a photoselection in the fluorophore absorption. The electric vector of the excitation light is oriented parallel to or in the same direction of the z-axis. Emitted light will be measured with a polarizer. When the emission is parallel to the excitation, the measured intensity is called $I_{||}$. When the emission is perpendicular to the excitation light, the measured intensity is called $I_{\perp}$. Light polarization and anisotropy are obtained according to Eq.5.1 and 5.2 :

$$P = \frac{I_{||} - I_{\perp}}{I_{||} + G\,I_{\perp}} \tag{5.1}$$

and

$$A = \frac{I_{||} - I_{\perp}}{I_{||} + 2\,G\,I_{\perp}} \tag{5.2}$$

G is the correction factor, it helps to take into account the differences in sensitivity of the detection system in the two polarizing directions I_{vv} and $I_{vh.}$ The G factor can be evaluated by measuring the fluorescence intensities parallel and perpendicular to the Z-axis with the excitation polarizer perpendicular to the Z-axis. G will be equal to:

$$G = I_{hv} \ / \ I_{hh} \tag{5.3}$$

P and A are interrelated with the following two equations:

$$P = 3\,A\,/\,(2 + A) \tag{5.4}$$

and

$$A = 2\,P\,/\,(3 - P) \tag{5.5}$$

From the definition of the polarization, one can conclude that a fluorophore free in solution can have a low polarization value while when it is bound to a macromolecule, its polarization increases. Therefore, upon protein denaturation, the polarization of the intrinsic or / and extrinsic fluorophores decreases as the result o the increased motions of the protein.

Polarization unit is a dimensionless entity, i.e., the value of P does not depend on the intensity of the emitted light and on the concentration of the fluorophore. However, this is the theory but the reality is quite different. In fact, measuring polarization at high fluorophore concentrations yield erroneous values and in many cases instead of reading the corrected values of P and A, values that neighbor the limiting ones are recorded.

2. Principles of polarization or of photoselection

In the space coordinate system, a, b and c, the exciting light is within direction a, polarized in the b-c plane (Fig. 5.1) (Albrecht, 1961). Fluorophore molecules which possess a transition moment parallel to the a direction will be excited. These molecules are shown in Figure 5.1 with shaded planes. An intensity vector can be projected into the three axis a, b and c. One intensity will be parallel to the a direction and thus will be called I_{II}. The other two intensities oriented parallel to the b and c axis and thus oriented perpendicular to the a axis. These two intensities are called $I_{\perp}$. Therefore, the global intensity I is equal to

$$I = I_{II} + 2\,G\,I_{\perp} \tag{5.6}$$

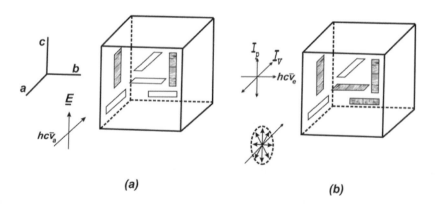

(a)   **(b)**

Figure 5.1. Photoselection of differently oriented molecules with plane-polarized (a) and unpolarized (b) light. Excited molecules are shaded. After Albrecht, A. C., 1961, J. Mol. Spectry. 6, 84. Figure published by Dörr, F., in Creation and Detection of the excited States. pp 53-122. Edited by Lamola, A. A. 1071, Marcel Dekker, Inc. New York.

Nevertheless, the global fluorescence intensity recorded in presence of palarizers is lower than that obtained in their absence. In fact, the presence of the excitation polarizer decreases the number of fluorophores absorbing the polarized excitation light, and the emission polarizer is going to decrease the number of emitting fluorophores whose emission are going to reach the detector.

Equation 5.1 clearly indicates that the values of P occur between 1 ($I_{\underline{\mathsf{I}}} = 0$) and – 1 ($I_{\mathsf{II}} = 0$). Natural or unpolarized light, where $I_{\underline{\mathsf{I}}} = I_{\mathsf{II}}$, yields a value of P equal to 0. These two extreme values of P occurred when the polarized absorption transition moment and that of the emission are colinear ($I_{\underline{\mathsf{I}}} = 0$) or perpendicular ($I_{\mathsf{II}} = 0$). However, in a rigid medium where motions are absent, the absorption and the emission transition dipoles can be oriented by an angle θ relative one to the other.

Therefore, the probability of light absorption by the molecules will be a function of $\cos^2 \theta$ and the polarization will be equal to

$$P_o = (3 \cos^2 \theta - 1) / (\cos^2 \theta + 3) \tag{5.7}$$

and

$$A_o = (3 \cos^2 \theta - 1) / 5 \tag{5.8}$$

θ	P_o	A_o
0	0.5	0.4
45	0.14	0.1
54.7	0.00	0.00
90	-0.33	-0.20

A value of θ equal to 0 indicates that the dipoles of the excitation and the emission are co-planar. This means that the excitation induces a $S_o \text{-→} S_1$ transition. One can notice also that for an angle equal to 54.7°, P and A are equal to zero. This angle is called the "magic angle". Measuring the fluorescence emission with the polarizers set up with an orientation equal to the magic angle, allows detecting an unpolarized light.

P_o and A_o are called respectively the intrinsic polarization and anisotropy. P_o and A_o are measured at temperatures equal to – 45°C or less. At this temperature, the fluorophores do not display any motions. Measurements of the intrinsic polarization are performed as follows:
- The sample is dissolved in the buffer.
- One to two drops of the sample solution are added to the fluorescence cuvette containing 1 ml of glycerol. The glycerol – buffer solution is mixed slowly to avoid bubbles formation.
- The cuvette in then placed in the optical compartment cooled with a 100% ethanol solution.
- Values of A or P are obtained from parallel and perpendicular fluorescence intensities after subtracting the Raman signal of the bufer measured in the same conditions.

Colinear excitation and emission vectors yield $S_o \text{-→} S_1$ and $S_1 \text{-→} S_o$ transitions. This means that high positive values of P would correspond to a $S_o \text{-→} S_1$ transition, while the negative values of P would correspond to the $S_o \text{-→} S_2$ transition.

Since the electronic transitions differ from one excitation wavelength to another, the value of P would change with the excitation wavelength. Emission occurs in general from the lowest excited state $S_1 v_0$, thus one can measure the anisotropy or the polarization along the absorption spectrum at a fixed emission wavelength. We obtain a

spectrum called the excitation polarization spectrum or simply the polarization
spectrum (Figures 5.2 and 5.3). The protein spectrum feature is characteristic of Trp
residues when this amino acid residue is the main emitting species. The value of
anisotropy will change from a protein to another depending on the local dynamics of the
Trp residues and of the global motions of the protein. In a highly rigid environment, the
values of the anisotropies along the excitation wavelengths will be more important than
when the protein and / or the fluorophore display global and local motions.

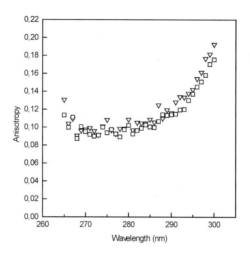

Figure 5.2. Fluorescence anisotropy spectra of *Lens culinaris* agglutinin LCA (squares) and
Vicia fava agglutinin VFA (triangles) obtained at 20°C. λ_{em} = 330 nm. The instrument used is a
Perkin-Elmer LS 5B.

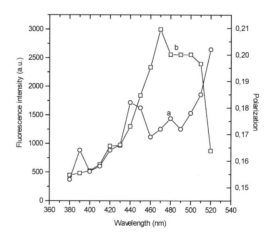

Figure 5.3. Polarization (a) and fluorescence excitation (b) spectra of fluorescein at − 17°C
dissolved in 6% water / 94% glycerol. Both spectra were obtained at λ_{em} = 530 nm.

As we can see from figures 5.2 and 5.3, the polarization spectrum is characteristic of the chemical structure and nature of the fluorophore.

Figure 5.4 displays the excitation polarization spectrum of protoporphyrin IX in propylene-glycol at –55°C (full line) and bound to the heme pocket of apohemoglobin recorded at 20°C (dotted line). One can notice that polarization at low temperature is higher than that observed when porphyrin is embedded in the heme pocket of apohemoglobin. This is the result of the local motions of the fluorophore within the pocket, independently of the global rotation of the protein.

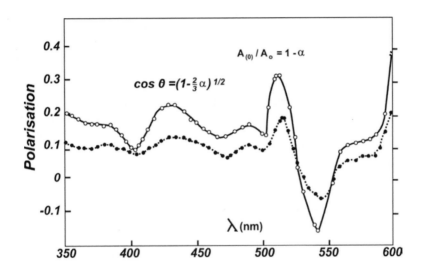

Figure 5.4. Excitation polarization spectrum of protoporphyrin IX in propylene-glycol at –55°C (full line) and bound to the heme pocket of apohemoglobin recorded at 20°C (dotted line). λ_{em} = 634 nm. Source: Thesis of Sebban Pierre, 1979, University of Paris 7. « Etude par fluorescence de molécule Hb^{desFe} ».

3. Absorption transitions and excitation polarization spectrum

An absorption spectrum could be the result of the overlapping of two or more absorption transitions. Separation of each band can be performed by means of the excitation polarization spectrum. If the spectrum consists of regions of constant polarization joined by regions of varying polarization, it is possible to consider that one region corresponds to a single electronic transition while the second region is the result of the overlap of contiguous transitions. Since anisotropy is additive, one can write:

$$<A> = f_a A_a + f_b A_b \qquad (5.9)$$

where $<A>$ is the mean anisotropy at a specific wavelength, A_a and A_b are the anisotropies of bands a and b and f the fractional contributions to the absorption of the two bands :

$$f_a + f_b = 1 \tag{5.10}$$

Combining equations 5.9 and 5.10 yields

$$f_a = \frac{<A> - A_b}{A_a - A_b} \tag{5.11}$$

and

$$f_a = \frac{A_a - <A>}{A_a - A_b} \tag{5.12}$$

Finally, the absorption spectrum of each band can be obtained by multiplying at each wavelength the absorption of the overall spectrum by the fractional contribution.

Figure 5.5 displays the excitation polarization and absorption spectra of FMN at – 50°C. One can notice that the anisotropy is constant between 310 and 370 nm and from 390nm. Therefore, one can easily separate the two absorption bands with maximum at 370 and 450 nm, respectively.

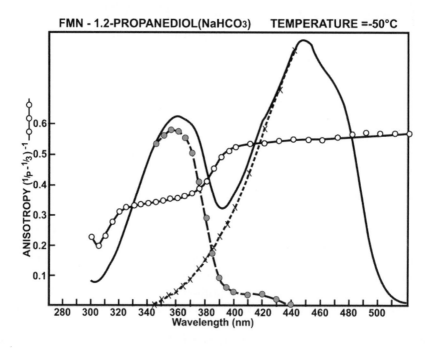

Figure 5.5. Separation of the 370 and 450 absorption bands of FMN in 1,2-propanediolat - 50°C by anisotropy measurements. Source: Spencer, R.D., 1970, PhD thesis. University of Illinois.

Tryptophan has two overlapping $S_o \longrightarrow S_1$ electronic transitions (^1L_a and ^1L_b) which are perpendicular one to each other. Both $S_o \rightarrow {}^1L_a$ and $S_o \rightarrow {}^1L_b$ transitions occur in the 260-300 nm range. Valeur and Weber used the polarization spectrum of indole to resolve the absorption spectra of the ^1L_a and ^1L_b. At wavelengths higher than 300 nm, the ^1L_a state is assumed to predominate. The authors determined the values of A_o for the two states and found values equal to 0.3 and -0.15 for the ^1L_a and ^1L_b states, respectively (Figure 5.6).

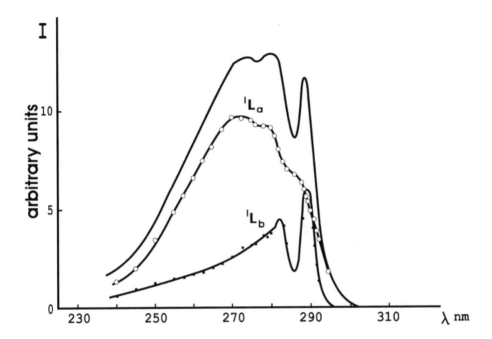

Figure 5.6. Resolution of the ^1La and ^1Lb absorption spectra of indole from the excitation polarization spectrum. Source:Valeur, B. and Weber, G. 1977, Photochem. Photobiol. 25, 441-444.

4. Fluorescence depolarization

4a. Principles and applications

In a vitrified solution and / or when the fluorophore molecules do not display any residual motions, the measured polarization is equal to the intrinsic one P_o. This value is obtained if during the excited state lifetime of the fluorophore, i.e., during the laps of time between light absorption and light emission, the orientation of the excitation and emission dipoles does not change.

However, during the excited state lifetime, energy transfer to neighboring molecules and / or local and global motions of the fluorophores can be observed. These two phenomena will induce a reorientation of the emission dipole and will provoke the depolarization of the system. Therefore, the value of the measured polarization P will be lower than the intrinsic one P_o.

The relation between P and P_o is given by the Perrin equation:

$$\frac{1}{P} - \frac{1}{3} = \left(\frac{1}{P_o} - \frac{1}{3}\right)\left(1 + \frac{RT\tau_o}{\eta V}\right) = \left(\frac{1}{P_o} - \frac{1}{3}\right)\left(1 + \frac{\tau_o}{\phi_R}\right) \qquad (5.13)$$

where R is the gaz constant = 2.10 Kcal. Mole^{-1} degré$^{-1}$ or $0,8 \times 10^3$ g cm^2 s^{-2} M^{-1} degré$^{-1}$. T, the temperature in Kelvin degree, η, the medium viscosity in centipoise (cp) or in g cm^{-1} s^{-1} (1 cp = 0.01 g cm^{-1} s^{-1}), V and ϕ_R the rotational volume and rotational correlation time of the fluorophore.

Plotting 1 / P as a function of T / η yields in principle a linear graph with a y-intercept equal to 1 / P_o and a slope equal to

$$\frac{R <\tau>}{V}\left(\frac{1}{P_o} - \frac{1}{3}\right).$$

The Perrin plot can also be written as :

$$\frac{1}{A} = \frac{1}{A_o}\left(1 + \frac{\tau_o}{\phi_R}\right) \qquad (5.14)$$

Decreasing the temperature or increasing the concentration of sucrose and / or glycerol in the medium, increases the viscosity of the medium.

Whether the fluorophores are intrinsic or extrinsic to the macromolecule (protein, peptide or DNA), depolarization is the result of two motions, the local motions of the fluorophore and the global rotation of the macromolecule.

The rotational correlation time of a spherical macromolecule can be determined with one of equations 5.15 and 5.16:

$$\phi_P = M(v + h)\eta / kTN \qquad (5.15)$$

where M is the molecular mass, v = 0,73 cm^3 / g characterizes the specific volume, h = 0,3 cm^3 / g is the hydration degree, η, the medium viscosity and N is the Avogadro number.

$$\phi_P(T) = 3.8 \, \eta(T) \times 10^{-4} M \qquad (5.16)$$

Equations 5.15 and 5.16 do not yield exactly the same value for ϕ_P. For example, for a protein with a molecular mass of 235 kDa such as flavocytochrome b_2 extracted from the yeast *Hansenula anomala,* the rotational correlation times calculated from Eq. 5.15 and 5.16 are 97 and 90 ns, respectively.

Perrin plot enables us to obtain information concerning the motion of the fluorophore. When the fluorophore is tightly bound to the protein, its motion will correspond to that of the protein. In this case, ϕ_R will be equal to the rotational correlation time of the protein ϕ_P and A_o obtained experimentally with the Perrin plot will be equal to that measured at $-45°C$ (Fig. 5.7).

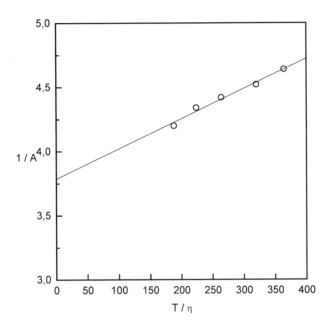

Figure 5.7. Steady-state fluorescence polarization versus temperature over viscosity ratio for Trp residues of human α_1 –acid glycoprotein prepared by acetonic precipitation. Data were obtained by thermal variations in the range 7-35°C. λ_{ex} = 300 nm. λ_{em} = 330 nm. Protein concentration is equal to 10 µM. The rotational correlation time determined from the Perrin plot is equal to 13 ns at 20°C is in the same range as that (17 ns) expected for the protein at the same temperature, indicates the presence of residual motions. Also, the extrapolated anisotropy (0.264) is equal to that measured at $-35°C$ (0.267). Source: Albani, J. R. 1998, Spectrochimica Acta, Part A. 54, 173-183.

When the fluorophore exhibits local motions, the extrapolated value of A, A(o) will be lower than the A_o value obtained at $-45°C$. The information obtained from the slope of the Perrin plot will depend on the fluorescence lifetime of the fluorophore and on the relative amplitudes of the motions of the fluorophore and of the protein.

When the fluorophore lifetime is equal to or lower than the rotational correlation time of the protein, the extrapolated anisotropy will be lower than the limiting one and the rotational correlation time obtained from the slope of the Perrin plot will correspond to an apparent rotational correlation time ϕ_A. ϕ_A is the result of two motions, that of the protein and of the segmental motion of the fluorophore (Fig. 5.8).

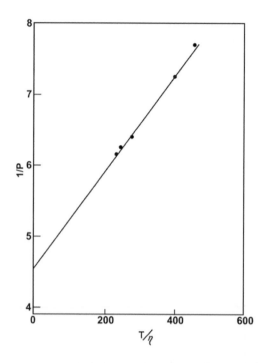

Figure 5.8. Steady-state fluorescence polarization versus temperature over viscosity ratio for Trp residues of human α_1 –acid glycoprotein. Data are obtained by thermal variations in the range 7-35°C. λ_{ex} = 295 nm. λ_{em} = 330 nm. Protein concentration is equal to 5 µM. The rotational correlation time determined from the Perrin plot is equal to 5.66 ns at 20°C and lower than that (17 ns) expected for the protein at the same temperature, indicates the presence of residual motions. Also, the extrapolated polarization (0.219) is lower than that measured at – 35°C (0.24). The same result is obtained when excitation is performed at 300 nm. Thus, the results observed are not the consequence of Trp→ Trp energy transfer. Source: Albani, J. 1992, Biophysical Chemistry 44, 129-137.

When the fluorescence lifetime is much higher than the rotational correlation time of the protein, the Perrin plot representation should enable us to obtain the value of ϕ_P. Since the fluorophore displays local motions, the extrapolated anisotropy will be lower than the limiting or intrinsic one (Fig. 5.9.)

It is possible to obtain experimentally the rotational correlation time of the fluorophore by increasing the medium viscosity in presence of glycerol and / or sucrose. In this case, the global rotation of the protein will be completely hindered and only the local motions of the fluorophore will be observed. 1 / P as a function of T / η yields a plot with two slopes. The rotational correlation times of the protein (ϕ_P) and of the fluorophore (ϕ_R) are calculated at high and low T / η, respectively (Fig. 5.10).

In the present case, extrapolation of the plot of high slope to T / η = 0, yields a value of P(o) or A(o) lower than the intrinsic ones (P_o or A_o) (Figures 5.10 and 5.11).

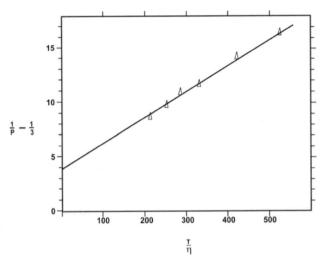

Figure 5.9. Steady-state fluorescence polarization versus temperature / viscosity ratio for protoporphyrin IX conjugate to horse apomyoglobin (Mb^{desFe}). Data are obtained by thermal variations in the range 10-45°C. Fluorescence lifetime of porphyrin is equal to 17.8 ns. The rotational correlation time (10 ns at 15°C) obtained from the Perrin plot is in the same range of that (9.2 ns) calculated theoretically for the protein. The extrapolated value of P(o) (=0.245) is lower than the limiting value P_o equal to 0.266 at λ_{ex} = 514 nm for porphyrin embedded in the apomyoglobin pocket. Thus, porphyrin displays free motion inside the heme pocket. λ_{em} = 630 nm. Source: Albani, J. and Alpert, B. 1986. Chem. Phys. Letters. 131, 147-152.

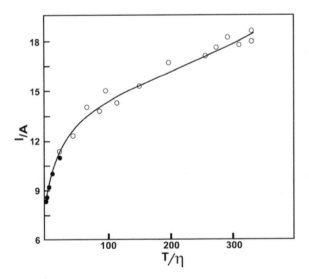

Fig. 5.10. Perrin plot for tryptophan fluorescence of apo α-chain (0.4 mg/ml) in 40 mM phosphate, pH 5.3, 25°C. The viscosity is varied by the addition of sucrose (0) or glycerol (●). The polarization (P) was measured as described in the text. The excitation and emission wavelengths were 285 and 345 nm, respectively. The units of T / η are degrees centipoise^{-1}. Source: Oton J, Franchi D, Steiner R. F., Martinez, C. F. and Bucci, E. 1984. Arch. Biochem. Biophys. 228, 519-524.

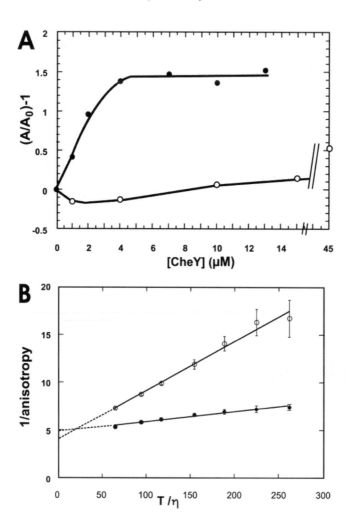

Figure 5.11. Fluorescence depolarization studies of fluorescein-labeled CheZ214FC.
The experiment was carried out in the absence (0) or presence (•) of 18 mM acetyl phosphate
(AcP). A: the effect of CheY concentration on the fluorescence anisotropy of fluorescein-
labeled CheZ214FC (0.2 µM). B: Perrin plot of the fluorescence anisotropy of fluorescein-
labeled CheZ2l4FC. The medium viscosity was increased by sucrose up to 4.6 cPoise. In this
viscosity range the change in fluorescence anisotropy reflects mainly the rotation of the whole
CheZ-fluorescein conjugate and to a much lesser extent the free rotation of the probe
(fluorescein). The latter rotation is not expected to be affected by the presence of Che Y and
AcP. Therefore, to a first approximation, the slopes of these lines, can be taken as the inverse of
the respective molecular volumes. CheY is a regulatory protein involved in chemotaxis
excitation. CheY interacts with three proteins: CheA protein phosphorylates CheY, CheZ
protein dephosphorylates CheY, and FliM, the protein of the flagellar motor switch to which
CheY binds. Source: Blat, Y.and   Eisenbach, M. 1996,   J. Biol. Chem. 271, 1226-1231.
Authorization of reprint accorded by The American Society for Biochemistry and Molecular
Biology.

Equation 5.14 indicates the possibility to calculate the rotational correlation time of the fluorophore not only by varying the T / η ratio but also by adding a collisional quencher. The interaction between the quencher and the fluorophore induces a decrease in the fluorescence lifetime and intensity of the fluorophore and an increase of the fluorescence anisotropy (Fig. 5.12). Plotting 1 / A as a function of τ_o yields a straight line of slope equal to ϕ_R. If the fluorophore is tightly bound to the macromolecule and does not exhibit any residual motions, the measured ϕ_R will be equal to ϕ_P and the extrapolated anisotropy equal to that measured at low temperature.

If however, the fluorophore exhibits free motions, the measured ϕ_R will be lower than that of the protein and the extrapolated anisotropy will be lower than the limiting anisotropy.

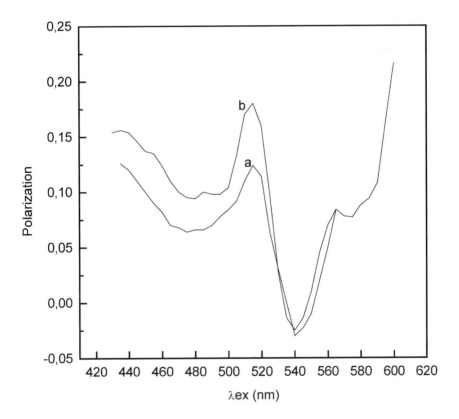

Figure 5.12. Fluorescence polarization spectrum of protoporphyrin IX in apoMb at 20°C in 3 psi (a) and 1200 psi (b) oxygen. λ_{em} = 634 nm.

4b. Measurements of rotational correlation time of tyrosine in small peptides

Rotational correlation time of tyrosine in small peptides can be performed with intensity or / and lifetime quenching.

Enkephalins are between several naturally occurring morphinelike substances called as endophins, released from nerve endings of the central nervous system and the adrenal medulla. They act as analgesics and sedatives in the body and appear to affect mood and motivation.

Dermenkephalin Tyr-DMet-Phe-His-Leu-Met-Asp-NH$_2$ (DREK) (also referred to as deltorphin or dermorphin gene-associated peptide) is a highly potent agonist at the opoid receptor. Whereas, the D-form is a highly potent opoid agonist, [L-Met^2] DREK is virtually inactive.

We are going to describe experiments of fluorescence intensity and anisotropy quenching with iodide to study the dynamics and the flexibility of the peptides and the tyrosine residue.

4b1. Fluorescence lifetime

The fluorescence intensity I(λ,t) of Tyr^1 residue of DREK obtained at λ_{ex} of 278 nm, can be adequately represented by a sum of two exponentials,

$$I(\lambda,t) = 0.54\ e^{-t/0.6} + 0.46\ e^{-t/2.12} \qquad (2.33)$$

where 0.54 and 0.46 are the normalized preexponential factors, and 0.60 ± 0.07 and 2.12 ± 0.03ns are the decay times. The emission wavelength (λ) is 303 nm. $\chi^2 = 1.2$. The mean fluorescence lifetime τ_o of DREK is 1.74 ns, respectively.

The mean fluorescence lifetime is used to calculate the rotational correlation time from the Perrin plot (quenching resolved emission anisotropy experiment).

4b2. Fluorescence intensity quenching of tyrosine residues by iodide

Fluorescence intensity quenching of tyrosine residues by iodide is analyzed with the Stern-Volmer formula

$$Io / I = 1 + Ksv\ [I^-] \qquad (4.6)$$

where Io, I, Ksv and (I$^-$) are the fluorescence intensities in absence and in presence of iodide, the Stern-Volmer constant and the concentration of added iodide.

The Stern-Volmer plots of the fluorescence quenching of the aminoterminal residue in dermenkephalin and [L-Met] dermenkephalin by iodide are shown in Fig. 5.13. Fluorescence of Tyr in (L-Met] DREK and that of free L-tyrosine are quenched identically by iodide (Ksv = 19.817 ± 0.025 and 19.298 ± 0.030 M^{-1} for L-tyrosine and [L-Met^2] DREK), respectively (Fig. 5.13a and b). Since iodide quenches tyrosine residue fluorescence within a spatial proximity, the similar values found for Ksv indicates the absence of any matrix surrounding the tyrosyl side chain. By contrast, the dynamic constant is lower for DREK (Ksv = 13.37 ± 0.02 M^{-1} (Fig. 5.13c.).

In the presence of 6 M guanidine, diffusion of iodide was found homogeneous for the two peptides and for L-tyrosine (Ksv = 19 M^{-1}). Thus, apparently, the low quenching efficiency of iodide in DREK relative to that in [L-Met^2] DREK would arise from differences in the conformation of the two compounds induced by the D configuration of the methionyl in position 2.

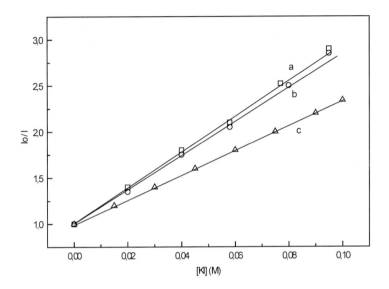

Figure 5.13. Fluorescence intensity quenching by iodide of L-Tyrosine (a), L-DREK (b) and D-DREK (c). λ_{ex}, 280 and λ_{em}, 303 nm. Values of K_{SV} are (a) : 19.82, (b) : 19.30 and (c) : 13.37 M^{-1}.

K_{SV} characterizes the accessibility of the fluorophore to the quencher while the bimolecular quenching constant k_q characterize the diffusion properties of the quencher. k_q is equal to 5.27 x 10^9 , 10.7 x 10^9 and 7.47 x 10^9 M^{-1} s^{-1} for L-tyrosine free in solution, L-DREK and DREK, respectively. We can notice that the bimolecular diffusion constant of KI is higher for the peptides than for free tyrosine in water. This result was also observed when fluorescence quenching experiments with acrylamide was performed on tyrosine residue of [Leu^5 enkephalin] (Lakowicz et al. 1993). In fact, the authors found for K_{SV} values equal to 7.2 and 10 M^{-1} for the tyrosine in the peptide and L-tyrosine free in solution, respectively. Since the mean fluorescence lifetime of the enkephalin is much shorter than that of free tyrosine in water, the bimolecular diffusion constant of acrylamide for the enkephalin fluorescence quenching is more important than that of L-tyrosine fluorescence quenching. The results obtained on both types of enkephalin indicate that quenching of tyrosine fluorescence intensity cannot be analyzed with the classical Stern-Volmer plot only.

4b3. Quenching emission anisotropy

In general, when anisotropy varies as a function of quencher concentration, the Perrin plot can be written as

$$1/A = 1/A_o + [\tau_o \; I/I_o]/\Phi_R A_o \tag{5.17}$$

where I and Io are the fluorescence intensities in the presence and absence of collisional quencher and Φ_R the rotational correlation time of the fluorophore. When the fluorescence lifetime τ_o and the intensity I_o decrease by collisional quenching, the anisotropy $A(\tau)$ increases and tends to the limiting anisotropy Ao. However, the presence of residual motions will lead to an extrapolated anisotropy A(o) different from A_o. Thus plotting $1/A$ vs I will yield information concerning the motion of the fluorophore.

Measurements of the emission anisotropy A as a function of added collisional quencher are made with the steady fluorescence intensity, which integrates the different weighted fluorescence lifetimes. Quenching emission anisotropy plot of $1/A$ vs I (Fig. 5.14) yields for A(o) a value of 0.246 and 0.243 for [L-Met2] DREK and DREK, respectively. These values, lower than that (0.278) measured at - 45 °C for tyrosine at 280 nm (Lakowicz and Maliwal, 1983), indicate that tyrosine residue in both peptides display residual motion independent of the global rotation of the peptide. It is possible to measure the relative importance of the mean residual motions of the tyrosine residues:

$$A(o)/A_o = 1 - \alpha \tag{5.18}$$

where α is the residual motion of the fluorophore. The average angular displacement θ of the fluorophore within the peptide is:

$$\cos^2\theta = (1 - 2\alpha/3) \tag{5.19}$$

The values of α and θ are 0.12 and 16.4° of arc, respectively.

The rotational correlation time Φ_R is equal to 0.49 ± 0.03 and 0.44 ± 0.04 ns for DREK and [L-Met2] DREK, respectively. These values are within the same range of the value (0.1 ns) found for free L-tyrosine in water (Rigler et al. 1990). Thus, the N-terminal tyrosyl in DREK and [L-Met2] DREK exhibits an internal motion independent of the global rotation of the whole molecules. However, the measured rotational time is 4 times higher than the rotational correlation time of free tyrosine in water, thus the motion of tyrosine in the morphins is inhibited by the neighboring amino acids.

The fluorescence anisotropy A of a fluorescent probe in a molecule can be given by

$$A(\tau) = \frac{\alpha \; A(o)}{1 + (1/\Phi_R + 1/\Phi_P)\tau_o} + \frac{(1-\alpha)A(o)}{1 + \tau_o/\Phi_P} \tag{5.20}$$

where τ_o is the fluorescence lifetime, Φ_R and Φ_P are the rotational correlation times of the fluorophore and the whole peptide, respectively.

The value of Φ_P found is equal to 1.1 ns, a value in good agreement with the 1 ns value suggested for these peptides from NMR experiments (Pastore et al. 1985).

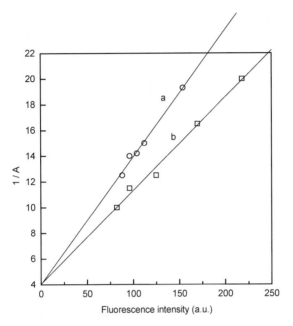

Figure 5.14. Fluorescence intensity quenching of L-and D-DREK with iodide. λ_{ex} = 280 nm and λ_{em} = 303 nm. A(o) = 0.2468 and Φ_R = 0.89 ns for the LDREK. A(o) = 0.243 and Φ_R = 0.97 ns for DREK.

Fluorescence anisotropy decay of [Leu] enkephalin tyrosine was measured using the frequency- domain up to 10 GHz. The data indicate a 44 ps correlation time for local tyrosine motions and a 219 ps correlation time for overall rotational diffusion of the pentapeptide (Lakowicz et al. 1993). Also a rotational correlation time of 26 ps was measured by ^{1}H NMR for H_α of tyrosine in position 1 of L-dermorphin (Simenel, 1990). These ps values determined by NMR and by fluorescence spectroscopy are the result of possible significant atomic fluctuations that occur in the picosecond time scale (Karplus and Mc Cammon, 1981). Since it was difficult in quenching experiments performed on DREK to measure such short correlation times we do not know whether these atomic fluctuations would depend on the conformation of the peptide or not. However, our results clearly put into evidence the presence of a local rotation within DREK.

4c. DNA-Protein interaction

Fluorescence anisotropy can be used to study complex formation between two interacting molecules. Anisotropy allows obtaining an idea on the motion of a molecule. A highly dynamic molecule yields a low anisotropy value.

Xeroderma pigmentation (XP) is an autosomal recessive human disease characterized by hypersensitivity to sunlight and a high incidence of skin cancer in sun-exposed area. XPA is a monomeric 31 kDa DNA-binding protein, known to be involved in the damage recognition step of nucleotide excision repair processes.

XPA binds to DNA-fluorescein complex recognizing the fluorophore as a lesion site. XPA binding induces a modification in the fluorescence observables of the external fluorophore. Depending on the fluorescence labeling position on DNA, binding of XPA can enhance (middle-labeled DNA) or decrease (Fig. 5.15) (end-labeled DNA) the fluorescence intensity of the fluorophore. The fluorescence intensity variation is accompanied by an increase in the fluorescence anisotropy (Fig. 5.16) indicating that binding of XPA to DNA induces a decrease in the rotational motion of DNA. At saturation, variation in both intensity and anisotropy cease.

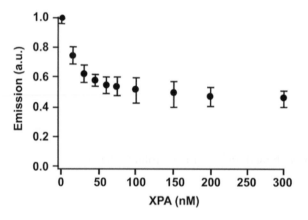

Figure 5.15. Fluorescence intensity titration curve. Fluorescence intensity decreases with recombinant, full-length *Xenopus* XPA; concentration increases due to XPA binding to the single fluorescein labeled at the 5′ -end of a ds 50-mer oligonucleotide. The titration curve is saturated (levels off) at 100 nM XPA. Curtosey from: William R. Wiley Environmental Molecular Sciences Laboratories.

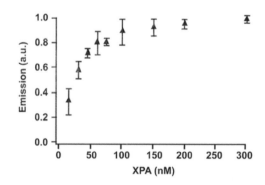

Figure 5.16. Fluorescence anisotropy titration curve. Anisotropy increases with XPA concentration, and is saturated at 100-nM XPA. Curtosey from: William R. Wiley Environmental Molecular Sciences Laboratories.

5. Fluorescence anisotropy decay time

Anisotropy measurements yield information on molecular movements that take place during the fluorescence lifetime. Thus, measuring the time-dependent decay of fluorescence anisotropy will give too interesting information regarding rotational and diffusive motions of macromolecules. Time-resolved anisotropy is determined by placing polarizers in the excitation and emission channels and measuring the fluorescence decay of the parallel and perpendicular components of the emission.

In the time-decay method, the fluorescence anisotropy can be calculated as follows:

$$A(t) = D (t) / S (t) \qquad\qquad (5.21)$$

where

$$D(t) = V_V - g\, V_H \qquad\qquad (5.22)$$

$$S(t) = V_V + 2\, g\, V_H \qquad\qquad (5.23)$$

V_v and V_H are the intensities of the vertically and horizontally polarized components, respectively, of the fluorescence elicited by vertically polarized light. g is the correction factor and is equal to H_V / H_H where H_V and H_H are the vertically and horizontally polarized components, respectively, of the fluorescence elicited by horizontally polarized light.

For a fluorophore bound tightly to a protein, the anisotropy decays as a single exponential,

$$A(t) =. A_o\, e^{-t/\Phi_P} \qquad\qquad (5.24)$$

where Φ_P is the rotational correlation time of the protein, and A_o is the intrinsic anisotropy of the fluorophore at the excitation wavelength.

When the fluorophore exhibits segmental motions, time-resolved anisotropy decay must be analyzed as the sum of exponential decays:

$$A(t) = A_o (\alpha\, e^{-t/\phi_S} + (1 - \alpha)\, e^{-t/\phi_L}) \qquad\qquad (5.25)$$

where ϕ_S and ϕ_L are the short and long rotational correlation times, α and $1-\alpha$ are the weighting factors for the respective depolarizing processes.

$$\frac{1}{\phi_S} = \frac{1}{\phi_P} + \frac{1}{6\, \phi_F} \qquad\qquad (5.26)$$

and

$$\frac{1}{\phi_L} = \frac{1}{\phi_P} \qquad\qquad (5.27)$$

Φ_F is the rotational correlation time of the fluorophore.

Time-resolved anisotropy decay can be used to study proteins folding. The fluorescence probe ANS was used to characterize dynamic properties of the transient partially folded forms that appear during the folding of the α–subunit of tryptophan synthase (αTS) from *Escherichia coli.*(Figures 5.17 and 5.18). Although the X-ray structure shows αTS to be a single α / β barrel structural domain (Hyde et al. 1988), the urea-induced equilibrium unfolding reaction involves two stable intermediates (Matthews and Crisanti, 1981). A small increase in the rotational correlation time was observed in the native baseline region, 0-2 M urea (Figure 5.18B). This, most likely, reflects the binding of urea to the surface of αTS. This conclusion is supported by the absence of changes in secondary and tertiary structure in this region

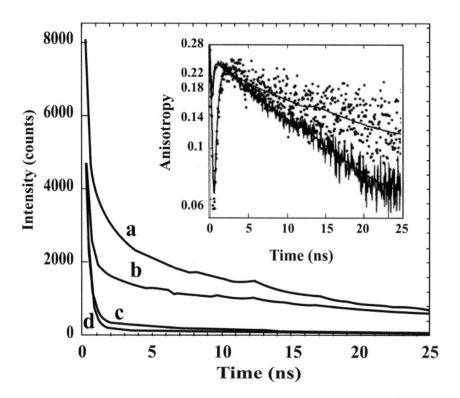

Figure 5.17. Time-correlated single photon counting traces acquired under equilibrium folding conditions with 10 μM αTS and 10 ΜM ANS. The vertical and horizontal polarization components of the ANS excited-state decay are shown at 0 M (a, vertical; b, horizontal) and 3 M urea (c, vertical; d, horizontal) upon excitation with 370 nm vertically polarized light. The fast component corresponds to the decay of unbound ANS. The corresponding calculated anisotropy curves at 0 and 3 M (solid and dotted trace, respectively) are also shown (inset). The smooth lines represent fits to the associative model. Source: Bilsel, O., Yang, L., Zitzewitz, J. A., Beechem, J. M. and Matthews, C. R. 1999, Biochemistry, 38, 4177 – 4187. Authorization of reprint accorded by the American Chemical Society.

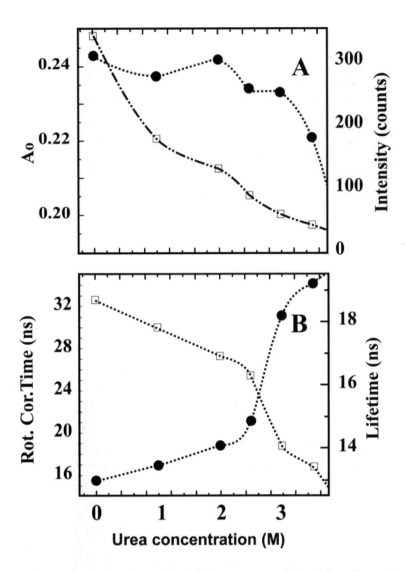

Figure 5.18. Urea concentration dependence of the parameters obtained from fits of the TCSPC traces under equilibrium folding conditions. The amplitude (panel A, □; right axis), excited-state lifetime (panel B, □; right axis), initial anisotropy (panel A, •, left axis) and rotational correlation time (panel B, •, left axis) associated with the longer lifetime component of ANS bound to αTS are shown. The behavior of the longer lifetime component was the focus of the present study because it makes the dominant contribution to the rotational correlation time of the bound ANS. The shorter lifetime component may reflect alternative ANS binding sites or excited-state dynamics of the bound ANS . αTS and ANS concentrations were each 10 μM. Source: Bilsel, O., Yang,, L., Zitzewitz, J. A., Beechem, J. M. and Matthews, C. R. 1999, *Biochemistry,* 38, 4177 – 4187. Authorization of reprint accorded by the American Chemical Society.

6. Depolarization and energy transfer

Energy transfer such as that observed between two or different chromophores, for examples between two tryptophan residues and / or from tryptophan to heme, is a source of depolarization. Energy transfer in hemoproteins between tryptophans and heme is commonly observed. The high energy transfer efficiency in hemoproteins can be put into evidence by plotting the Perrin plot by varying the polarization or the anisotropy as a function of the temperature.

Cytochrome b_2 core from the yeast *Hansenula anomala* has a molecular mass equal to 14 kDa and its sequence shows the presence of two tryptophan residues. Their fluorescence intensity dcay can be adequately described by a sum of three exponentials. The lifetimes obtained from the fitting are equal to 0.054, 0.529 and 2.042 ns, with fractional intensities equal to 0.922, 0.068 and 0.010. The mean fluorescence lifetime τ_o is equal to 0.0473 ns.

The main fluorescence lifetime ($\tau = 54$ ps) and its important fractional intensity ($f_i = 92\%$) indicates that an important energy transfer occurs between the Trp residues and the heme. The protein fluorescence is described by three lifetimes, since one Trp residue may have multiple fluorescence lifetimes, it is difficult to assign the fluorescence lifetimes to any particular Trp residue (s). Nevertheless, some assumptions may be made to explain the origin of the three lifetimes:

(1) The two Trp residues are located in different microenvironments, one near the surface and the second in the core of the protein with a completely quenched fluorescence. In this case, the Trp residue located near the surface of the protein would be responsible of the observed fluorescence. The lifetimes would originate from the different relative orientations between the emission dipole of the Trp residue and the absorption dipole of the heme. The residual motion of the heme and / or that of the Trp residue would generate the different orientations. We notice from the values of the fractional intensities that the relative orientation between the dipoles is favorable to that where the energy transfer is the most efficient.

(2) The two Trp residues are located in the same microenvironment. However, the relative orientations of their emission dipoles to that of the absorption dipole of the heme are not identical, giving rise to different energy transfer efficiencies.

Some authors attributed the nanosecond fluorescence lifetimes in hemoproteins to the presence of an apo form and / or to some impurities. The reader should go to chapter seven for more information concerning this matter.

In an attempt to measure the rotational correlation time of the protein, we have measured the anisotropy of the Trp residues of cytochrome b_2 core at different temperatures and thus for different viscosities. The results are described with the classical Perrin plot (1 / A as a function of T / η) (Fig. 5.19). The data yield a rotational correlation time equal to 38 ps instead of 5.9 ns calculated theoretically for the cytochrome b_2 core, with an extrapolated value A(o) equal to 0.208, lower than that (0.265) usually found for Trp residues at $\lambda_{ex} = 300$ nm at $-45°C$. The fact that the extrapolated anisotropy is lower than the limiting one means that the system is depolarized due to the global and local motions within the protein. In this case, the value of the apparent rotational correlation time (Φ_A) calculated from the Perrin plot will be lower than the global rotational time of the protein (Φ_P). However, the fact that Φ_A is 1000 times lower than Φ_P indicates that a third process different than the global and local rotations is participating in the depolarization of the system. This process is the high energy transfer that is occurring from the tryptophans to the heme.

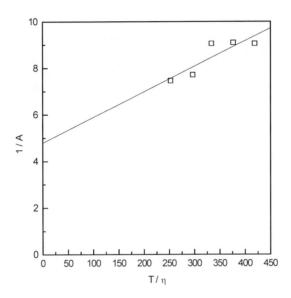

Figure 5.19. Steady-state fluorescence anisotropy versus temperature/viscosity ratio for tryptophan residues of cytochrome b_2 core. Data are obtained by thermal variations in the range 10-36°C.

The extrapolated value A(o) is equal to

$$A(o) = A_o \; x \; d_P \; x \; d_F \; x \; d_T \tag{5.28}$$

Where A_o, d_P, d_F and d_T are the intrinsic anisotropy, the depolarization factor due to the global rotation of the protein, the depolarization factor due to the local motions of the fluorophore and the depolarization factor due to the energy transfer and Brownian motion, respectively.

In the absence of energy transfer, Eq. 5.28 is equal to

$$A(o) = A_o \; x \; d_P \; x \; d_F \tag{5.29}$$

Zentz et al (2003) applied equation 5.29 to investigate the protein elastic forces responsible for the Trp angular displacement inside the apoHb molecule. In fact, at each temperature (T), a certain amount of the energy (W) from the thermal bath is converted into mechanical energy

$$W = (1/2) \; k_B \; T \tag{5.30}$$

where k_B is the Boltzmann constant. This mechanical energy produces a change in the Trp orientation and is related to the Trp mean square angular motion, θ^2, by a torque spring constant C

$$W = (1/2)\, k_B\, T = (1/2)\, C\, \theta^2 \qquad (5.31)$$

The value of C was found equal to 1.3×10^{-20} and 0.8×10^{-20} Nm / rad in the absence and the presence of 32% glycerol, respectively. Also, the variation of θ^2 with the temperature is not the same whether glycerol is absent or present. These results indicate that protein – solvent interaction modifies the intramolecular mobility of the protein.

Chapter 6

FORSTER ENERGY TRANSFER

1. Principles

Förster energy transfer or energy transfer at distance occurs between two molecules, a donor, the excited fluorophore, and an acceptor, a chromophore or a fluorophore. The energy is transferred by resonance, i.e., the electron of the excited molecule induces an oscillating electric field that excites the electrons of the acceptor. The latter will reach an excited state. If the acceptor is fluorescent, its deexcitation will occur mainly by a photon emission. However, if it does not fluoresce, it will return to the fundamental state as the result of its interaction with the solvent. The efficiency of the energy transfer depends on three parameters, the distance R between the donor and the acceptor, the spectral overlap between the fluorescence spectrum of the donor and the absorption spectrum of the acceptor (Fig. 6.1) and the orientation factor κ^2 which gives an indication on the relative alignment of the dipoles of acceptor in the fundamental state and of donor in the excited state.

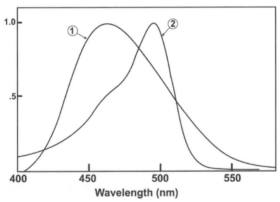

Figure 6.1. (1) Normalized corrected emission spectrum of apoHb-ANS complex upon excitation at 350 nm. (2) Normalized absorption spectrum of apo FIA. Source: Sassaroli, M., Bucci, E., Liesegang, J., Fronticelli, C. and Steiner, R. F. 1984, Biochemistry. 2487 – 2491. Authorization of reprint accorded by the American Chemical Society.

The distance that separates the two molecules goes from 10 to 60-100 Å. Below 10 Å, electron transfer may occur between the two molecules inducing an energy transfer from the donor to the acceptor. κ^2 values hold from 0 to 4. For aligned and parallel transition dipoles (maximal energy transfer) κ^2 is equal 4 and if the dipoles are oriented perpendicular to one another (very weak energy transfer), κ^2 is equal to 0. When κ^2 cannot be known, the value of 2/3 is taken. This value corresponds to a random relative orientation of the dipoles.

Energy transfer mechanism can be described with the diagram below :

$$D + h\nu_0 \text{ --------> } D^* \qquad \text{absorption}$$
$$D^* \qquad \text{--------> } D + h\nu_1 \qquad \text{fluorescence}$$
$$D^* \qquad \text{--------> } D \qquad \text{non radiative deexcitation}$$
$$D^* + A \text{ --------> } D + A^* \qquad \text{energy transfer}$$
$$A^* \qquad \text{--------> } A \qquad \text{non radiative deexcitation}$$
$$A^* \qquad \text{--------> } A + h\nu_2 \qquad \text{induced fluorescence}$$

Figure 6.2 displays fluorescence intensity quenching of DAPI complexed to DNA in presence of two concentrations of acridine orange. In fact, one can notice that while the fluorescence intensity of DAPI decreases, that of acridine orange increases. This type of experiments is one way between others to show the presence of energy transfer mechanism.

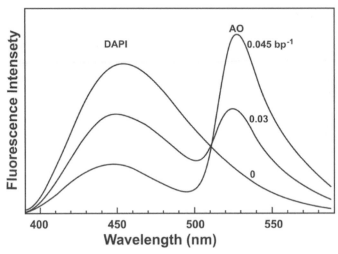

Figure 6.2. Fluorescence emission spectra of DNA-bound DAPI in the presence of 0.03 and 0.045 bp^{-1} of AO. The excitation wavelength is 360 nm. Source: Maliwal, B.P., Kusba, J. and Lakowicz, J. R 1995, Biopolymers, 35, 245-255. Authorization of reprint accorded by John Wiley & Sons, Inc.

Also, one can put into evidence energy transfer mechanism by recording the fluorescence excitation spectrum of the complex (donor-acceptor) (λ_{em} is set at a wavelength where only the acceptor emits) and comparing it to the absorption spectrum of the donor alone. In presence of energy transfer between the two molecules, a peak characteristic of the absorption of the donor will be displayed in the fluorescence excitation spectrum.

Figure 6.3 shows the absorption spectrum of TNS in buffer (a) and the excitation spectrum of TNS-BSA complex ($\lambda_{em} = 460$ nm) (b). We can observe in the excitation spectrum a peak at 280 nm characteristic of the tryptophan residues of BSA. Thus, energy transfer from tryptophans to TNS occurs within the protein-ligand complex.

Excitation of tryptophan residues in BSA at 295 nm in presence of increased concentrations of TNS induces a decrease in the fluorescence intensity of the tryptophans accompanied by an increase in the fluorescence intensity of TNS (data not shown). TNS absorbs at 295 nm (see figure 1.12), thus the increase of its emission intensity upon excitation of the TNS-protein complex is the result of its absorption and of the energy transfer mechanism. The absence of the latter phenomenon will yield emission spectra of TNS with intensities much lower than those observed in its presence.

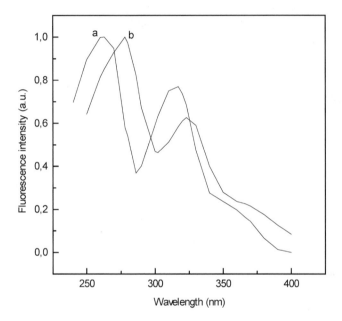

Figure 6.3. Absorption spectrum of TNS in phosphate buffer at pH 7 (a) and fluorescence excitation spectrum of TNS in presence of bovine serum albumin (b) λ_{em} = 460 nm.

Energy transfer is sensitive to the modifications occurring within the structure and the protein dynamics. For example, Staphylococcal nuclease is a small protein of 149 amino acids residues with no cysteine residues and only a single tryptophan residue at position 140 close to the C-terminus (Fig. 6.4). Staphylococcal nuclease called also thermonuclease, is an enzyme that hydrolyses both single- and double-stranded DNA and RNA, although it appears to be more active on single-stranded substrates, yielding 3'-phosphomononucleotide and 3'-phospholigonucleotide end-products. Cleavage of DNA or RNA occurs preferentially at AT or AU-rich regions.

Energy transfer studies were performed on Staphylococcal nuclease between the single tryptophan residue and a cysteine, labeled with the fluorescent probe IAEDANS, incorporated on the surface of the native conformation at position 64 in place of lysine (Fig. 6.4).

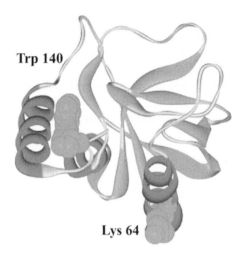

Figure 6.4. Ribbon model of the structure of Staphylococcal nuclease showing the location of the two fluorescent probes, Trp140 and Lys64 (Cys-IAEDANS). The distance between the tryptophan and the cysteine residues is equal to 21 Å. Source : Nishimura, C., Riley, R., Eastman, P. and Fink, A. L, 2000, Journal of Molecular Biology. 299, 1133-1146.

Energy transfer was studied in both the native and acid-unfolded state. Figure 6.5 clearly indicates that energy transfer between tryptophan and IAEDANS is much more important in the native than in the unfolded state. Energy transfer efficiency was found equal to 66% and 18%, respectively, and the distance between tryptophan and the cysteine residues was found equal to 20.4 and 29.3 Å in the native and acid-unfolded states, respectively.

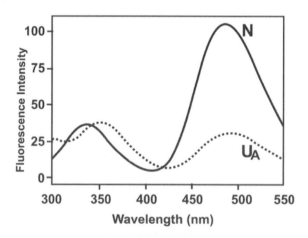

Figure 6.5. Fluorescence emission spectra of native (N) (pH 7.5) and acid-unfolded (U_A) (pH 2.0) staphylococcal nuclease K46C labeled with IAEDANS. Excitation of tryptophan residue ($\square_{ex}$ = 295 nm) increases the emission at 485 nm of IAEDANS as the result of Trp → IAEDANS energy transfer. This increase is less important in the unfolded state than in the folded one. Source: Nishimura, C., Riley, R., Eastman, P. and Fink, A. L, 2000, Journal of Molecular Biology. 299, 1133-1146.

Green fluorescent protein (GFP) is commonly used for energy transfer experiments. The fluorescent moiety of GFP protein is the ser-tyr-gly derived chromophore. GFP can be expressed in variety of cells where it becomes fluorescent, it can be fused to a host protein and it can be mutated so that the mutants have different fluorescence properties and can be used in energy transfer studies. Some examples, the blue fluorescent protein (BFP), which is a GFP mutant with a Tyr66His mutation, absorbs at 383 nm and emits at 447 nm. In energy transfer experiments, BFP is a donor molecule to GFP. In a system with randomly oriented chromophores, the distance Ro at which 50% energy transfer occurs is found equal to 4 nm (Heim, 1999). Cyan fluorescent protein (CFP) contains a Tyr66Trp mutation and absorbs and emits at 436 and 476 nm, respectively. Yellow fluorescent protein (YFP) is a Thr203Tyr mutant with excitation and emission peaks equal to 516 and 529 nm, respectively. In the CFP-YFP pair, CFP is the donor molecule and the YFP is the acceptor one. The value of Ro found for randomly oriented chromophores is 5.2 nm. CFP-YFP pair fused to proteins has been used in resonance energy transfer studies along with multifocal multiphoton microscopy to measure transport phenomena in living cells (Majoul et al. 2002) and protein-protein interaction and structural changes within a molecule (Truong and Ikura, 2001).

Also, the Phe-64→Leu and Ser-65→Thr mutant shows also higher fluorescence parameters comparative to the wild green fluorescent protein. Excitation at 458 nm yields a fluorescence spectrum with two peaks at 512 and 530 nm The fluorescence properties of this enhanced green fluorescent protein (EGFP) were found similar to the recombinant glutathione S-transferase–EGFP (GST–EGFP) protein, expressed in Escherichia coli (Cinelli et al. 2004).

Imaging fluorescence resonance energy transfer studies between two GFP mutants in living yeast were also performed. The donor was the GFP mutant P4.3, bearing the Y66H and Y145F mutations and the acceptor was the GFP mutant S65T (Sagot et al, 1999). The authors constructed a concatemer where the two GFP mutants are linked by a spacer containing a protease-specific recognition site. The tobacco etch virus (TEV) protease was added in order to cleave the covalent bond between the two fluorophores. Experiments performed in vitro show that excitation of the donor at 385 nm leads to the emission of both the donor and the acceptor at emission peaks equal to 385 and 445 nm, respectively. Addition of TEV protease induces an increase of the donor emission and a decrease of the emission acceptor. Experiments performed in vivo reveal that, in presence of protease, energy transfer loss is around 57%.

Membranes fusion can be studied with energy transfer mechanism. In fact, membrane vesicles labeled with both NBD and rhodamine probes are fused with unlabeled vesicles. In the labeled vesicles, upon excitation of NBD at 470 nm, emission from rhodamine is observed at 585 nm as the result of energy transfer from NBD to rhodamine. The average distance separating the donor from the acceptor molecules increases with vesicules fusion, decreasing by that energy transfer efficiency (Struck et al. 1981).

Interaction between DNA and protein can also be performed with energy transfer studies. One generate two DNA fragments each containing half of a protein binding site, each fragment being labeled with two different fluorophores that form a donor / acceptor pair. Protein binding induces the annealing of the two DNA fragments bringing the fluorescence donor and acceptor into proximity. This will induce an increase in the energy transfer mechanism between the two fluorophores (Heyduk and Heyduk, 2002).

Energy transfer is considered as a type of dynamic quenching. In general, dynamic quenching can be described with three different processes: electron transfer, electron exchange and Förster or coulombic energy transfer (Fig. 6.6).

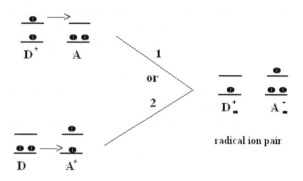

Electron transfer

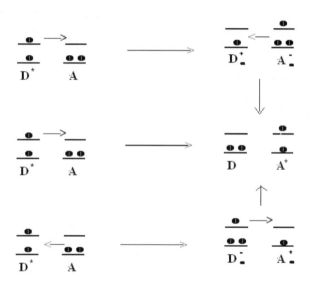

Electron exchange

Dipole-dipole coupling

Figure 6.6. Different mechanisms of dynamic quenching

Electron exchange and Coulombic modes are part of what we can call in general energy transfer mechanism. In the Coulombic mode an electron transfer does not occur between the donor and the acceptor molecules, however two transitions are observed simultaneously, the rate of the dipole-dipole transfer depends on the oscillator strength of the radiative transitions, which is not the case for the electron-exchange mechanism. Also, in the Förster mechanism there is a dipole-dipole coupling mechanism, which is effective in the singlet-singlet energy transfer only, as the result of the large transition dipoles.

Both Förster and electron exchange rate constants depend on the spectral overlap integral J between the emission spectrum of the donor and the absorption spectrum of the acceptor. A high value of J induces a high electron rate

The rate constant of the electron exchange (k_E) is equal to

$$k_E = KJ\, e^{-2\, r/\, L} \tag{6.1}$$

where K is a parameter that is dependent on orbital interactions, J, the normalized spectral overlap integral, r the distance that separates the donor from the acceptor and L the van der Waals radii.

Thus, the electron exchange rate constant decreases to very small values with the increase of the intermolecular distance. The electron exchange is observed up to distances equal to 10 Å.

In the Förster mechanism, the distance r is sufficiently large that dipole-dipole interactions are the main generating potential of the system. Let us remind that we have an energy transfer between an excited molecule denoted D^* and a molecule in the ground state denoted A. The interacting potential P between the dipoles of the two molecules is equal to

$$P = \mu_A\, \mu_D\, r^{-3}\, \kappa\, n^{-2} \tag{6.2}$$

where

$$\kappa = -\, 2\, \cos\theta_D \cos\theta_A + \sin\theta_D \sin\theta_A \cos(\phi_D - \phi_A) \tag{6.3}$$

μ is the magnitude of the dipole moment, n^{-2} is the result of the reduction of the interaction by polarization of the medium, r, θ and ϕ can be considered as the usual spherical polar coordinates (Fig. 6.7).

The rate constant k_{ET} of dipole-induced energy transfer can be calculated from the Fermi golden rule formula

$$k_{ET} = 2\, \pi\, \hbar^{-1}\, |P|^2\, \rho \tag{6.4}$$

where ρ is the number of states coupled by P per energy interval for which deexcitation energy of D equals excitation energy of A.

Combining equations 6.2 and 6.4 shows that k_{ET} decreases as r^{-6}. In the next sections, r will be noted R.

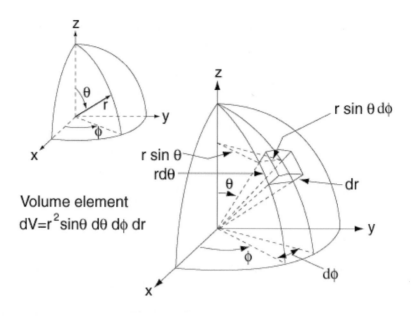

Figure 6.7. Spherical polar coordinates.

2. Energy transfer parameters

Let us consider k_t as the rate constant of the excited state depopulation via energy transfer mechanism, the measured fluorescence lifetime is thus equal to

$$1 / \tau_\tau = 1 / \tau_0 + k_t \qquad (6.5)$$

wher $1 / \tau$ is the deexcitation rate constant in the absence of energy transfer. The efficiency E of deexcitation by means of energy transfer mechanism is equal to:

$$E = \frac{k_t}{k_t + 1 / \tau_o} = \frac{1 / \tau_t - 1 / \tau_o}{1 / \tau_t} = 1 - \frac{1 / \tau_o}{1 / \tau_t} = 1 - \frac{\tau_t}{\tau_o} \qquad (6.6)$$

Energy transfer mechanism appears to affect fluorescence lifetime, intensity and quantum yield.

$$E = 1 - \frac{I}{I_o} = 1 - \frac{\tau_t}{\tau_o} = 1 - \frac{\phi_t}{\phi_o} \qquad (6.7)$$

where τ and I and ϕ_t are the mean fluorescence lifetime and intensity in the absence (τ_o I_o and ϕ_o) and in the presence of the quencher (τ_t, I and ϕ_t). .

The distance R that separates the donor from the acceptor is calculated using equation 6.8:

$$R = R_0 \left(\frac{1-E}{E} \right)^{1/6} \qquad (6.8)$$

The Forster distance R_0 (in Å) at which energy transfer efficiency is 50% is calculated with equation 2:

$$R_0 = 9.78 \times 10^3 [\kappa^2 n^{-4} Q_D J(\lambda)]^{1/6} \qquad (6.9)$$

where κ^2 is the orientation factor (= 2/3), n the refractive index (= 1.33) and Q_D the average quantum yield of the donor in the absence of the acceptor.

From the overlap of the fluorescence spectrum of the donor and the absorption spectrum of the acceptor, we can calculate the overlap integral J ($M^{-1} cm^3$)

$$J(\lambda) = \frac{\int_0^\infty F_D(\lambda).\varepsilon_A(\lambda).\lambda^4 d\lambda}{\int_0^\infty F_D(\lambda)d\lambda} \qquad (6.10)$$

ε_A is the extinction coefficient of the acceptor expressed in $M^{-1} cm^{-1}$.

F_D the fluorescence intensity of the donor expressed in arbitrary units.

λ the wavelength expressed in cm.

The distance R that separates the donor from the acceptor can be known thanks to the X-rays diffraction studies. In this case, from the energy transfer experiments equations 6.4 to 6.6 allow to calculate precisely the value of κ^2. For example, the value of κ^2 for the five tryptophan residues of cytochrome P-450 has been determined. Cytochrome P-450, a superfamily of heme proteins, catalyses monoxygenation of a wide variety of hydrophobic substances of xenobiotic and exogenous origin (Guengerich, 1991; Porter and Coon, 1991). Cytochrome P-450 from *Bacillus megaterium* strain ATCC 14581 called cytochrome P-450$_{BM3}$ is a monomer of molecular mass of 119 kDa, comprising an N-terminal P-450 heme domain linked to a C-terminal reductase domain (Narhi and Fulco, 1986). The heme domain of cytochrome P-450$_{BM3}$ contains five tryptophan residues (Fig. 6.8). The distances of the tryptophans from the heme are 1.224 (Trp96), 1.834 (Trp90), 1.85 (Trp325), 1.968 (Trp367) and 2.365 nm (Trp130). Fluorescence intensity decay yields three fluorescence lifetimes equal to 0.2, 1 and 5.4 ns with fractional contributions of 61%, 33% and 6%, respectively. The longest lifetime was assigned to possible impurities and the two shortest lifetimes to the three Trp residues.

Figure 6.9 displays the plot of Trp to heme distance obtained by fluorescence energy transfer as a function of κ^2 (from 0 to 4) for the two shortest fluorescence lifetimes (solid curve). The five dotted lines in the figure show the crystallographic distances between tryptophan residues and heme. The intersection between the crystallographic lines and the energy transfer plots is equal to the κ^2 correspondig to the specificic tryptophan.

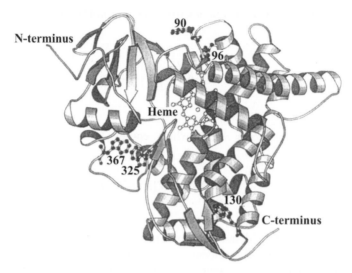

Figure 6.8. Location of five tryptophan residues (shown in a ball and stick representation) and the heme prosthetic group (white ball and stick) in the heme domain of cytochrome P-450$_{BM3}$.

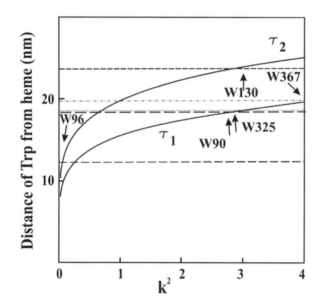

Figure 6.9. Plot of κ^2 against the distance of tryptophan residues from the heme in cytochrome P-450. The dotted lines in the figure show the crystallographic distances between tryptophan residues and lines labelled with τ_1 and τ_2 show the calculated distances between the heme and different tryptophan residues using all possible value of κ^2 (0.0-4.0). The point where these two lines intersect shows the κ^2 value for the particular residue and is indicated by the arrows. Source of figures 6.8 and 6.9: Khan, K.K., Mazumdar, S., Modi, S., Sutcliffe, M., Roberts, G. C. K. and Mitra, S. 1997, Eur. J. Biochem. 244, 361-370. Authorization of reprint accorded by Blackwell Publishing.

3. Relation between energy transfer and static quenching

We have seen in chapter 4 that static quenching occurs when there is a complex between the two interacting molecules. This complex is observed by following the fluorescence intensity variation (increase or decrease) of the fluorophore. Binding of a ligand to a protein induces in general a decrease of the fluorescence intensity of the tryptophan residues of the protein. When energy transfer occurs between the tryptophans and the ligand, the fluorescence intensity decrease of the typtophans is accompanied by an enhancement of the fluorescence intensity of the ligand if the latter fluoresces.

The binding parameters (stoichiometry and association constant) are usually determined from the intensity decrease of the tryptophan residues. However, when energy transfer occurs, errors in the determination of the association constant can be observed.

The following example describing hemin binding to human α_1-acid glycoprotein illustrates the latter two paragraphs.

α_1-Acid glycoprotein contains three Trp residues that all contribute to the fluorescence of the protein. Performing anisotropy and quenching studies on the tryptophan residues allowed us to suggest a three-dimensional model for α_1-acid glycoprotein. In this model, the protein would contain a pocket with two main domains, a hydrophobic domain where ligands such as progesterone and TNS can bind and a hydrophilic one formed mainly by the glycosylation site of the protein (Albani, 1999; De Ceukeleire and Albani, 2002). This model is based on conclusions drawn from dynamics and structural experiments, and none of the experiments performed is straightforward.

TNS binds to α_1-acid glycoprotein with a dissociation constant of 60 μM and a stoichiometry of 1:1. The nature of the binding site is hydrophobic although contacts with a polar environment (solvent) do exist (Albani et al. 1995).

Binding studies are performed between α_1-acid glycoprotein and hemin, a co-factor that binds to the hydrophobic pocket of hemoproteins. If α_1-acid glycoprotein contains a hydrophobic pocket, hemin will bind to this pocket with a well defined stoichiometry. Also, if TNS and hemin bind to the same binding site and if the affinity of hemin to α_1-acid glycoprotein is more important than that of TNS, binding of hemin to a protein-TNS complex will remove the TNS from its binding site.

Addition of hemin to a solution of a fixed amount of α_1-acid glycoprotein - TNS complex induces a complete abolishment of TNS fluorescence (λ_{ex}, 320 nm and λ_{em}, 420 nm) by 1 mol hemin / mol α_1-acid glycoprotein (Fig. 6.10). Thus, TNS and hemin bind to the same site on α_1-acid glycoprotein.

The decrease of TNS fluorescence intensity was analyzed by plotting the ratio of the fluorescence intensities in the absence and presence of hemin as a function of hemin concentration (Fig. 6.11)

$$I_0 / I = 1 + K_a [hemin] \qquad (4.21)$$

where K_a is the association constant of α_1-acid glycoprotein – hemin complex.

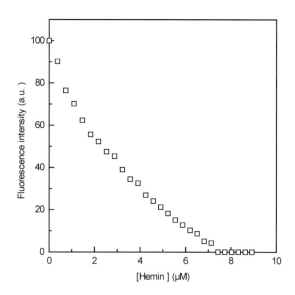

Figure 6.10. Displacement of TNS (4 μM) bound to 7.5 μM α_1-acid glycoprotein with hemin. λ_{ex}, 320 nm and λ_{em}, 420 nm.

Fluorescence intensity decrease of TNS was corrected for the dilution and for absorption at 320 and 442 nm. At the highest concentration of hemin added (9 μM), the corrected intensity for dilution was 1.18 times higher than that recorded before correction. Also, at the same concentration of hemin, the intensity after correction for the inner filter effect was 1.5 times higher than that obtained before correction.

At the lowest concentration of hemin (0.37 μM), the intensity after correction for the inner filter effect was 1.11 times higher than that obtained before correction. Source: Albani, J. R. 2004, Carbohydrate Rsearch, 339, 607-612.

The value of K_a was found equal to 0.44 μM^{-1}, i.e., a dissociation constant equal to 2.3 μM. Looking closely to figure 6.10, one can notice that 50% decrease in the fluorescence intensity of TNS corresponds to a concentration equal to 2.2 μM. This value is in good agreement with the dissociation constant obtained with Eq. 4.21 (Fig. 6.11).

TNS binds to α_1-acid glycoprotein on a hydrophobic site with a dissociation constant of 60 μM (Albani et al. 1995). TNS bound to α_1-acid glycoprotein is displaced by hemin as shown by the complete abolishment of its fluorescence (Fig. 6.10).

Since, hemin is known to bind to the hydrophobic pockets of hemoproteins, one should expect that hemin binds to the only hydrophobic site of α_1-acid glycoprotein, its pocket.

Data of figure 6.10 can be analyzed by plotting the fluorescence intensity as a function of the logarithm of added hemin concentration. This will yield a curve going from 100% to 0% TNS labeling (Fig. 6.12). At 50% labeling, the corresponding concentration of ligand is called the IC50 or 50% inhibitory concentration and is equal to 2.43 μM.

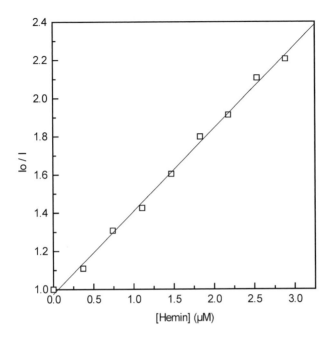

Figure 6.11. Plot of the fluorescence intensity decrease of TNS bound to α_1-acid glycoprotein by hemin. Data are from figure 6.10. Io and I are the fluorescence intensities of TNS before and after addition of hemin (see Eq.4.21). The dissociation constant obtained for the hemin - α_1-acid glycoprotein is equal to 2.3 µM.

Cheng and Prusoff (1973) derived an equation allowing calculating the dissociation constant (K_i) of the inhibitor-protein complex:

$$K_i = \frac{IC50}{1 + L / K_d} \qquad (6.11)$$

where in our case K_i is the dissociation constant of the hemin - α_1-acid glycoprotein complex, L the concentration of TNS (4 µM) and K_d (= 60 µM) the dissociation constant of TNS-α_1-acid glycoprotein (Fig. 6.12). Eq. 6.11 yields a value for K_i equal to 2.27 µM, a value in good agreement with that obtained from figure 6.10 (K_d = 2.2 µM) or from figure 6.11 (K_d = 2.3 µM).

Addition of hemin to a solution of 7.5 µM α_1-acid glycoprotein induces a decrease of the fluorescence intensities of the Trp residues (λ_{ex} = 280 nm and λ_{em} = 332 nm) (Fig. 6.13).

Binding constant of the α_1-acid glycoprotein - hemin complex was analyzed using Eq. 6.12 and 6.13, where free ligand concentration was taken into consideration:

$$Flu = \frac{Flu_0 \times (L_0 - L_b) + Flu_1 \times L_b}{L_0} \qquad (6.12)$$

where Flu is the observed fluorescence, Flu_0 and Flu_1 are the fluorescence of free and bound hemin, respectively, L_0 and L_b are the concentrations of total and bound hemin.

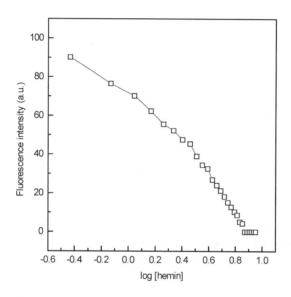

Figure 6.12. Semi-logarithmic plot of the displacement of TNS bound to α_1-acid glycoprotein by hemin. Data are from Figure 6.10. Source: Albani, J. R. 2004, Carbohydrate Rsearch, 339, 607-612.

The intensity decrease is clearly hyperbolic and therefore a mathematical binding analysis can be performed using the following quadratic equation obtained from the definition of the equilibrium constant:

$$L_b = 0.5 \left[(nP_0 + L_0 + K_d) - \{(P_0 + L_0 + Kd)^2 - 4\,n\,P_0\,L_0\}^{1/2} \right] \qquad (6.13)$$

where P_0 is the protein concentration. The parameter Flu_1 was found equal to 57. The dissociation constant Kd of the α_1-acid glycoprotein-hemin complex determined from Eq. 6.13 is equal to 0.67 ± 0.67 μM. This value is different from that (2.3 μM) determined from figures 6.10, 6.11 and 6.12. The small value of K_d with its important relative uncertainty means that the concept of calculating K_d from Eq. 6.13 is not correct. The reason for this comes from the fact that the decrease of the fluorescence intensity of the Trp residues in presence of hemin is the result of a high Förster energy transfer from the Trp residues to the hemin and not only because of binding of hemin to the protein such as when TNS displacement was followed (Fig. 6.10) (see also discussion).

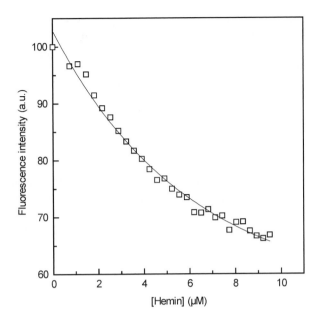

Figure 6.13. Titration curve of 7.5 µM of α_1-acid glycoprotein with hemin. λ_{ex}, 280 nm and λ_{em} 332 nm. Fluorescence intensity decrease was corrected for the dilution and for absorption at 280 and 332 nm. At the highest concentration of hemin added (9.5 µM), the corrected intensity for dilution was 1.15 times higher than that recorded before correction. Also, at the same concentration of hemin, the intensity after correction for the inner filter effect was twice that obtained before correction.

At the lowest concentration of hemin (0.74 µM), the intensity after correction for the inner filter effect was 1.2 times higher than that obtained before correction. Source: Albani, J. R. 2004, Carbohydrate Rsearch, 339, 607-612.

From the overlap of the emission spectrum of Trp residues in α_1-acid glycoprotein and absorption spectrum of hemin in presence of α_1-acid glycoprotein (Fig. 6.14), we have calculated the overlap integral J

$$J(\lambda) = \frac{\int_0^\infty F_D(\lambda).\varepsilon_A(\lambda).\lambda^4 d\lambda}{\int_0^\infty F_D(\lambda) d\lambda} \qquad (6.10)$$

J was found equal to $4.33 \times 10^{-14} M^{-1} cm^3$.
The Forster distance R_o (in Å) at which the efficiency of energy transfer is 50% was calculated with equation 6.12:

$$R_0 = 9.78 \times 10^3 [\kappa^2 n^{-4} Q_D J(\lambda)]^{1/6} \qquad (6.9)$$

where κ^2 is the orientation factor (= 2/3), n the refractive index (= 1.33) and Q_d the average quantum yield (= 0.064). R_o is equal to 25.6 Å. This large distance suggests the presence of an efficient energy transfer between Trp residues and hemin that occurs specifically inside the formed complex.

The efficiency of quenching (E) is equal to

$$E = 1 - \frac{I}{I_o} \qquad (6.7)$$

where I and I_o are the fluorescence intensity in the presence and the absence of hemin.

The value of E calculated at infinite concentrations of hemin was obtained by plotting 1/E as a function of 1/[hemin] (Fig. 6.15). E was found equal to 0.80.

The distance that separates the donor from the acceptor was calculated with equation 6.8 :

$$R = R_0 \left(\frac{1-E}{E} \right)^{1/6} \qquad (6.8)$$

R is equal to 20.3 Å.

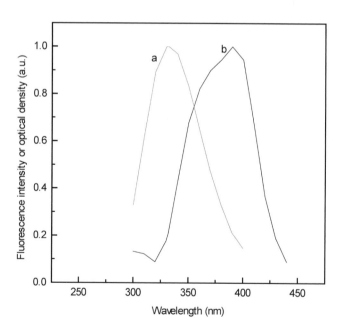

Figure 6.14. Spectral overlap between Trp residues fluorescence emission of α_1- acid glycoprotein (a) and absorption spectrum of hemin bound to the protein (b). Spectrum b was obtained by subtracting (protein) from (protein-hemin complex). Source: Albani, J. R. 2004, Carbohydrate Rsearch, 339, 607-612.

The constant rate of the energy transfer (k_t) can be calculated from equation 6.14:

$$k_t = (1 / \tau_o) (R_o / R)^6 \qquad (6.14)$$

where τ_o (= 2.285 ns) is the fluorescence lifetime of Trp residues in α_1-acid glycoprotein. k_t is equal to 2.8×10^9 s^{-1}, a value equal to 6.5 times that calculated for Trp→ calcofluor energy transfer.

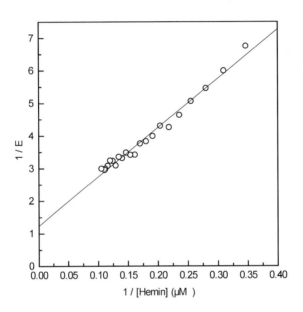

Figure 6.15. Determination of energy transfer efficiency from Trp residues of α_1-acid glycoprotein to hemin. The value of E determined at the extrapolation is equal to 0.80. Source: Albani, J. R. 2004, Carbohydrate Rsearch, 339, 607-612.

Discussion

The stoichiometry of α_1-acid glycoprotein - hemin complex (1:1) indicates that the hemin binds to a specific site. Displacement of TNS by hemin gives the nature of this site: a hydrophobic one. Therefore, hemin does not bind to the surface of α_1-acid glycoprotein, a hydrophilic area. It binds to a hydrophobic domain of the pocket of the protein inducing a decrease of the fluorescence intensity of Trp residues of the protein via a Förster energy transfer mechanism.

Important energy transfer occurs between the Trp residues and hemin as it is revealed by the values of E (0.80) and R_o (25.6 Å). It is interesting to compare these results with those obtained for the energy transfer between the Trp residues of α_1-acid glycoprotein and TNS and calcofluor white. The two extrinsic fluorophores bind to the pocket of α_1-acid glycoprotein in two different domains, hydrophobic (case of TNS) and hydrophilic (case of calcofluor white).

Energy transfer studies between Trp residues and TNS and calcofluor white gave values for R_o equal to 28 Å (Albani et al. 1995) and 18 Å (Albani, 2003), respectively. The very close values of Ro for Trp –> TNS and Trp –> hemin energy transfer is also an indication of the fact that TNS and hemin binds to the same hydrophobic site within α_1-acid glycoprotein pocket.

Displacement of TNS by hemin is a direct proof that the two chromophores bind to the same site. Also, the structural nature of hemin prohibits it to bind specifically to the surface of α_1-acid glycoprotein. In general, hemin interacts with the hydrophobic domain of hemoproteins pocket (Dickerson and Geis, 1983). Also, it is important to indicate that the fluorescence intensity decrease of Trp residues in presence of saturated concentrations of hemin is not equal to 100%, such as it is the case for TNS fluorescence decrease. Trp residues are part of the protein and they are not displaced by the binding of hemin. Therefore, the intensity decrease of Trp residues is the result mainly of their energy transfer to the hemin. However, TNS is displaced by hemin. TNS does not fluoresce when it is free in solution and each time a defined concentration of hemin binds to the pocket, it displaces a defined concentration of TNS. The fluorescence that results occurs from TNS still bound to the protein. Upon reaching saturation (1 hemin for 1 α_1-acid glycoprotein), all the TNS will be free in solution and the fluorescence intensity recorded will be equal to zero. We plotted 1 / E vs 1 / [hemin] from the data of figure 6.10, we found a value for E equal to 1.44 (Fig. 6. 16).

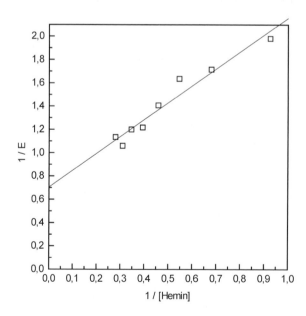

Figure 6.16. Determination of efficiency of TNS fluorescence bound to α_1-acid lipoprotein by hemin. The value of E determined at the extrapolation is equal to 1.44. This excludes the presence of any energy transfer from TNS to heme upon binding of hemin to the protein-TNS complex.

This simply means that energy transfer between TNS and hemin does not occur since efficiency of quenching cannot be higher than 100%. TNS is displaced by hemin as the result of the binding of hemin to the same site. Therefore, the dissociation constant of the hemin-α_1-acid glycoprotein complex calculated from the data of figure 6.10 is the correct one since it was obtained from binding experiments only and not by a combination of binding and high energy transfer phenomena. Equation 6.13 yields a value for the dissociation constant equal to 0.67 ± 0.67 µM. If this value is correct, this means that around this concentration, we should have a decrease by 50% of the fluorescence intensity of Trp residues. Looking closely to the data of figure 6.13, one can notice that at 0.7 µM hemin, the intensity decrease is only 4% of the initial fluorescence intensity. Thus, this value cannot be considered as the dissociation constant for the α_1-acid glycoprotein - hemin complex. Let us consider now the lowest fluorescence intensity value calculated by the computer with Eq. 6.13. This value was found equal to 57. For simplicity, let us consider it equal to 60. In other terms, the fluorescence intensity of Trp residues begins at 100% in absence of hemin and reaches 60 at saturation concentrations of hemin. Thus the decrease is equal to 40. In this case, the K_d calculated in figure 6.13 corresponding to a 50% intensity decrease and thus to an intensity equal to 80, will be equal to 4 µM. This value is twice that found when displacement of TNS was performed (Fig. 6.10). This difference is due simply because the intensity at saturation is not equal to 60 as generated by the computer. Then how can we analyze the data of figure 6.13? If we consider that binding of hemin to α_1-acid glycoprotein induces energy transfer from Trp residues to the heme and thus fluorescence intensity quenching of Trp residues, the decrease in the observed fluorescence intensity should stop when the stoichiometry reaches 1 mole of hemin for 1 mole of protein. The same analysis applies for TNS fluorescence intensity quenching. When we reach 1 mol of hemin for 1 mol of protein, the fluorescence intensity of TNS is equal to zero (Fig. 6.10). Quenching experiments in figure 6.13 were performed on 7.5 µM of α_1-acid glycoprotein. Thus, saturation should be reached at 7.5 µM of hemin. The corresponding intensity is around 70. This means that the dissociation constant of the complex will be calculated for a fluorescence intensity of 85.

For this value, the plot in figure 6.13 yields a K_d equal to 2.8 µM, a value close to that (2.3 µM) obtained from figures 6.10 and 6.11 or from Equation 6.11.

Thus, the decrease of the fluorescence intensity of Trp residues beyond the stoichiometry concentration is the result of a non specific quenching mechanism by aggregated hemin binding nonspecifically to the protein (Albani, 1985).

In hemoproteins such as myoglobin and hemoglobin, the iron of the heme interacts with two histidine residues of the pocket so that stability is maintained (Dickerson and Geis, 1983). Studies concerning hydrophobic domain of α_1-acid glycoprotein pocket does not show any histidine residue (Kute and Westphal, 1976) and thus binding of hemin to the pocket does not allow a real docking on the wall of the pocket. For this reason also, interaction between hemin and pocket of α_1-acid glycoprotein could be of a nature different from that we know on hemoproteins.

The present data show that hemin can bind to a non-hemo protein on a specific site. The presence of a pocket within α_1-acid glycoprotein facilitates this binding. The diameter of the hemin is estimated to be 12 Å (Dickerson and Geis,1983). Thus, the volume that hemin occupies within the α_1-acid glycoprotein pocket is approximately equal to 904 Å^3. X-ray diffraction studies show that α_1-acid glycoprotein is oblongue and occupies a volume equal to 81680 Å^3 (see chapter 8, paragraph 14a). Therefore, hemin occupies a very small volume compared to that of the protein. Since the pocket of α_1-acid glycoprotein is formed by two domains, one hydrophobic and the other hydrophilic, its volume should have a size higher than the volume occupied by the hemin.

The carbohydrate residues possess a well defined structure when bound to α_1-acid glycoprotein (Albani and Plancke, 1999; Albani et al. 1999; 2000), their presence in the pocket confers to them their specific structure. However, we do not know the volume occupied by the carbohydrate residues within the pocket although some fluorescence data tend to indicate that the carbohydrate residues cover most of the interior surface of the pocket (De Ceukeleire and Albani, 2000). X-ray diffraction and electron microscopy studies have shown that carbohydrate residues cover 91% of α_1-acid glycoprotein surface (see chapter 8, paragraph 14c.)

Chapter 7

ORIGIN OF PROTEINS FLUORESCENCE

Literature contains an important number of papers dealing with proteins fluorescence and especially with the origin of the Tryptophan fluorescence. All these studies correlate the origin of the fluorescence to the primary and tertiary structures of the proteins. In peptides and proteins, rotation of the polypeptide chain is observed at the level of the α - carbon atom of each amino acid. This rotation is characterized by two well defined angles of torsion or dihedral ϕ and ψ (Figure 7.1).

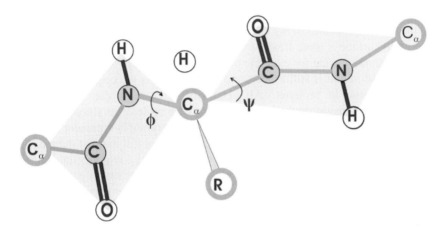

Figure 7.1. Illlustration of the peptide plane (gray area) and ϕ-Ψ angles. The line formed by the repeating -Cα-C-N-Cα- is the backbone of the peptide chain. Mathematically, phi (ϕ) and psi (Ψ) are the dihedral angle (also known as torsional angle) which is defined as the angle between the point (e.g., Cα) at the end of a 4-point sequence and the plane (e.g., peptide plane) occupied by the other three points. In a peptide, phi-psi angles are restricted to certain ranges. A plot of their distribution is called the Ramachandran plot. Source: Web Book publications.

The values of the two angles are illustrated in the Ramachandran diagram which defines the limits of conformational freedom for each peptide bond unit and hence for the entire polypeptide chain. However, even if limited conformations are allowed for each amino acid, the number of possible conformations for the polypeptide is high. For example, a 100-residue protein with three conformations for each amino acid would have 3^{100} possible conformations. Therefore, the number of conformations that a protein can possess is very important.

A folded protein could have a set of different conformations, thus we have here a first definition of a protein structure: the global structure is a combination of sub-structures or conformations. The interconversion between them is not too fast. Each conformation is rigid and has a definite specific structure. This model is known as the rotamers model.

However, even if this model is correct, one should not forget that a protein displays motions. Therefore, each conformation or sub-structure possesses local motions, i.e., a folded protein would display very complex dynamics. Therefore, when we perform spectroscopic measurements, we could monitor the mean dynamics of the different conformations.

A fast interconversion between different sub-species induces a more complex system. In this case, fluorescence measurements would be a mean value of all the different possible sub-species.

Interactions that occur between different amino acid residues of a protein generate local dynamics that should be analyzed differently from the interconvesions between the different conformations. Therefore, proteins and macromolecules in general possess local flexibilities allowing the diffusion of small molecules such as solvent or oxygen through the macromolecules. This flexibility varies from a domain to another and in general protein surface is more flexible than its interior. This local flexibility can be put into evidence by monitoring for example oxygen diffusion as a function of temperature.

The main question that is still asked up to now is what is the origin of the tryptophan fluorescence lifetime in proteins? All scientists agree with the fact that when a tryptophan is buried inside the hydrophobic core of a protein, its fluorescence is blue-shifted compared to the fluorescence observed from a tryptophan present at the protein surface. However, the origin of the fluorescence lifetime is still in great debate.

Tryptophan in peptides and proteins exhibits a bi or multi – exponential fluorescence decay. An explanation of this decay is the presence of conformers in equilibrium in the folded state. Each conformer exhibits one specific fluorescence lifetime. The origin of the conformers arises from the rotation of the indole ring within the C_α - C_β bond and / or the C_β - C_γ bond, the interconversion between the rotamers being slow relative to the fluorescence time scale. The rotamers are considered also as rigid entities.

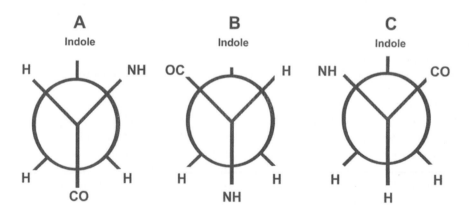

Figure 7.2. The three possible rotamers of tryptophan in proteins.

The rotamers model was used to explain the origin of the biexponential decay of tryptophan free in solution. In polypeptides, lifetime of each rotamer is explained as the result of the quenching interactions between the indole and quenching groups in the fluorophore.

For example, the unique tryptophan residue of the human serum albumin displays a biexponential decay with lifetimes equal to 3.14 and 0.51 ns. The long and short lifetimes are attributed to rotamers A and C of figure 7.2, respectively. In rotamer A, the quenching group near the indole is the amino-group. Fluorescence quenching is weak and thus fluorescence lifetime is long. In rotamer C, the carbonyl and amino groups surround indole group. Thus, fluorescence quenching is important and lifetime is weak. However, the rotamer model does not explain why rotamer B does not have any fluorescence!!! Since this rotamer is not fluorescing, can we say that fluorescence quenching by the carbonyl group is highly efficient so that fluorescence of indole is equal to zero? If this is the case, we do not understand why rotamer C still has an observable fluorescence lifetime? So, just for a simple fluorophore free in solution, we have at least three questions without any answer. Can we imagine the quiz we have when we are dealing with proteins?

Finally, the rotamers model was introduced to explain the biexponential decay of the tryptophan in solution. If we have to apply the rotamers model to tryptophan in polypeptides, what will be the contribution of the protein structure and dynamics to the fluorescence lifetime?

We are going to detail now different examples concerning the fluorescence of Trp residues in proteins. We shall notice that it is difficult to draw one single rule in the interpretation of the origin of the Trp fluorescence lifetime.

Antithrombin (AT) is a serine protease inhibitor that inhibits thrombin, factor Xa, IXa, XIa, XIIa, tPA, urokinase, trypsin, plasmin, and kallikrein (Lahiri et al. 1976; Travis and Salvesen, 1983; Menache, 1991; Menache et al. 1992). Human Antithrombin of molecular weight equal approximately to 58,000 Da, contains 432 amino acids, three disulfide bridges, and four carbohydrate side chains, which account for 15% of the total mass (Magnusson, 1979; Franzen et al. 1980). Human antithrombin is synthesized in the liver and present in plasma at levels of 14 to 20 mg/dl (Murano et al. 1980; Rosenberg et al. 1986). One cause of decreased level of antithrombin is pathological conditions.

Human antithrombin, a member of the serpin family of protein proteinase inhibitors, contains four tryptophan residues (Fig. 7.3) that belong to two classes, buried and solvent exposed. Fluorescence lifetime measurements of wild type protein and of four variants containing single Trp→Phe mutations, allow obtaining for each protein three lifetimes, 0.23 ± 0.04, 1.6 ± 0.13 and 6 ± 0.9 ns.

In this example, it is difficult to assign a specific fluorescence lifetime to a specific Trp residue. Also, the rotamers model can in no way explain the origin of the fluorescence lifetimes. One can consider the fluorescence lifetime as being the result of the Trp residues interaction with their microenvironments. However, since fluorescence lifetimes do not vary significantly with the different mutants, one should consider the possibility of having, around the Trp residues, a common identical protein structure responsible of the three measured fluorescence lifetimes.

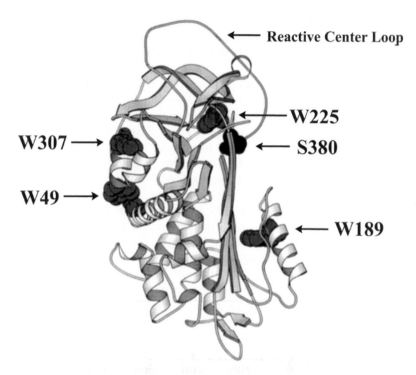

Figure 7.3. Structure of human antithrombin showing the location of the four tryptophans, based on the crystal structure of (Schreuder, H. A., de Boer, B., Dijkema, R., Mulders, J., Theunissen, H. J. M., Grootenhuis, P. D. J., and Hol, W. G. J. 1994. Nat. Struct. Biol. 1, 48-54). Also shown is Ser-380, the P14 residue that is initially present inserted into β-sheet A, but is expelled by the heparin-induced conformational change. The reactive center loop is indicated at the top of the molecule. Note that the two-dimensional representation makes Trp-307 and Trp-49 appear closer than they are. Source: Meagher, J. L., Beechem, J. M., Olson, S. T. and and Gettins, P. G. W. 1998, J. Biol. Chem. 273, 23283-23289. Authorization of reprint accorded by The American Society for Biochemistry and Molecular Biology.

Barnase, a 110-residue extracellular protein found in bacillus amyloliquefaciens, is a ribonuclease with potentially lethal functions within the cell. The RNase activity of barnase is inhibited by barstar, a 90-residue polypeptide, which binds tightly to barnase through salt bridges and hydrogen bonds.

A double mutant of the single-domain protein barstar having a single tryptophan (W53) is obtained by mutating the remaining two tryptophans (W38 and W44) into phenylalanines. W53 is buried in the core of the protein. Fluorescence intensity decay of the mutant follows a single exponential behavior with a lifetime equal to 4.9 ns. Figure 7.4 displays the Stern-Volmer plots of the lifetime quenching of Trp53 by KI and acrylamide. Accessibility of the tryptophan to the acrylamide is very weak and inexistent to the KI. These experiments agree with the fact that Trp53 is buried in the hydrophobic core of barstar. Therefore, the environment of the tryptophan could be highly rigid. In fact, fluorescence anisotropy decay shows the presence of one rotational correlation time equal to 4.1 ns. This value corresponds to the global rotation of the barstar of molecular mass equal to 10.1 kDa.

Therefore, Trp53 is devoid from any residual motion. In this case, the large fluorescence lifetime could be the result of the absence of any segmental motion of the Trp residues, reducing by that the local dynamic interaction with the neighboring amino acids.

Wild type barstar shows two fluorescence lifetimes, 4.1 and 1.5 ns. The long lifetime is attributed to two of the three Trp residues including Trp53. Thus, comparing lifetime data of wild type and of double mutants, one found difficult any interpretation of the results with the conformers model, especially if we consider that the tertiary structure of barstar is identical in the mutant and wild type forms.

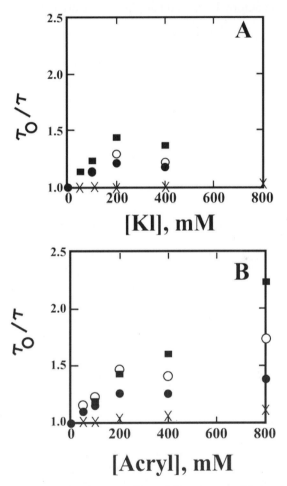

Figure 7.4. Stern-Volmer plots for the quenching of fluorescence lifetime(s) of W38FW44F in the N-state (pH 7) and in the A-form (pH 3) of barstar, by either KI (A) or acrylamide (B). τ_o and τ are the lifetimes in the absence and in the presence of quencher, respectively: (×) pH 7 (N-state); (■) pH 3, mean lifetime; (○) pH 3, 2.1 ns component; (●) pH 3, 5.2 ns component. In the case of quenching by KI, the medium also contained $Na_2S_2O_3$ (100 μM) to avoid the formation of I_3^-. Source: Swaminathan, R., Nath, U., Udgaonkar, J. B., Periasamy, N. and Krishnamoorthy, G. 1996, Biochemistry, 35, 9150 –9157. Authorization of reprint accorded by The American Chemical Society.

Iron-sulfur proteins such as High – potential iron proteins (HiPIPs) belong to a group of proteins possessing only the iron-sulfur complex as the prosthetic group. These proteins play an important role in electron carrier in almost all living organisms. They function in anaerobic electron transport chains, contribute to plant photosynthesis, nitrogen fixation, steroid metabolism and oxidative phosphorylation.

HiPIP from *Chromatium vinosum* contains three tryptophan residues (W60, W76 and W80) (Fig. 7.5). The distance between tryptophan 80 and the Fe_4S_4 cluster (7.5 Å) and their relative orientation contribute to a high energy transfer Förster type. Fe-S bonds covalently bind the Fe4S4 cluster to the protein matrix at cysteine residues 43, 46, 63 and 77. The polypeptide chain conformation may be described as a sequence of α helical or extended conformations and hairpin turns (Carter et al. 1974).

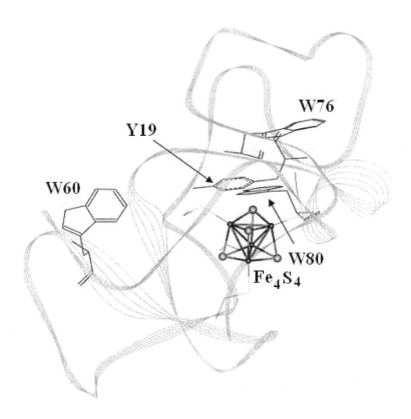

Figure 7.5. Location of the tryptophan residues, the tyrosine residue, and the iron sulfur cluster (shown in ball and stick) of high potential iron protein (HiPIP) from *Chromatium vinosum*. This schematic representation of the protein (monomer unit) was generated from the crystal structure (Protein data bank), accession number 1CKU using the Weblab Viewer program (MSI). The coordinates were obtained from the Protein data bank (Parisini, E., F. Capozzi, P. Lubini, V. Lamzin, C. Luchinat, and G. M. Sheldrick. 1999. Acta Crystallogr. D. 55:1773-1784). Ab *initio* solution and refinement of two high potential iron protein structures at atomic resolution). Source: Sau, A. K., Chen, C. A., Cowan, J. A., Mazumdar, S. and Mitra, S. 2001, Biophys. J. 81, 2320-2330. Authorization of reprint accorded by the American Biophysical Society.

Fluorescence intensity decay of the Trp residues can be described with four lifetimes (Table 7.1).

Table 7.1. Fluorescence intensity decay parameters of holoHiPIP from *Chromatium vinosum*.

τ_i (ns)	0.04	0.5	2.1	5.4	
α_i	0.75	0.13	0.10	0.02	$\tau_0 = 2.56$ ns.
f_i	0.07	0.157	0.508	0.263	

Source: Sau, A. K., Chen, C. A., Cowan, J. A., Mazumdar, S. and Mitra, S. 2001, Biophys. J. 81, 2320-2330. Authorization of reprint accorded by the American Biophysical Society.

The mean fluorescence lifetime of the Trp residues in the apoform of HiPIP decreases compared to the holo form. This decrease is the result of the modification in the fractional contribution (f_i) of the two longest lifetimes to the total emission (Table 7.2).

Table 7.2. Fluorescence intensity decay parameters of apoHiPIP from *Chromatium vinosum*.

τ_i (ns)	0.14	0.5	1.8	5.5	
α_i	0.54	0.28	0.16	0.02	$\tau_0 = 1.96$ ns.
f_i	0.123	0.228	0.469	0.18	

Source: Sau, A. K., Chen, C. A., Cowan, J. A., Mazumdar, S. and Mitra, S. 2001, Biophys. J. 81, 2320-2330. Authorization of reprint accorded by the American Biophysical Society.

Also, in the apo form, only the shortest fluorescence lifetime increases accompanied with an increase in its fractional contribution to the total fluorescence.

Constructing W80N mutant leads to the abolishment of the shortest lifetime (Table 7.3). These results indicate that the shortest lifetime originates from tryptophan 80 and is the consequence of the high energy transfer Förster type from the tryptophan to the Fe_4S_4 cluster.

However, it is difficult to assign to any of the other two Trp residues a specific fluorescence lifetime. Also, the rotamers model appears here inappropriate.

Table 7.3. Fluorescence intensity decay parameters of the HiPIP mutant W80N

τ_i (ns)	-	0.4	2.1	5.6	
α_i	-	0.37	0.424	0.207	$\tau_0 = 3.8$ ns.

Source: Sau, A. K., Chen, C. A., Cowan, J. A., Mazumdar, S. and Mitra, S. 2001, Biophys. J. 81, 2320-2330. Authorization of reprint accorded by the American Biophysical Society.

In many cases, it is very easy to assign fluorescence spectra to specific tryptophan residues. For example, in luminescent bacteria, a lux – specific acyl – ACP thioesterase (Lux D) is responsible for providing myristic acid to the luminescent system. The Lux D protein interacts with a multienzyme fatty acid reductase complex that reduces the fatty acid to aldehyde for the light emitting reaction.

The thioesterase of *Vibrio harveyi* contains four tryptophan residues at positions 23, 99, 186 and 213 that are conserved in the lux-specific thioesterases from all species. The three dimensional structure of thioesterase (Fig. 7.6) obtained by X-ray diffraction studies, reveals the position of each of the Trp residues of the protein. Trp-23 and 99 residues are partially buried while Trp-186 is deeply buried inside the molecule. Finally, Trp-23 is at the protein surface and thus is exposed to the solvent.

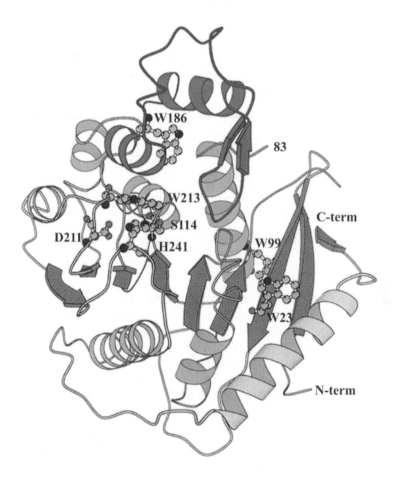

Figure 7.6. Schematic representation of the three-dimensional structure of the thioesterase from *V. harveyi*. The atomic coordinates were taken from the X-ray study of (Lawson D.M, Derewenda, U., Serre, L., Ferri, S., Szittner, R., Wei, Y., Meighen, E. A. and Derewenda, Z. S. 1994, Biochemistry. 33, 9382-9388). The amino and carboxyl termini are denoted by the letters N and C, respectively. Residues 73 and 83 flank the surface loop that is invisible in the electron density map. Source: Li, J., Szittner, R. and Meighen, E. A. 1998, Biochemistry, 37, 16130 – 16138. Authorization of reprint accorded by the American Chemical Society.

In order to put into evidence the contribution of each Trp residue to the global fluorescence of the protein, construction of mutants was performed. In each mutant, one Trp residue was replaced by a phenylalanine. Figure 7.7a shows the fluorescence emission spectra of the wild type and the difference spectra of the tryptophan mutants calculated by subtraction of their respective fluorescence emission from that of the wild type enzyme. Therefore, the results described in the figure give the fluorescence emission spectra of the specific tryptophan residue missing in each of the mutants. The spectra obtained indicate that Trp 186 contribute the most to the fluorescence intensity of the wild type. The position of the emission maximum indicates the relative position in the protein of each Trp residue. In fact, the maximum of the fluorescence spectrum of Trp 186 (λ_{max} = 332 nm) indicates that the Trp residue is buried in the protein core surrounded by hydrophobic environment. The other three Trp residues display fluorescence spectra at 338 and 340 nm. Therefore, the environment of these residues is not completely hydrophobic. The emission maximum of the wild type is located at 337 nm indicating that contribution from Tryptophan 23, 213 and 99 to the global fluorescence is significant. The sum of the fluorescence spectra of the four tryptophan residues was identical to that of the wild type enzyme (Fig. 7.7b) showing that the performed mutations do not induce a significant change in the protein conformation. Also, this reveals that energy transfer between the Trp residues is not significant.

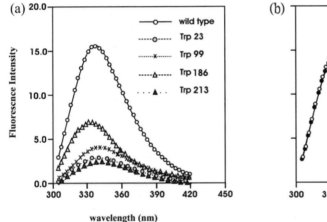
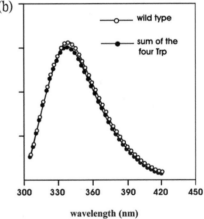

Figure 7.7. Contributions of each tryptophan to the fluorescence emission spectrum of the wild type thioesterase. Fluorescence spectra of 1 μM enzyme in 50 mM phosphate (pH 7.5) were recorded at 20°C on excitation of 296 nm. (a) Spectrum of the wild type enzyme and the difference spectra between the wild type and the tryptophan mutants. The difference spectra were calculated by subtracting the spectrum of a particular tryptophan mutant (e.g., W23Y) from that of the wild type enzyme to obtain the spectrum for a specific tryptophan (e.g., Trp23). (b) Comparison of the spectra of the wild type and the sum of the difference spectra of the individual tryptophan residues. Source: Li, J., Szittner, R. and Meighen, E. A. 1998, Biochemistry, 37, 16130–16138. Authorization of reprint accorded by the American Chemical Society.

In some proteins such as actin, for example, where 4 tryptophan residues are present, the fluorescence intensity decay can be described with two lifetimes 1.99 and 5.41 ns ($\lambda_{ex,}$ 297 and $\lambda_{em,}$ 350 nm) (Turoverov et al. 1999). In such a case, we find it difficult to explain that the two fluorescence lifetimes are originating from two rotamers.

P13[suc1] acts in the fission yeast cell division cycle. Addition of the protein to the kinase assay in vitro was able to rescue the defect in the Cdc2 mutant kinase activity. P13[suc1] contains two tryptophans (Trp-71 and Trp-82) and seven tyrosine residues. The maximum of the fluorescence emission spectrum is located at 336 nm ($\lambda_{ex,}$ 295 or 275 nm). Thus, tryptophan residues are responsible for the protein fluorescence. Three lifetimes are found for the intensity decay, 0.6, 2.9 and 6.1 ns, with fractional intensities of 0.04, 0.32 and 0.64, respectively.

Decay associated spectra (Fig. 7.8) show that fluorescence lifetime components display a characteristic emission spectrum with a maximum located at 325, 332 and 340 nm, respectively. The results reveal that a very weak emission spectrum corresponds to the short lifetime component.

In this example too, we do not see how we can link the fluorescence lifetimes to the presence of conformers. Since we have two tryptophan residues, we should have at least four lifetimes. Attributing the fluorescence lifetimes to conformers, means that we have a combination of conformers that yields three fluorescence lifetimes.

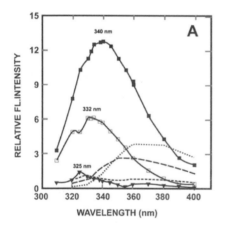 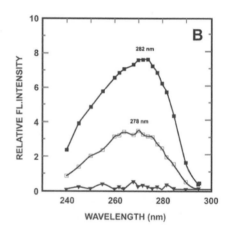

Figure 7.8. DAS (emission) and EDAS (excitation) of p13[suc1]. A, DAS of native p13[suc1] (0.2 mg/ml in 50 mM Tris-HCl, pH 8.0, 2 mM EDTA, 20 °C). Data were collected using an excitation wavelength of 295 nm. The DAS were obtained from the global analysis of a multiple emission wavelength experiment (16 data sets). The filled square symbols refer to the long lifetime DAS, the open square symbols refer to the medium lifetime DAS, and the filled downward triangle symbols refer to the short lifetime DAS. With the dotted line, the short dashed line, and the dashed line are reported the long, the medium, and the short lifetime DAS obtained in 4.0 M GdnHCl, respectively. B, EDAS of native p13[suc1] obtained under the same experimental conditions of Fig. 7.8A Data were collected using an emission wavelength of 336 nm. The EDAS were obtained from the global analysis of a multiple excitation wavelength experiment (18 data sets). The same symbols used in Fig. 7.8A for each lifetime DAS are used here to indicate each lifetime EDAS. Source: Neyroz, P., Menna, C., Polverini, E. and Masotti, L. 1996, Journal of Biological Chemistry. 271, 27249-27258. Authorization of reprint accorded by the American Society for Biochemistry and Molecular Biology.

Time resolved anisotropy decay performed at different temperatures, reveals the presence of two rotational correlation times, one Φ_P, corresponding to the global rotation of the protein and the second Φ_A, a shorter one, reveals the presence of local residual motions around and / or near the two tryptophan residues. Φ_A is an apparent rotational time that is a mathematical combination of the global rotation of the protein and the segmental motion of the fluorophore.

The values of Φ_P is temperature dependent, it decreases when the temperature increases. However, Φ_A appears to be in certain way independent of the temperature (Table 7.4). This results from the complexity of the local rotation around Trp residues.

Table 7.4. Values of the rotational correlation time Φ_P of native p13^{sucl} and of the apparent rotational correlation time Φ_A at different temperatures.

Temperature (°C)	Φ_P (ns)	Φ_A (ns)
6	15.6	0.54
10	12.3	0.33
14	11.5	0.49
19	9.3	0.37
21	10.5	0.64
25	9.3	0.71

Source: Neyroz, P., Menna, C., Polverini, E. and Masotti, L. 1996, Journal of Biological Chemistry. 271, 27249-27258.

The amino acids themselves display residual motions that can originate from the motions of the backbone of the Trp residues itself and from the motion of the surrounding amino acids.

These motions are complex and do no not allow to find a simple correlation between the temperature and the short correlation time. Calculating the value of the rotational correlation time of the fluorophore itself Φ_F from equation 5.27

$$\frac{1}{\phi_S} = \frac{1}{\phi_P} + \frac{1}{6\,\phi_F} \qquad (5.27)$$

yields values equal to 95, 56, 84 and 113 ps at 6, 10, 14 and 21°C, respectively. Most of these values are close to that (100 ps) usually found for free fluorophores in solution. Thus, it is difficult to find a simple correlation between the temperature and the value of Φ_F. The latter seems to be independent of the temperature and lies around 100 ps. This value is common to many fluorophores and thus it can be assigned to motions that correspond to a switch of the molecule around its axis by a certain angle. When the fluorophore is intrinsic to a protein or to a macromolecule or covalently bound to it, the value of Φ_F does not correspond always to a complete rotation of the fluorophore around itself.

Human immunoefficiency virus (HIV) is the agent responsible for the acquired immunodeficiency syndrom (AIDS). HIV-1 protease is among the targets identified for chemotherapy. It is a homodimeric aspartyl protease of 99 amino acid residues per monomer. The protein contains two Trp residues (W6 and W42) per monomer located in different environment. The fluorescence emission spectrum of the protein shows a peak at 341 nm (λ_{ex}, 275 nm) indicating that the Trp residues are responsible for the protein emission. Also, the position of this peak is blue-shifted compared to the fluorescence maximum of Tryptophan in water and thus the average exposure of the Trp residues to the solvent is not total. Trp residues are partly buried within the hydrophobic core of the protein.

Fluorescence intensity decay of the Trp residues (λ_{ex}, 300 nm) yields four fluorescence lifetimes, 0.22, 0.44, 2.06 and 4.46 ns. The authors considered the two shortest lifetimes as the result of the excited state quenching of the tryptophan emission by the protein backbone, while the two longest lifetimes are characteristics of the emission of the Trp residues from two different environments (Kung et al. 1998). Tryptophan free in solution emits with two lifetimes, 0.5 and 3.1 ns. In this case, the short lifetime cannot be assigned to the presence of a backbone protein. The two lifetimes are considered to be the result of the different side chain rotamers conformations. Thus, a tryptophan in presence of different environments, water and protein matrix, can have the same or identical fluorescence lifetimes. This example shows clearly that it is not obvious to explain the origin of the fluorescence lifetime of tryptophan residues in proteins by the rotamers model.

Nevertheless, let us discuss now the HIV –1 protease fluorescence in terms of rotamers model. The purpose of this discussion is to find out how far we can go in the application of this model. Therefore, let us consider the rotamers model as correct and thus the two fluorescence lifetimes (0.5 and 3.1 ns) of free Trp in solution originates from two rotamers. If this model is correct in proteins also and thus in HIV-1 protease, we have two tryptophans with four lifetimes, two short (close to 0.5 ns) and two long (close to 3.1 ns). Thus, we can attribute to each tryptophan, long and short fluorescence lifetimes. In other terms, the presence of the protein matrix around Trp residues does not play any fundamental role in the fluorescence lifetimes of the Trp residues. In this case, the protein backbone has no effect or non significant effect on the fluorescence lifetime of HIV-1 protease. In simple words, the fluorescence lifetime of tryptophan would be in dependent of the tertiary structure of HIV-1 protease.

Fluorescence emission lifetime of one tryptophan residue located in five different positions within an 18-residue amphiphatic peptide (Table 7.5) was also analyzed with the rotamers model. These peptides are unstructured in aqueous solution and become structured when associated to a lipid. A blue shift in the emission maximum of the Trp residue is observed, indicating that the fluorophore is in contact with a hydrophobic environment when the lipid is bound to the peptides. Only peptide 18D-12 shows an emission maximum located at 354 revealing that in this peptide tryptophan is in contact with aqueous environment. These data are consistent with the fact that association of lipid with peptide yield an α-helix where the tryptophan residues located on the hydrophobic surface of the α-helix are in the hydrophobic environment provided by the lipid surface, whereas tryptophans located on the hydrophilic face of the helix are directed toward the aqueous phase. Also, the helix axis of the peptides is parallel to the lipid surface (Fig. 7.9).

Table 7.5. Sequence of amphiphatic peptides and position of the emission maximum of the Trp residue in the absence and presence of unilamellar bilayer vesicles composed of egg phosphatidylcholine.

Peptide	Sequence	$\lambda_{solution}$	$\lambda_{bilayer}$
18A-2	DWLKAFYDKVAEKLKEAF	353 nm	341 nm
18B-3	DRWKAFYDKVAEKLKEAF	353 nm	340 nm
18C-7	DRLKAFWDKVAEKLKEAF	353 nm	339 nm
18D-12	DRLKAFYDKVAWKLKEAF	354 nm	354 nm
18E-14	DRLKAFYDKVAEKWKEAF	350 nm	344 nm

Source: Clayton, A. H. A. and Sawyer, W. H. 1999, Biophysical Journal, 76, 3235-3242. Authorization of reprint accorded by the American Biophysical Society.

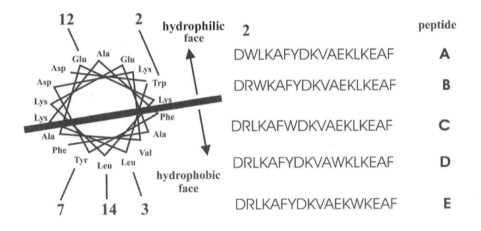

Figure 7.9. Helical wheel representation of 18-residue amphiphatic peptides. The figures in bold refer to the positions of tryptophan substitution in peptides A-E. The solid line represents the putative boundary between the hydrophilic and hydrophobic faces of the α-helix. Source: Clayton, A. H. A. and Sawyer, W. H. 1999, Biophysical Journal, 76, 3235-3242. Authorization of reprint accorded by the American Biophysical Society.

Fluorescence intensity decay of Trp residue in the peptides is described by means of three exponentials in the absence of lipid and by means of two lifetimes in its presence (Table 7.6). The data show that the fluorescence lifetimes are dependent on the position of the Trp residue in the sequence of the peptide. Therefore, the lifetime would be the result of the interaction of the tryptophan with its environment. Dipolar relaxation and environment heterogeneity seems to be the major factors responsible for the fluorescence lifetimes.

Table 7.6. Lifetimes (τ) and prexponential factors (α) of the 18-residue amphiphatic peptides in solution (-v) and bound to unilamellar bilayer vesicles (+v).

Peptide	environment	$\tau_{1 (ns)}$	$\tau_{2 (ns)}$	$\tau_{3 (ns)}$	$\alpha_1 (\%)$	$\alpha_2 (\%)$	$\alpha_3 (\%)$
18A-2	-v	0.200	2.091	4.870	38	54	8
	+v	–	2.582	6.160	–	70	30
18B-3	-v	0.174	1.995	4.144	30	40	30
	+v	–	2.640	6.973	-	72	28
18C-7	-v	1.266	3.365	5.765	35	49	16
	+v	–	2.270	6.454	-	64	36
18D-12	-v	1.137	3.698	8.964	39	59	3
	+v	–	3.167	9.174	-	75	25
18E-14	-v	1.251	4.010	6.123	31	53	16
	+v	–	2.494	6.000	-	71	29

Source: Clayton, A. H. A. and Sawyer, W. H. 1999, Biophysical Journal, 76, 3235-3242. Authorization of reprint accorded by the American Biophysical Society.

In case of the preexponential terms are characteristics of the percentage of rotamers present in solution, the corresponding values do not indicate that we are in presence of three conformers since the preexponential factor of each lifetime varies from one peptide to another.

The authors analyzed the data with the global analysis method. In this method, fluorescence lifetimes are considered to be shared between the data sets whereas the preexponential factors are designated nonglobal. Global analysis is used in general to recover the spectral distribution of α_i (known also as decay associated spectra). By treating the fluorescence intensity decays with the global analysis method, the authors considered the presence of rotamers as evidence and as a fact. In other words, they decided that rotamers are responsible for the fluorescence lifetimes. Therefore, with the global analysis method, τ_1, τ_2 and τ_3 will be identical for the five peptides.

Other authors (Ladokhin and White, 2001) considered that the global analysis method as a non-convincing one for the peptides described by Clayton and Sawyer. In fact, the tryptophan in the different peptides is in contact with different lipid chemical groups and the water environment. This is put into evidence by the important variation of the emission peak (up to 15 nm) of the steady-state spectra. Therefore, the heterogeneity of the surrounding medium of tryptophan was not taken into consideration by applying the global analysis method. Ladokhin and White assigned the multiple fluorescence decay to a dipolar relaxation phenomenon of the tryptophan environment. In fact, they showed that dipolar relaxation of indole in a bulk phase could produce similar-looking decay-associated spectra in the absence of rotamers (Figure 7.10).

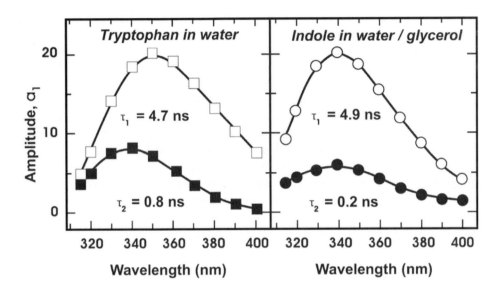

Figure 7.10. Decay-associated spectra obtained by global analysis of heterogeneous fluorescence kinetics of tryptophan in water and of indole in a water/glycerol mixture (3/2, V/V) measured at 10°C. Two components in the case of tryptophan (left) are related to rotameric forms. The structural relaxation in indole's environment in the absence of any rotamers, can lead to an apparently similar heterogeneity (right). Source: Ladokhin, A. S. and White, S. H, 2001, Biophysical Journal. 181, 1825-1827. Authorization of reprint accorded by the American Biophysical Society.

If we consider the rotamers model as the ideal one to describe the results obtained by Clayton and Sawyer, thus we can draw the following conclusion: the modification of the environment of a tryptophan affects fluorescence parameters such as fluorescence intensity and fluorescence emission peak only, the fluorescence lifetime being independent of the Trp environment. This conclusion is short but is a real interpretation and definition of the rotamers model. In a certain way, why not? Therefore, this implies that spatial structure of a protein does not affect the fluorescence lifetime of tryptophan residues.

In this case, what is the real significance of the measured fluorescence lifetimes?

Sillen et al (2003) measured the fluorescence lifetimes of the three Trp residues (4, 57 and 170) present in the calcium binding protein isolated from sacroplasm of the muscles of the sand worm *Nereis diversicolor* (NSCP). In three measured forms, apo, Mg^{2+} and Ca^{2+}, the authors measured four fluorescence lifetimes (Table 2.7). Measuring the fluorescence lifetimes in three forms of the mutant W4F/W170F, the authors found also four fluorescence lifetimes in the same range of those measured in the wild type. Therefore, a question should be asked here, can we describe the fluorescence lifetimes of the tryptophan in NSCP with the rotamers model?

In order to understand the origin of the fluorescence of the Trp residues in a protein, we compared the fluorescence parameters of α_1-acid glycoprotein prepared in two different ways: 1) by a successive combination of ion displacement chromatography, gel filtration and ion exchange chromatography (α_1-acid glycoprotein[c]) and 2) by ammonium sulfate precipitation (α_1-acid glycoprotein[s]).

Excitation at 295 nm yields a fluorescence spectrum with a peak equal to 332 nm (α_1-acid glycoprotein[c]) and to 337 nm (α_1-acid glycoprotein[s]). The positions of the peaks are typical for proteins containing tryptophan residues in a hydrophobic environment and result from the contribution of both solvent-exposed and buried Trp residues. The fluorescence maximum of α_1-acid glycoprotein[c] is blue shifted compared to that of α_1-acid glycoprotein[c]. Thus, the two Trp residues embedded in the protein core of α_1-acid glycoprotein[c] are less exposed to the solvent than those embedded in α_1-acid glycoprotein[s].

Steady-state fluorescence anisotropy of Trp residues in both preparations of α_1-acid glycoprotein was performed at different temperatures. The Perrin plots (Fig. 5.7 and 5.8) reveal that Trp residues of α_1-acid glycoprotein[c] display free motions while those of α_1-acid glycoprotein[s] follow the global rotation of the protein.

Fig. 7.11 reports the fluorescence anisotropy spectra of α_1-acid glycoprotein[s] (a) and α_1-acid glycoprotein[c] (b) recorded at 20°C between 265 and 305 nm (λ_{em}, 330 nm). The general qualitative feature resembles that obtained for Trp residue in proteins.

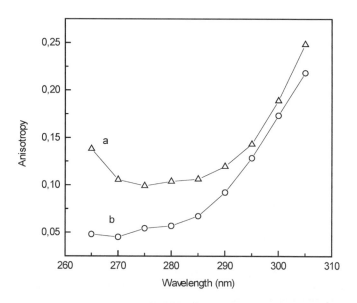

Figure 7.11. Fluorescence excitation anisotropy spectra of α_1-acid glycoprotein[s] (a) and α_1-acid glycoprotein[c] (b) at 20°C. $\lambda_{em} = 330$ nm. Source: Albani, J. R. 1998, Spectrochimica Acta Part A. 54, 175-183.

The shape of the spectra is almost identical indicating that the Trp residues in both proteins have the same microenvironments. However, the values of the anisotropies are higher for α_1-acid glycoprotein^s, i.e. the Trp residues in α_1-acid glycoprotein^s show a more restricted mobility than those in α_1-acid glycoprotein^c.

We notice that the fundamental anisotropy in α_1-acid glycoprotein^s is not reached despite the absence of motion of the Trp residues. This may be explained by the fact that the global rotation of the protein and the energy transfer between the Trp residues and the energy transfer from tyrosine to tryptophan residues induce a depolarization of the system. Trp residue in a fluid medium or in a denatured protein does not display an increase in the anisotropy with the excitation wavelength. However, in a vitrified solution or when Trp residues are surrounded by a rigid microenvironment, the anisotropy increases with the excitation wavelength. Fig. 7.11 shows that in both α_1-acid glycoprotein preparations, the anisotropy of the Trp residue increases. This increase has the same shape for both proteins. Trp residues in α_1-acid glycoprotein^c exhibit residual motions, while those present in α_1-acid glycoprotein^s are rigid. Therefore, one may expect to observe an increase in the anisotropy with the excitation wavelength for α_1-acid glycoprotein^s only. As it is not the case, one may say that the general feature of the excitation anisotropy spectra of Trp residues in proteins will depend on the structure and the dynamics of the protein and of the fluorophore microenvironment.

Indeed, the Trp residues in α_1-acid glycoprotein^c exhibit residual motions but not free ones, because of the surrounding amino acids that constrained the motion of the fluorophore. Therefore, the increase in the anisotropy with the excitation wavelength will depend on the importance of these restrictions.

Tryptophan has two overlapping $S_o \longrightarrow S_1$ electronic transitions ('L_a and 'L_b) which are perpendicular to each other (Fig. 5.6). Both $S_o \rightarrow {}^1L_a$ and $S_o \rightarrow$ 'L_b transitions occur in the 260-300 nm range. In polar solvents, 'L_a has lower energy than 'L_b, and emission from this lowest state will be observed. When the fluorophore environment exhibits restricted motions, absorption at the red-edge photoselects the lowest energy (S_1) ('L_a in a polar solvent). In this case, the anisotropy will be higher than when absorption occurs at the blue-edge, since only depolarization due to small angular differences between the absorption and emission transition moments and solvent reorientation, if any, occurs. Therefore, the increase in the anisotropy toward the red edge of the absorption band is due to the photoselection of the predominantly 'L_a transition and to the importance of the restricted mobility of the fluorophore.

Fig. 7.12 displays the variation of the steady-state anisotropy of tryptophan residues of α_1-acid glycoprotein^s (a) and α_1-acid glycoprotein^c (b) as a function of the emission wavelength. Experiments were performed at excitation wavelength of 300 nm in order to avoid any depolarization process that may occur as the result of an energy transfer between the Trp residues. The anisotropy in α_1-acid glycoprotein^c decreases from 0.178 at 320 nm to 0.148 at 380 nm, a decrease of 18%, while it decreases in α_1-acid glycoprotein^s from 0.215 at 320 nm to 0.160 at 380 nm, a decrease of 25%. The anisotropy variation across the emission spectrum was observed for fluorophores bound tightly to the proteins or to the membranes. However, Fig. 7.12 indicates that the decrease in the anisotropy along the emission wavelength can be observed also when the Trp residues in proteins display residual motions.

The variation of the anisotropy across the emission spectrum or with the excitation wavelength was observed for fluorophores that are bound tightly to the membranes or to the proteins. However, Figures 7.11 and 7.12 indicate that this variation occurs also for the Trp residues that exhibit motions. We notice also that more the motions of the fluorophore are constrained by the surrounding microenvironment, more the decrease in the anisotropy along the emission wavelength or its increase with the excitation wavelength is important.

A conclusion can be drawn from the above results is that the dynamics and the structure of the microenvironment of the Trp residues and the tertiary structure of the protein have an important effect on the fluorescence intensity, the position of the maximum, the center of gravity of the spectrum and finally on the dependence of the anisotropy on the emission and excitation wavelengths.

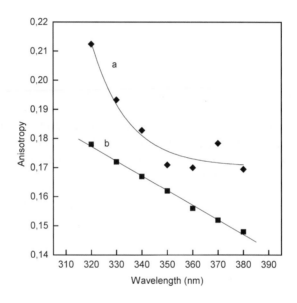

Figure 7.12. Fluorescence anisotropy of α_1-acid glycoprotein[s] (a) and α_1-acid glycoprotein[c] (b) at 20°C as a function of the emission wavelength. λ_{ex} = 295 nm. Source: Albani, J. R. 1998, Spectrochimica Acta Part A. 54, 175-183.

The time decay of fluorescence intensity measurements of α_1-acid glycoprotein[s] was performed with an Edinburgh Analytical Instruments CD 900 fluorometer. The technique used was time correlated single photon counting. The sample was excited with series of pulses at a frequency of 20 kHz. The time decay of α_1-acid glycoprotein[c] was measured by the phase method.

Table 7.7 gives the fluorescence lifetimes values for α_1-acid glycoprotein measured by the two methods. One would expect to obtain different fluorescence lifetimes for the two types of preparation if the dynamics has an influence on the lifetime data. Instead, we notice that the two preparations give very close fluorescence lifetimes.

Table 7.7. Fluorescence lifetimes data of α_1-acid glycoprotein[c] and α_1-acid glycoprotein[s]

α_1-Acid glycoprotein[c]

τ_1	f_1	τ_2	f_2	τ_3	f_3	$<\tau>$
0.354	0.101	1.664	0.66	4.638	0.238	2.285
± 0.034	± 0.05	± 0.072	± 0.03	± 0.342	± 0.01	

α_1-Acid glycoprotein[s]

0.26	0.0955	1.75	0.7	5.15	0.203	2.29
± 0.07	± 0.02	± 0.03	± 0.05	± 0.09	± 0.06	

Sources : Albani, J., Vos, R., Willaert, K. and Engelborghs, Y, 1995, Photochem. Photobiol. 62, 30-34 and Albani, J. R, 1998, Spectrochimica Acta Part A 54, 175-183.

It has been shown that α_1-acid glycoprotein exhibits genetic polymorphism and that amino acid sequences may vary within a single individual. Thus, one should not exclude the possibility of having three lifetimes originating from Trp residues that emit from three α_1- acid glycoprotein isoforms. However, multiexponential decay is frequently observed for numbers of proteins that do not exhibit genetic polymorphism. Also, since all the α_1- acid glycoprotein preparations always give three lifetimes that are identical from one preparation to another, the possibility of an emission from two or three α_1- acid glycoprotein isoforms is to be excluded. This possibility was also excluded by studying the interaction between α_1- acid glycoprotein and TNS (Albani et al. 1995).

The multiexponential decay of α_1- acid glycoprotein may have different origins such as the emission from different relaxed states, the emission from different rotamer conformations, the dynamics of the Trp residues within their microenvironments and the presence of two protein isoforms. The rotamers model proposes that the Trp side chain may adopt low-energy conformations, due to rotation around the C_α-C_β and/or the C_β-C_γ bonds of the side chains, with each conformation displaying a distinct decay time. The two α_1- acid glycoproteins possess three tryptophan residues emitting with three fluorescence lifetimes that are identical for both proteins. In one of the protein, the tryptophan microenvironment is rigid and in the other it is mobile. Therefore, one could exclude the environment dynamics of the tryptophan residues as the main origin of the fluorescence lifetimes in α_1- acid glycoprotein. The rotamers model by itself cannot explain the tri-exponential decay, we should at least have six fluorescence lifetimes, two for each tryptophan residues. Or there could be a combination of the fluorescence lifetimes of the rotamers so that the final result would be three fluorescence lifetimes. Also, energy transfer between the tryptophan residues and from tyrosines to tryptophans cannot be excluded. Thus, even in the absence of motions, different internal processes can contribute to the origin of the fluorescence lifetime.

Hemoproteins are also interesting to investigate because of the efficient energy transfer from tryptophans to heme. Although fluorescence parameters such as intensity, lifetime and polarization of tryptophans in hemoproteins are weak, they still can be measured. Energy transfer rate between a tryptophan and heme depends on the dipole – dipole orientation and the distance that separates the two chromophores. Thus, in a certain way, energy transfer will be affected by the internal dynamics of the protein. Anyway, residual motions will always affect energy transfer between a donor and an acceptor, independently of the chemical nature of the two molecules.

Despite the important energy transfer between Trp residues and heme, different fluorescence lifetimes, going from the ps to the ns range, can be measured. For example, fluorescence lifetimes of the two Trp residues of horse heart myoglobin are measured at different pH from 7.2 to 4.42. Four lifetimes are obtained at all pH. From one pH to another, only the fractional contribution of each lifetime is modified (Table 7.8).

Table 7.8. Fluorescence lifetime parameters of the intensity decay of horse heart myoglobin

pH	τ_1	f_1	α_1	τ_2	f_2	α_2	τ_3	f_3	α_3	τ_4	f_4	α_4
7.20	35	0.142	0.550	130	0.407	0.425	1491	0.198	0.018	4894	0.253	0.007
	40	0.149	0.509	116	0.394	0.463	1363	0.210	0.021	4822	0.247	0.007
4.95	43	0.188	0.603	127	0.342	0.371	1505	0.207	0.019	5167	0.262	0.007
	40	0.160	0.544	116	0.365	0.428	1363	0.191	0.019	4822	0.284	0.008
4.72	47	0.133	0.514	121	0.299	0.449	1599	0.220	0.025	5254	0.347	0.012
	40	0.103	0.455	116	0.335	0.508	1363	0.178	0.023	4822	0.384	0.014
4.62	56	0.122	0.519	113	0.200	0.420	1304	0.209	0.038	4842	0.469	0.023
	40	0.075	0.435	116	0.256	0.510	1363	0.189	0.032	4822	0.480	0.023
4.42	46	0.086	0.522	126	0.184	0.411	1679	0.245	0.041	5236	0.485	0.026
	40	0.065	0.450	116	0.203	0.480	1363	0.188	0.038	4822	0.544	0.031

Source : Gryczynski, Z., Lubkowski, J. and Bucci, E. 1995, J. Biol. Chem. 270, 19232-19237. Authorization of reprint accorded by the American Society for Biochemistry and Molecular Biology.

The authors attribute the lifetimes to three different emitting species of myoglobin (Fig. 7.13). (1) **Species I** with normal heme as shown in the crystal structure; (2) **species II** where heme is inverted, i.e rotated 180° around the α-γ-*meso*-axis of the porphyrin ring; and (3) **species III** in reversible dissociation equilibrium with heme. **Species I** with normal hemes have the shortest lifetimes, up to 150 ps, **species II** with disordered hemes have longer, 'intermediate' lifetimes of a few hundred ps, **species III** with dissociated hemes have the longest lifetimes near 5000 ps.

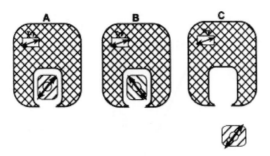

Figure 7.13. The three emitting species of myoglobin. Namely, going from the left, **species I** with normally oriented heme, **species II** with inverted heme, **species III** with reversibly dissociated heme. Note the *quasi* orthogonal orientation of heme and tryptophan transition moments produced by inverted heme, which much reduces the excitation energy transfer from tryptophan to heme. Source: Gryczynski, Z and Bucci, E. 1998. Biophys. Chem. 74, 187-196.

The authors assigned the lifetime of 40 ps to Trp-14 in the presence of normal heme, and the lifetime of 116 ps to Trp-7 in the presence of normal hemes and to Trp-14 in the presence of inverted hemes. The lifetime at 1.3 ns was assigned to Trp-7 in the presence of inverted hemes. The lifetime at 4.8 ns could be assigned only to reversibly heme-dissociated myoglobin and its value used as that of nonquenched tryptophan in the system.

The assignment of the longest lifetime to **species III,** i.e. to the nonquenched tryptophan in the system, is based on the fact that decreasing the pH and thus denaturing the myoglobin, induces an increase in the value of the fractional intensity of the longest lifetime and a decrease of the value of the fractional intensity of the two shortest lifetimes. In fact, decreasing the pH destabilizes the myoglobin complex inducing a release of the heme from its pocket. This will increase the fraction of apoprotein in the system and its corresponding fluorescence intensity relative to the holoform (Figure 7.14).

In the present work, the rotamers model was not taken into consideration as if it cannot be applied. The high energy transfer from tryptophans to heme induces lifetimes in the picoseconds range. In the absence of this energy transfer, the lifetimes increase to reach the nanosecond scale. However, if we look closely to the data obtained, the two longest values are in the same range of those assigned in general to the presence of rotamers. The shortest two picosecond lifetimes are the results of tryptophan → heme energy transfer process.

We can see from this example, that hemoprotein such as myoglobin is analyzed differently from a non-hemo protein.

Fluorescence lifetimes of the tryptophan residues in flavocytochrome b_2 extracted from the yeast *Hansenula anomala*, measured at 9°C, were found equal to 0.073 ± 0.019, 1.118 ± 0.024, 1.472 ± 0.235 and 4.894 ± 0.450 ns with preexponential factors equal to 0.125 ± 0.023, 0.761 ± 0.154, 0.083 ± 0.017 and 0.031, respectively. One can notice that the two shorter lifetimes in the picoseconds range, corresponding to the high energy transfer between tryptophan residues and heme, contribute the most to the global fluorescence.

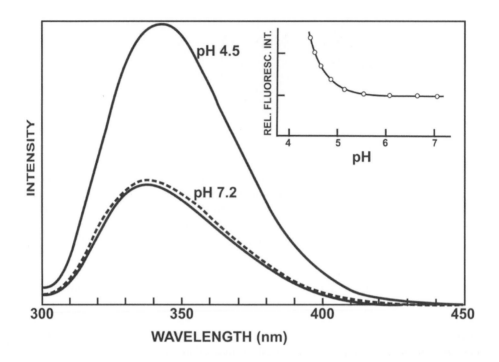

Figure 7.14. Emission spectrum of horse heart myoglobin at pH 7.2 (lower curve) and at pH 4.5 (upper curve). The dashed curve was obtained upon neutralization of the sample previously at pH 4.5. The inset shows the pH dependence of the intensity. Source: Gryczynski, Z., Lubkowski, J. and Bucci, E. 1995, Journal of Biological Chemistry. 270, 19232-19237. Authorization of reprint accorded by the American Society for Biochemistry and Molecular Biology

Another example on fluorescence lifetimes of hemoproteins; peroxidases (donor: H_2O_2, oxidoreductase: E.C. 1.11.1.7) are heme enzymes that catalyze oxidative reactions that use hydrogen peroxidase as an electron acceptor. The seed coat soybean peroxidase (SBP) belongs to class III of the plant peroxidase super family, which includes horseradish (HRP), barely (BP1) and peanut (PNP) peroxidases. All the enzymes of this class contain a protoheme located within a pocket that plays an important role in the catalytic cycle. Soybean peroxidase (SBP) is a glycoprotein of molecular mass equal to 37 kDa. It is a monomer composed of 326 amino acids with a single tryptophan at position 117 (Figure 7.15).

Fluorescence studies are performed on both the holo and the apo forms of the protein. In the holo form, the authors measured four fluorescence lifetimes in the same range of those observed for myoglobin, while in the apoform three fluorescence lifetimes are obtained (Table 7.9) The shortest one (35 ps) measured in the holo form is considered as the result of the energy transfer between the tryptophan and the heme, while the longest three ones have been considered by the authors as emanating from the apoform of the protein.

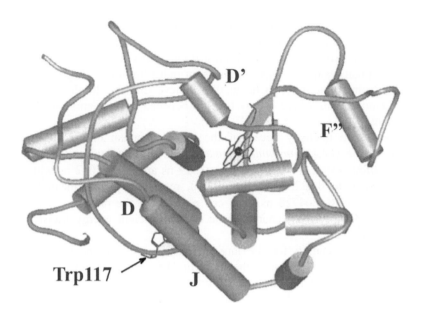

Figure 7.15. Structure of SBP obtained from crystallographic results. Herne and Trp 117 are indicated as stick models. Helices are named according to nomenclature used by (Gajhede, M., Schuller, D. J., Henriksen, A., Smith, A. T. and Poulos, T. L. 1997. Nat. Struct. Biology. 4 : 1032-1038). Source: Amisha Kamal, J. K. and Behere, D. V. 2001. Biochem. Biophys. Res. Commun. 289, 427-433.

Table 7.9. Steady-state and time-resolved decay analysis of SBP, HRP-C and their apoforms. and apo-SBP. Lifetimes are in ns and wavelengths in nm.

	Φ_F	λ_{em}	τ_1	τ_2	τ_3	τ_4	α_1	α_2	α_3	α_4
SBP	2×10^{-3}	335	6.3	2.0	0.3	0.035	0.2	0.8	2	97
ApoSBP	20×10^{-3}	340	5.9	1.73	0.3	-	7.0	27	66	-
HRP-C	1×10^{-3}	326	4.6	1.4	0.045	-	1.0	2.0	97	-
ApoHRP-C	20×10^{-3}	342	4.4	1.32	0.25	-	6.0	15.0	79	-

Sources: Amisha Kamal, J. K. and Behere, D. V. 2001. Biochem. Biophys. Res. Commun. 289, 427-433 ; Das, T. K. and Mazumdar, S. 1995. Eur. J. Biochem. 227, 823-828 ; Das, T. K. and Mazumdar, S. 1995. J. Phys. Chem. 99, 13283-13290.

In the present two cases, the results described in Table 7.9 indicate that the rotamers model is not adequate to describe the origin of the fluorescence lifetime of the tryptophans in the two proteins. The shortest fluorescence lifetime (35 or 45 ps) found also for tryptophan residues in myoglobin is an indication of the high energy transfer Förster type from tryptophans to heme. We believe that this lifetime is common or almost all hemoproteins where energy transfer between tryptophan(s) and the heme is very important.

The fluorescence intensity decay of the two tryptophan of cytochrome b_2 core extracted from the yeast *Hansenula anomala,* can be described as the sum of three fluorescence lifetimes: $\tau_1 = 0.054$ ns, $\tau_2 = 0.529$ ns and $\tau_3 = 2.042$ ns with fractional intensities f equal to 0.922, 0.068 and 0.010, respectively. The mean fluorescence lifetime τ_o is equal to 0.04728 ns. One can notice also the presence of a very short fluorescence lifetime (54 ps) characteristic of the efficient energy transfer from the tryptophans to the heme.

We have no explanation for the largest fluorescence lifetime since we did not measure the fluorescence lifetime of the apoform. Different possibilities can explain the origin of the two longest fluorescence lifetimes: protein impurities, heme-dissociated protein and internal motions that induce different relative heme-tryptophan orientations.

Finally, it is interesting to report data obtained with a tyrosine residue. Dynamics of the inactivating Shaker B (ShB) peptide (H_2N-MAAVAGLYGLGEDRQHRKKQ) from the Shaker B potassium channel were studied in the absence and presence of anionic phospholipid membrane. The peptide contains one tyrosine residue and no tryptophan (Poveda et al. 2003). Fluorescence intensity decay of tyrosine in ShB peptide can be described with three fluorescence lifetimes: 0.4, 1.4 and 3.0 ns with preexponential factors equal to 0.26, 0.55 and 0.19, respectively, with a mean lifetime of 2.0 ns. Time resolved anisotropy decay for the peptide in solution shows the presence of two rotational correlation times, 0.2 and 0.9 ns, corresponding respectively to the segmental motion in proximity of the tyrosine and to the global rotation of the peptide. When incorporated into the phosphatidic acid, fluorescence lifetimes were found equal to 0.4, 1.5 and 3.7 ns with preexponential factors equal to 0.32, 0.51 and 0.17, respectively with a mean fluorescence lifetime of 2.3 ns. Time resolved anisotropy decay shows the presence of two rotational correlation times 0.2 and 4.3 ns. ShB peptide inserts deeply enough into the vesicle bilayer in contact with the hydrophobic region, although electrostatic interactions between the C-terminal and the membrane surface do exist. Thus, the 0.2 ns will reflect internal fluctuations of the tyrosine residue and the 4.3 ns component will correspond to the fluctuations of the whole peptide segment incorporated into the membrane. These data reveal that incorporation of the peptide in the membrane does not modify significantly the fluorescence lifetimes and does not affect the local flexibility of the tyrosine residue. Also, the peptide fragment incorporated in the membrane vesicles is not rigid. The authors explained the triexponential behavior of the fluorescence intensity decay as the consequence of the existence of ground-state rotamers sensing different chemical environments. However, since the tyrosine residue and the whole peptide displays continuous motions whether the peptide is free in solution or incorporated in the membrane vesicles, we do not see why the fluorescence lifetime cannot be interpreted as the consequence of the internal dynamics of the tyrosine residue along with different types of interactions with the neighboring amino acids.

Chapter 8

DESCRIPTION OF THE STRUCTURE AND DYNAMICS OF α_1-ACID GLYCOPROTEIN BY FLUORESCENCE STUDIES

1. Introduction

α_1-Acid glycoprotein (orosomucoid) is a small acute-phase glycoprotein (Mr = 41000) that is negatively charged at physiological pH. It consists of a chain of 183 amino acids (Dente et al. 1987), contains 40% carbohydrate by weight and has up to 16 sialic acid residues (10-14 % by weight) (Kute and Westphal, 1976). Five heteropolysaccharide groups are linked via an N-glycosidic bond to the asparaginyl residues of the protein (Schmid et al. 1973) (Fig. 8.1). The protein contains tetra-antennary as well as di- and tri-antennary glycans.

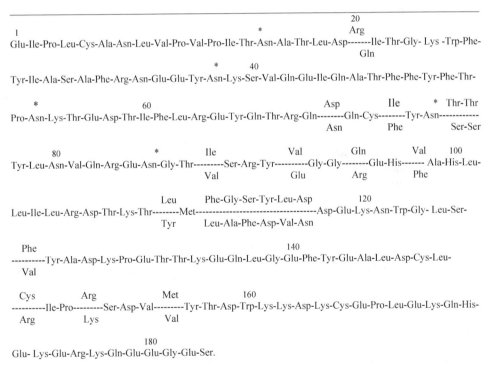

Figure 8.1. Primary structure of α_1-acid glycoprotein. The five heteropolysaccharide units are linked N-glycosidacilly to the asparagine residues that are marked with a star. Sources: Schmid, K Kaufmann H, Isemura S, Bauer F, Emura J, Motoyama T, Ishiguro M and Nanno S, 1973, Biochemistry. 12, 2711-2724 and Dente L, Pizza MG, Metspalu A, Cortese R, 1987, Structure and expression of the genes coding for human α_1-acid glycoprotein. EMBO J. 6, 2289-2296.

Although the biological function of α_1-acid glycoprotein is still obscure, a number of activities of possible significance have been described such as, the ability to bind the β-drug adrenergic blocker, propranolol (Sager et al. 1979) and certain steroid hormones such as progesterone (Kute and Westphal, 1976). Many of these activities have been shown to be dependent on the glycoform of α_1-acid glycoprotein (Chiu et al. 1977). As the serum concentration of specific glycoforms of α_1-acid glycoprotein changes markedly under acute or chronic inflammatory conditions, as well as in pregnancy and tumor growth, a pathophysiological dependence change in the carbohydrate-dependent activities of the protein may occur. Therefore, the relation between the function of α_1-acid glycoprotein and pathophysiological changes in glycosylation was extensively studied (Van Dijk et al. 1995; Mackiewicz and Mackiewicz, 1995; Brinkman-Van Der Linder et al. 1996). However, the relation between structure and dynamics of the protein part of α_1-acid glycoprotein and its function was not studied sofar.

Dynamics of proteins are in general important to their functions (Frauenfelder et al. 1979; Karplus and McCammon, 1983, Capeillère-Blandin and Albani, 1987). It was shown for example, that binding of cytidine 5'-triphosphate (CTP) and N-(phosphonoacetyl)-L-aspartate (PALA) to aspartate transcarbamylase (ATCase) from *Escherichia coli* induces a modification in the dynamics behavior of the two Trp residues of the protein (Royer et al. 1987).

α_1-Acid glycoprotein shows a chemical nature identical to many serum components that have the specific ability of interacting with hormones such as progesterone, to form dissociable complexes. α_1-Acid glycoprotein displays a high affinity for progesterone, but it appears to bind only a small portion of the circulating ligand, the major part is associated with serum albumin and corticosteroid binding globulin. Interaction between α_1-acid glycoprotein and progesterone is temperature and pH dependent (Canguly and Westphal, 1968; Kirley et al. 1982), thus binding of progesterone to α_1-acid glycoprotein is dependent on the pathophysiological condition. Binding of drugs to circulating plasma proteins such as α_1-acid glycoprotein can decrease the plasma concentration to below the minimal effective concentration thus inhibiting efficacy and abolishing therapeutic effect. Paterson et al showed that purification process of α_1-acid glycoprotein could alter the binding properties of the protein. Studies were carried out with imatinib mesylate following fluorescence intensity quenching of tryptophan residues of the protein. Significant binding will decrease the fluorescence intensity of the intrinsic fluorophores while in the absence of any interaction, one will not observe any fluorescence decrease. Applying fluorescence method overcomes the major disadvantages of the commonly used dialysis method (namely that binding of drugs to α_1-acid glycoprotein can be influenced by the presence of lipids in the dialysis medium, and bound drug can leak through the membrane). Additionally, as reported here, it can also be utilized to identify the source of differences in binding between the same drug and glycoprotein in different laboratories. Imatinib (Glivec®) is a signal transduction inhibitor used in the treatment of chronic myeloid leukaemia (CML), a haematopoietic disorder of stem cell origin. Interaction between imatinib and α_1-acid glycoprotein (isolated using a non-acidic low pressure chromatography technique) was found inexistence at physiological concentrations (Jørgensen et al, 2002). Other studies using commercially source of α_1-acid glycoprotein have argued the opposite (Gambacorti-Passerini et al; 2000) and implicated α_1-acid glycoprotein as a mechanism of resistance.

The following data re-evaluate the interaction of α_1-acid glycoprotein with imatinib within and above the physiological concentration range of the drug in the plasma using the same source of α_1-acid glycoprotein and spectrofluorimetry. Since there is disagreement regarding the ability of α_1-acid glycoprotein to bind to, and thus reduce the efficacy of imatinib, the fluorescence study of α_1-acid glycoprotein -imatinib interactions described in the next two figures was conducted using a commercial source of the protein previously intimated as binding to the drug.

Figure 8.2 displays the fluorescence emission spectra of α_1-acid glycoprotein (0.5mg/mL) in the absence and the presence of increased concentrations of imatinib.

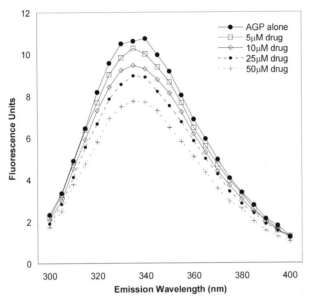

Figure 8.2. Fluorescence emission spectrum of α_1-acid glycoprotein and α_1-acid glycoprotein - imatinib solutions. λ_{ex} = 280 nm. Unpublished results by Sarah Paterson BSc(Hons) and Kevin Smith PhD, Department of Bioscience, University of Strathclyde, Royal College Building, 204 George Street, Glasgow, G1 1XW, Scotland. A work supported by the Caledonian Research Foundation.

The quenching, or percentage decrease in α_1-acid glycoprotein fluorescence following addition of drug, can be plotted as a function of drug concentration, and the data fitted using the quadratic binding equation (Parikh et al, 2000)

$$\% \text{ Quenching} = C_1 \frac{S - \sqrt{S^2 - 4\,D_T\,P_T}}{2} - C_2\,D_T \qquad (4.25)$$

where D_T = total drug concentration; P_T = total protein concentration; $S = D_T + P_T + K_d$; C_1, C_2 and K_d are unknown constants, determined by least squares curve fitting

The resulting quenching curve shown in figure 8.3 indicates that there is a limited interaction between the protein and imatinib, and that the extent of interaction increases with imatinib concentration.

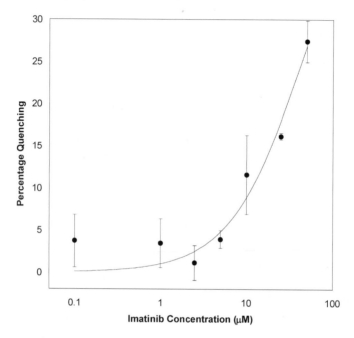

Figure 8.3. Quenching curve showing the interaction of imatinib with α_1-acid glycoprotein. Unpublished results by Sarah Paterson BSc(Hons) and Kevin Smith PhD, Department of Bioscience, University of Strathclyde, Royal College Building, 204 George Street, Glasgow, G1 1XW, Scotland. A work supported by the Caledonian Research Foundation.

The findings confirm the inability of α_1-acid glycoprotein to bind to imatinib at physiological concentrations of both glycoprotein (0.5mg/mL) and drug (0-10µM), as previously reported by Jørgensen et al (2002). Accordingly, it has been demonstrated that the percentage quenching within the physiological range of imatinib is negligible, with only 10 per cent quenching of the α_1-acid glycoprotein spectrum at 10 µM imatinib. At lower imatinib concentrations, quenching approaches 0 per cent, indicating very weak interaction between the drug and α_1-acid glycoprotein.

The method of isolation has been proven not to result in any structural alteration (e.g. desialylation and denaturation) (Smith et al. 1994), thus it could be argued that the difference in the binding experiments may result from structural changes in the conformation of the protein as a result of the purification process used. Additionally, isoforms of α_1-acid glycoprotein (same protein, different glycosylation patterns) could be obtained under several physiological and pathological conditions. This structural alteration is likely to affect the degree of drug binding. Therefore the observed differences between two or many α_1-acid glycoprotein preparations might be due to differences in glycosylation.

Also, one should mention that altering the carbohydrate conformation could modify the global and / or the local structure of the protein. Isoforms of α_1-acid glycoprotein are in general considered as the results of the modifications of the nature of the carbohydrate residues bound covalently to the protein. However, upon structural modification of the carbohydrate residues, its dynamics could be altered and its interaction with the protein matrix modified. A new carbohydrate – protein interaction can modify the structure of the protein matrix. Studying the structure, dynamics and physiological properties of a protein with fluorescence spectroscopy requires the use of intrinsic and/or extrinsic fluorophores. Also, it is important to indicate that there are no crystallographic data on α_1-acid glycoprotein. The reason for this is that no one up to our days succeeded to obtain crystals of the protein that are possible to be analyzed with X-ray diffraction studies and / or with electron microscopy diffraction. The crystallization of α_1-acid glycoprotein proved to be somewhat difficult, probably because of the high solubility and the large carbohydrate moiety of the protein. The first time, the protein has been crystallized was in 1953 (Schmid, 1953). Then in 1980 and 1984 other authors reported the possibility of obtaining crystals of proteins (McPherson et al. 1980; 1984). However, no crystallographic data have been reported concerning the spatial structure of the protein, probably because of the instability of the crystals themselves.

Therefore, we have here an important reason to focus on fluorescence studies since this method can help to obtain information on the structure and the dynamics of a macromolecule.

We performed our studies using three fluorescent probes, Trp residues, 2,p-toluidinylnaphthalene-6-sulfonate (TNS) and calcofluor white. Also, we studied the binding of hemin to α_1-acid glycoprotein. Hemin is non-fluorescent and we found that it binds specifically to α_1-acid glycoprotein. This experiment allowed us to prove that the glycoprotein contains a pocket where the different ligands bind (The results are detailed in chapter six).

α_1-Acid glycoprotein contains three Trp residues, one residue, Trp-160, is at the surface of the protein and two, Trp-122 and Trp-25, are located in the protein matrix (Kute and Westphal, 1976; Schmid et al. 1973; Hof et al. 1996)

TNS shows a weak fluorescence when dissolved in a polar solvent. Its fluorescence intensity increase significantly when it binds to proteins (McClure and Edelman,, 1966) or to membranes (Easter et al. 1976).

Calcofluor white is a fluorescent probe capable of making hydrogen bonds with β–(1–>4) and β-(1-3) polysaccharides (Rattee and Greur, 1974). The fluorophore shows a high affinity for chitin, cellulose and succinoglycan, forming hydrogen bonds with free hydroxyl groups (Maeda and Ishida, 1967). In the presence of succinoglycan, a polymer of an octasaccharide repeating unit, consisting of galactose, glucose acetate, succinate and pyruvate in a ratio of 1:7:1:1:1 (Aman et al. 1981), calcofluor fluoresces brightly as the result of its binding to the oligosaccharide (York and Walker, 1998).

Calcofluor is commonly used to study the mechanism by which cellulose and other carbohydrate structures are formed *in vivo* and is also widely used in clinical studies (Sridhar and Sharma, 2003; Grossniklaus et al. 2003).

The different studies we are describing here allowed us to obtain a global picture of the structure and the dynamics of α_1-acid glycoprotein.

2. Methods

Fluorescence parameters such as the intensity and the position of the emission maximum are sensitive to the modifications occurring in the microenvironment of the fluorophore.

The fluorescence intensity decay of a fluorophore in its environment could be analyzed with one, two or more lifetimes. The fluorescence lifetimes may originate from the structural heterogeneity surrounding the fluorophore, the dynamics of this environment and the energy transfer to the neighboring amino acids. The amplitudes, the values and the number of these lifetimes may vary with the modification of the surrounding environment of the fluorophore or / and with any important structural modification that can occur in the protein near the fluorophore (Libich et al. 2003; Kuppens et al. 2003).

The red-edge excitation spectra method is largely used as a tool for monitoring motions around a fluorophore. The method is very sensitive to the changes that occur in the microenvironment of a fluorophore (Demchenko, 2002).

Quenching of the fluorescence intensity by cesium ion of a fluorophore is the result of the dynamic interaction between the two molecules. Such as when oxygen is used as a quencher, two constants can be obtained, the Stern-Volmer constant K_{SV} and the bimolecular diffusion constant k_q. The value of K_{SV} will give an idea on the accessibility of the fluorophore to the solvent and thus to the quencher while the value of k_q will give information on the dynamics of the system (Lakowicz and Weber,1973).

When a protein contains two classes of intrinsic fluorophore, one at the surface of the protein and the second embedded in the protein matrix, fluorescence intensity quenching with cesium or iodide allows obtaining the spectra of these two classes. A selective quenching implies that addition of quencher induces a decrease in the fluorescence observables (intensity, anisotropy and lifetime) of the accessible class. At high quencher concentration the remaining observables measured will reflect essentially those of the embedded fluorophore residues. In this case, one can determine the fraction of fluorescence intensity that is accessible (f_a). Knowing f_a along the emission spectrum will allow us to draw the spectrum of each class of fluorophore (Lehrer, 1971).

Upon binding of a ligand on the protein and if this binding induces structural changes inside the protein, the intensities of the spectrum of the classes can be modified and in the case of the Trp residues, the emission maximum can shift to the higher or lower wavelengths.

Steady-state fluorescence anisotropy of fluorophore residues, measured as a function of temperature (5 to 35°C) (Perrin plot), will give us information concerning the motion of the fluorophore within the protein (Weber, 1952). When the fluorophore is tightly bound to the protein, its motion will correspond to that of the protein. In this case, the rotational correlation time will be equal to that of the protein. When the fluorophore exhibits significant motions when bound to the protein, the rotational correlation time determined from the Perrin plot will be lower than that of the protein and will be the results of two motions, that of the protein and of the segmental motion of the fluorophore.

When a protein contains fluorophore residues located at the surface and in the core as it is the case for α_1-acid glycoprotein, the results obtained from the Perrin plot contain contributions from all the residues. In order to obtain information on the motion of each class of fluorophore residues, one may follow anisotropy and intensity variations as a function of quencher concentration (Quenching Resolved Emission Anisotropy) (Eftink, 1983) or / and anisotropy and lifetime variations as a function of temperature (-50 to +35°C) (Weber method) (Weber et al. 1984).

The QREA method allows determining the fraction f_b and the limiting anisotropy of the non-accessible residues. Knowing the value of the anisotropy in absence of quencher, one can determine the intrinsic anisotropy of the accessible residues.

The Weber method provides parameters characteristic of the environment of the rotating unit such as the limiting anisotropy and the thermal coefficient to resistance to a rotation. Also, the method permits to separate the motion of different classes of fluorophore.

All these methods have their limits and their advantages. Also, many problems arise due to a misinterpretation of the results obtained. The most important thing is to be sure of the data obtained, and then one has to deal with them and find the best way to analyze them. When we study a protein for the first time, performing more and more experiments allows understanding better the results obtained and helps to have clearer ideas on the interpretation of all the data obtained.

Working on $α_1$-acid glycoprotein was challenging since literature was poor on information concerning its structure and dynamics. In fact, $α_1$-acid glycoprotein is studied since now more than 50 years. Many aspects have been studied, whether in biochemistry, genetics, physiology and medicine. From the results obtained up to these days, we notice that the protein has not only one specific function but it acts as a probe that allows monitoring physiological changes. For example, the serum concentration of specific glycoforms of $α_1$-acid glycoprotein changes markedly under acute or chronic inflammatory conditions, as well as in pregnancy and tumor growth. Also, it has been reported that concentration of $α_1$-acid glycoprotein increases with age particularly in depressed female patients (Young et al. 1999). The relation between the function of $α_1$-acid glycoprotein and pathophysiological changes in glycosylation was extensively studied. However, the relation between structure and dynamics of $α_1$-acid glycoprotein and its function was not studied.

3. Fluorescence properties of TNS bound to sialylated $α_1$-acid glycoprotein

3 a. Binding parameters

In phosphate buffer, free TNS presents a very weak fluorescence. Once $α_1$-acid glycoprotein is added to the solution, an enhancement in the TNS fluorescence appears with a maximum at 425 nm ($λ_{ex}$ = 320 nm) (Fig. 8.4). The stoichiometry of the TNS - $α_1$-acid glycoprotein complex was determined according to Job method. In this method, the concentrations of TNS and $α_1$-acid glycoprotein are varied, but the sum of their molar concentrations is held constant. A plot of relative fluorescence *vs* mole fraction of enzyme will give a maximum value at the mole fraction corresponding to the binding ratio. Figure 8.5 indicates that the stoichiometry of the complex is 1: 1.

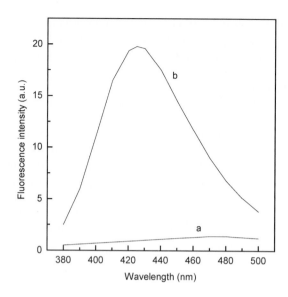

Figure 8.4. Fluorescence emission of TNS alone in solution (a) and in presence of α_1-acid glycoprotein (b).

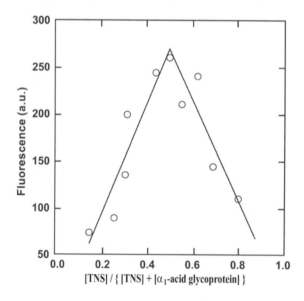

Figure 8.5. Determination of the stoichiometry of the TNS - α_1-acid glycoprotein complex at 20°C with the Job plot ([TNS] + ([([α_1-acid glycoprotein]) = 4.5 μM. Source: Albani, J., Vos, R., Willaert, K and Engelborghs, Y. 1995. Photochem. Photobiol. 62, 30-34. Authorization of reprint accorded by the American Society for Photochemistry.

A titration of a fixed amount of TNS (0.85 µM) with α_1-acid glycoprotein was performed in order to determine the association constant of the complex. Figure 8.4 displays the intensity increase of TNS fluorescence at 20°C. The association constant of the TNS- α_1-acid glycoprotein complex obtained at three temperatures were found equal to by fitting the data of Fig. 8.6 to Eq.8.1

$$\frac{1}{I} = \frac{1}{[P]\,K_a\,I_{max}} + \frac{1}{I_{max}} \tag{8.1}$$

where [P] is the protein concentration, I the fluorescence intensity at a certain concentration of protein, and I_{max} the fluorescence intensity at infinite concentration of protein, K_a is the association constant. The K_a obtained is equal to 19.6 ($\pm$4) x 10^3 M^{-1}, 16.7 ($\pm$3) x 10^3 M^{-1} and 11 ($\pm$2) x 10^3 M^{-1} at 10, 20 and 33°C, respectively. The maximal fluorescence of the complex decreased with increasing temperature (respectively: 301 $\pm$ 60; 201 $\pm$ 30; 156 $\pm$ 30 in arbitrarily scaled units), as to be expected from the known temperature dependence of nonradiative decay processes.

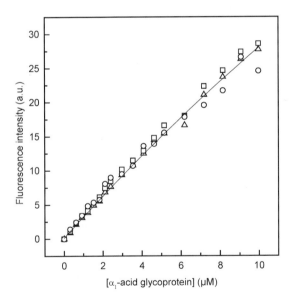

Figure 8.6. Titration curve of 0.85 µM of TNS with α_1-acid glycoprotein. λ_{ex} = 320 nm and λ_{em} = 430 nm.

The standard enthalpy change for the binding is ΔH^o = -18 $\pm$ 3 kJ mol^{-1} and ΔS^o = 19 $\pm$ 12 J K^{-1} mol^{-1}. The negative value of the enthalpy change indicates that the interaction between TNS and α_1-acid glycoprotein is exothermic and is not of a pure nonhydrophobic nature. This means that while TNS binds to a hydrophobic site, it is in close contact with the buffer.

3b. Fluorescence lifetime

The fluorescence intensity decay of TNS bound to α_1-acid glycoprotein can be adequately described by a sum of three exponentials. The lifetimes obtained from the fitting are 0.4 ± 0.1, 4.3 ± 0.1 and 11 ± 0.1 ns and the corresponding amplitude fractions are 0.21 ± 0.05, 0.34 ± 0.05 and $\pm 0.45 \pm 0.06$ ($\lambda_{em} = 425$ nm). The weighted average fluorescence lifetime is 9.33 ns. The short lifetime 0.4 ns correspond to free TNS while the two long ones correspond to bound TNS. The increase of the lifetime in presence of a structure around the TNS is the result of the decrease of the interaction between the fluorophore and the solvent. The Brownian motions are less important upon the interaction of the fluorophore with the surrounding amino acids of the protein. In solution, almost all the energy of the excited state is dissipated in the medium as a thermal quenching. When bound to a protein, the thermal quenching decreases and the fluorophore will stay much longer at the excited state giving by that a longer fluorescence lifetime. Most intriguing are the values of the fluorescence lifetimes of TNS in presence of a protein. In fact, the 4.3 and 11 ns lifetime values have been always found for TNS bound to different proteins such as apomyoglobin (Lakowicz et al. 1984), mellitin (Demchenko, 1985), troponin c (Steiner and Norris, 1987) and *Lens culinaris* agglutinin (Albani, 1996). In all these proteins, TNS was found to have restricted motions with a binding site that is more or less hydrophobic. Therefore, the two lifetimes (4 and 10-11 ns) observed for the TNS-protein complexes are common features of TNS bound to a protein. Thus, these values are more specific to the TNS structure itself within the protein, rather to the protein itself, although they are observed only when TNS is bound to a protein. Therefore, structure of free TNS in solution could be different from that of TNS bound to a protein. The values of the fluorescence lifetimes do not seem to be dependent on the nature of the macromolecule, i.e., specific and same type of interactions occur between TNS and its binding site. However, one should add the possibility of having local structure of TNS that varies slightly with the nature of the interacting macromolecule. In this case, lifetime measurements are not sufficiently sensitive to these small structural differences.

3c. Dynamics of TNS bound to α_1-acid glycoprotein

This study was performed using three methods, the red-edge excitation spectra, the static Perrin plot and the anisotropy decay with time. In the red-edge excitation spectra, it is the fluorophore environment that is relaxing around the excited fluorophore. In the fluorescence anisotropy experiments, on the other hand, the displacement of the emission dipole moment of the fluorophore is monitored. In the first approach, it is the environment that is either fluid or rigid. In the second approach, the restricted reorientational motion of the fluorophore is followed

Figure 8.7 shows the fluorescence spectra of TNS bound to α_1-acid glycoprotein obtained at four excitation wavelengths. At 360 nm, the emission maximum is located at 425 nm. It shifts to higher wavelength (430 nm and 440 nm) when excitation wavelengths are 380 and 400 nm, respectively. The red-edge excitation shift is significant (15 nm). This is taken as direct evidence that TNS has a restricted mobility on α_1-acid glycoprotein Thus, fluorescence emission occurs before the dipole relaxation.

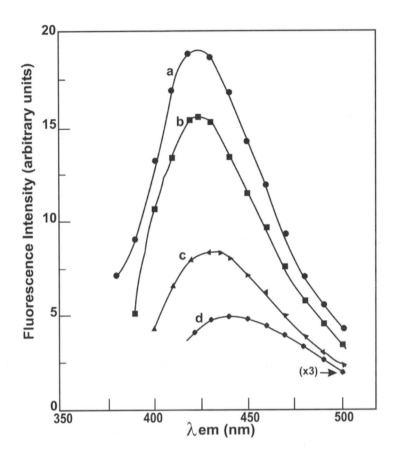

Figure 8.7. Steady-state fluorescence spectra of the 2,6-TNS-α_1-acid glycoprotein complex recorded at four excitation wavelengths, 360 nm (a), 370 nm (b), 380 nm (c) and 400 nm (d). [TNS] = [α_1-acid glycoprotein] = 5 µM, temperature = 20°C. Source: Albani, J. 1992, Biophys. Chem. 44, 129-137.

The dependence of the emission maximum on the excitation wavelength is plotted in Fig. 8.8 at three temperatures (8, 25 and 40°C). The temperature does not seem to affect the behavior of the plot. If an increase in temperature accelerates the relaxation time τ_r, the magnitude of the red-edge excitation effect will be lower at high temperatures than at low temperatures, and the behavior of the plot obtained in Fig. 8.8 will differ from one temperature to another.

The behavior of the plot (λ_{em} = f(λ_{ex})) is temperature independent. This result indicates that the red-edge excitation effect is not a dynamic process but a static one, i.e. TNS molecules are bound tightly to the protein. In this case, emission observed at any excitation wavelength occurs from a non-relaxed state and is temperature independent. Otherwise, increasing the temperature will increase the mobility of the binding site, affecting at the same time the fluorescence of the probe.

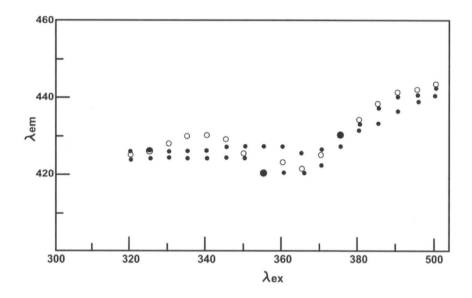

Figure 8.8. Dependence of the maximum of the emission spectra on excitation wavelength of the 2,6-TNS-α_1-acid glycoprotein   complex. [TNS] = [α_1-acid glycoprotein] = 5 μM. Open circles, 40°C. Closed circles, 8°C and stars, 25°C. Source: Albani, J. 1992, Biophys. Chem. 44, 129-137.

We notice that emission maximum does not present the same position at the three temperatures. However, this change is not regular along the excitation wavelengths. For instance, at 335 nm and 340 nm, the difference of emission maxima at 8°C and 40°C is more than 30% of full range emission maxima changes, while at 320 and 395 nm, the difference is much less important.

This means that the difference change observed is not the effect of optical phenomenon that interferes with fluorescence such as scatter. Otherwise, we should observe the same difference change along the plot. Also, scatter would be observed in the Perrin plots affecting the extrapolated value of P.

A possible explanation of the difference in the emission maximum is: the continuous irradiation of TNS-protein complex may induce a partial photodestruction of the probe site altering the position of the maximum. This photodestruction is less important at 8°C than at 25°C or 40°C. Also we used the same sample for the three temperatures, a fact that makes the TNS-protein complex more sensitive to irradiation than a sample that has been irradiated for a shorter time.

The data in Fig. 8.8 are affected by different experimental errors such as irradiation of the same sample for a long period of time. Reading off the intensities from the fluorescence spectra to determine the maximum can also be a source of error, especially at 35°C because of the broadness of the peaks. We may have three to five subsequent emission wavelengths around the real maximum at the same fluorescence intensity.

However, since the behavior of the plot λ_{em} = f(λ_{ex}) is identical at all three experimental temperatures, then, the red-edge excitation effect is a static process.

Thus, red-edge excitation spectra experiments indicate that TNS microenvironment has motions that are dependent of the global rotation of the protein, i.e., TNS will follow the motion of α_1-acid glycoprotein.

The rotational correlation time Φ_p of a hydrated sphere can be obtained for example from equation 5.16

$$\Phi_p (T) = 3.8 \ \eta(T) \times 10^{-4} \ M \tag{5.16}$$

where M is the protein molecular weight and η the viscosity of the medium (Leenders et al. 1993). Eq. 5.16 yields a calculated rotational correlation time of 16 ns at 20°C for α_1-acid glycoprotein if the protein can be considered as spheric.

The fluorescence polarization of the TNS-α_1-acid glycoprotein complex ($\lambda_{ex} = 320$ nm and $\lambda_{em} = 430$ nm) was measured as a function of temperature. The perrin plot representation (Fig. 8.9) yields values for P_o and Φ_R equal to 0.431 and 19 ns, respectively. The value of P_o is equal to that measured for TNS at $-35°C$ and is in the same range of that (0.415) found for TNS in propylene glycol at $-65°C$. The value of Φ_R (= 20 ns) found from the Perrin plot is identical to the theoretical value of Φ_P (16 ns). Thus, TNS binding site is rigid and TNS motions will follow the global rotation of α_1-acid glycoprotein. The polarization results are in good agreement with the results obtained in red-edge excitation shift experiments, i.e., the binding site of TNS is rigid on the protein.

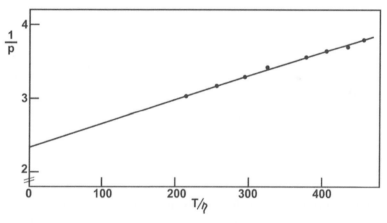

Figure 8.9. Steady-state fluorescence polarization versus temperature / viscosity ratio for 2.6-TNS- -α_1-acid glycoprotein complex. Data were obtained by thermal variation of temperature. λ_{ex}, 320 nm and λ_{em}, 430 nm. Source: Albani, J. 1992, Biophys. Chem. 44, 129-137.

The time decay of fluorescence anisotropy of the TNS-α_1-acid glycoprotein complex plotted as ln A as a function of time is presented in Fig. 8.10. A single correlation time of 16 ns is observed. This value is equal to that calculated theoretically for α_1-acid glycoprotein. This result indicates clearly that TNS is bound tightly to α_1-acid glycoprotein. The fluorescence anisotropy decay experiment results are in good agreement with those obtained for static polarization (Perrin plot) and the red-edge excitation shift.

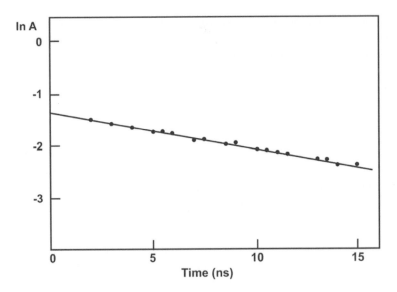

Figure 8.10. Fluorescence anisotropy time decay of TNS-α_1-acid glycoprotein complex. Source : Albani, J. R. 1994. J. Biochem. 116, 625-630.

The fluorescence emission maximum of TNS bound to α_1-acid glycoprotein is located at 425 nm upon excitation at 320 and 360 nm. This wavelength is short and is very close to those observed for TNS dissolved in solvents of low polarity such as ethanol and butanol. However, a red-edge excitation shift does not occur in these solvents. Thus, the spectral properties of TNS bound to α_1-acid glycoprotein and in liquid isotropic solvents are not equivalent. The solvent polarity scale is insufficient to describe the spectral properties of TNS when bound to a protein. Emission from the non-relaxed state, i.e., the binding site is rigid, yields a fluorescence spectrum with a maximum located at short wavelengths. When the fluorophore is excited at the red-edge of the absorption spectrum, the fluorescence maximum will depend on the excitation wavelength.

4. Fluorescence properties of calcofluor bound to α_1-acid glycoprotein

4a. Fluorescence parameters of calcofluor bound to α_1-acid glycoprotein

We report here studies on the nature of the interaction between calcofluor and carbohydrate residues of α_1-acid glycoprotein by comparing the fluorescence spectra of free calcofluor with those of the following complexes: calcofluor- α_1-acid glycoprotein, calcofluor-human serum albumin (HSA) and calcofluor - *Lens culinaris* agglutinin (LCA) (a lentil lectin). HSA and α_1-acid glycoprotein have similar isoelectric points, however, serum albumin does not contain any glycan, while α_1-acid glycoprotein is highly glycosylated. The lentil lectin is a tetramer of molecular weight of 52,570 Da, which is not glycosylated (Foriers et al. 1977).

Fig. 3.32 exhibits the normalized fluorescence emission spectra of calcofluor free in phosphate-NaCl buffer (b), in the presence of a saturated concentration of α_1-acid glycoprotein (c) and of HSA at a concentration close to saturation (a). The emission maximum of the fluorophore shifts from 435 nm in buffer to 448 or to 415 nm, respectively, in the presence of α_1-acid glycoprotein or HSA. The red shift (13 nm) of the maximum observed in presence of α_1-acid glycoprotein compared with the maximum obtained in water indicates that the microenvironment of the excited state of calcofluor on α_1-acid glycoprotein is hydrophilic. The blue shift (20 nm) observed in presence of the serum albumin corresponds to an emission from a hydrophobic environment. Thus, the interaction of calcofluor with serum albumin arises from molecular energy transitions different from those present in the interaction between the fluorophore and α_1-acid glycoprotein.

An increase in the fluorescence intensity of calcofluor was observed upon its interaction with both proteins. The fluorescence intensity is three times higher when the fluorophore is in isobutanol than in water. Thus, calcofluor emission is sensitive to the polarity of the solvent. The emission maximum of calcofluor on HSA (415 nm) is close to that (413 nm) observed when the fluorophore is dissolved in isobutanol.

Fluorescence sensitivity of calcofluor to the medium is common to many fluorophores such as TNS (McClure and Edelman, 1966), Trp residues (Burstein et al. 1973) and flavin (Weber, 1950). However, the fluorescence emission maxima of the above fluorophores are also viscosity dependent. Thus, the solvent polarity scale is insufficient to describe the spectral properties of a fluorophore in a protein (case of Trp residues) or bound to a protein (case of TNS). In fact, when the fluorophore is surrounded by a rigid or viscous environment, or when it is bound tightly to a protein, its fluorescence emission will be located at short wavelengths. In this case, the emission occurs from a non-relaxed state, and the spectrum obtained will be identical to that obtained when the emission occurs from a hydrophobic environment such as isobutanol. Therefore, emission of calcofluor on HSA may be the result of an emission from a hydrophobic binding site and/or a highly rigid binding site.

When the binding site or when the microenvironment of the fluorophore displays free motions, the emission will occur from a relaxed state, yielding a fluorescence spectrum with a maximum located at high wavelengths compared to that observed in a rigid or a hydrophobic environment. Therefore, the emission of calcofluor bound to α_1-acid glycoprotein could be the result of an emission from a hydrophilic binding site and / or a highly dynamic one.

The shift observed for calcofluor from 432 nm in water to 447 nm when bound to α_1-acid glycoprotein can in no way be the result of increasing dynamics of the fluorophore. Otherwise, calcofluor would have a higher degree of freedom when bound to α_1 - acid glycoprotein than when it is free in solution. This is not logical. Therefore, the only explanation of this red shift is that the microenvironment of calcofluor on α_1-acid glycoprotein is highly polar. This polarity is the result of the spatial organization of the carbohydrates of the protein, providing a complex and a highly polar microenvironment. Therefore, calcofluor interacts preferentially with the carbohydrates, although an interaction with the protein moiety cannot be excluded.

Interaction between calcofluor and α_1-acid glycoprotein differs from that observed between TNS and α_1 -acid glycoprotein. TNS binds to the protein on a rigid binding site. Also, the fluorescence maximum of TNS shifts to shorter wavelengths upon binding to α_1-acid glycoprotein. Comparing the results obtained with TNS and calcofluor, one may conclude that the microenvironment of the binding site of calcofluor on α_1-acid glycoprotein differs from that of the binding site of TNS. Calcofluor binds to the carbohydrate residues while TNS binds to a hydrophobic region of the protein.

At 0.9 μM of calcofluor and in the presence of a saturated concentration of α_1-acid glycoprotein (45 μM) we observed an increase of the fluorescence intensity. However, this enhancement was not accompanied by a shift in the emission maximum.

Upon increasing the concentration of calcofluor to 5 μM, and when calcofluor is in excess compared to α_1-acid glycoprotein (for example, 200 μM of calcofluor in presence of 5 μM of protein), the shift to 447 nm is obtained (Fig. 3.32). Thus, it seems that a minimum concentration of calcofluor is necessary to observe the red shift.

The anisotropy A, at 20°C, of calcofluor free in solution increases from 0.1830 ± 0.0007 to 0.1982 ± 0.0007 and 0.2094 ± 0.0014 in the presence of α_1-acid glycoprotein and HSA, respectively. The increase in anisotropy indicates that the calcofluor interacts with both proteins (Le Tilly and Royer, 1993).

4b. Binding parameters

The interaction between calcofluor and α_1-acid glycoprotein or HSA induces a decrease in the fluorescence intensity of the Trp residues of the proteins (λ_{ex}, 295 nm). In order to determine the binding parameters (stoichiometry and the association constant) of the two complexes, a titration of a constant amount of each protein with calcofluor was performed following the intensity decrease of the Trp residues (Fig. 8. 11).

The dissociation constants of the calcofluor-protein complexes were determined by fitting the data to Eq. 6.12 and 6.13

$$Flu= \frac{Flu_0 \times (L_0 - L_b) + Flu_1 \times L_b}{L_0} \tag{6.12}$$

where Flu is the observed fluorescence, Flu_0 and Flu_1 are the fluorescence of free and bound calcofluor, respectively, L_0 and L_b are the concentrations of total and bound calcofluor.

The intensity decrease is clearly hyperbolic and therefore a mathematical binding analysis can be performed using the following quadratic equation obtained from the definition of the equilibrium constant

$$L_b = 0.5 \left[(P_0 + L_0 + K_d) - \{(P_0 + L_0 + Kd)^2 - 4 \, P_0 \, L_0\}^{1/2} \right] \tag{6.13}$$

where P_0 is the protein concentration. The parameter Flu_1 was obtained by extrapolation from a reciprocal plot and found to be 0.3348 and 0.78 for the calcofluor- α_1-acid glycoprotein and calcofluor – HSA complexes, respectively.

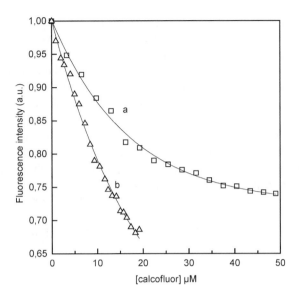

Figure 8.11. Titration curves of α_1-acid glycoprotein (17 µM) (a) and of HSA (10.5 µM) (b) with calcofluor. Source: Albani, J. R. and Plancke, Y. 1999. Carbohydrate Research. 318, 193-200.

The dissociation constants were found equal to 6.6 ± 0.5 µM, i.e., an association constant of 0.15 μM^{-1} and to 25 ± 1 μM^{-1}, i.e., an association constant of 0.04 μM^{-1} for the calcofluor - α_1-acid glycoprotein and calcofluor-HSA complexes, respectively. Thus, the affinity of calcofluor to α_1-acid glycoprotein is higher than its affinity to HSA.

The number of interacting sites was obtained from the slope of $\Delta F / \Delta F_{max}$ versus the calcofluor concentration. The equation used is that derived by (Weber and Young, 1964):

$$x = n L / (P + K_d) \tag{8.2}$$

where x = $\Delta F / \Delta F_{max}$ and n is the number of sites. The value of n was found equal to 0.8 ± 0.1 and 2.1 ± 0.1 for the calcofluor-α_1 acid glycoprotein and calcofluor-HSA complexes, respectively.

4c. Nature of the interaction of calcofluor with α_1 -acid glycoprotein and HSA

Addition of LCA to calcofluor does not modify the fluorescence observables (intensity, position of the maximum and anisotropy) of calcofluor. Thus, the fluorophore does not bind to the LCA, indicating that it is not just binding to any hydrophobic site. Nevertheless, one cannot exclude that binding of the fluorophore to HSA is nonspecific, since serum albumin has the characteristic of binding many types of ligands.

Addition of carbohydrate residues extracted from α_1-acid glycoprotein to a solution of calcofluor induces a decrease in the fluorescence intensity of the fluorophore (Fig. 8.12). This decrease clearly indicates that interaction between the carbohydrates and calcofluor is occurring. The same result was obtained when the carbohydrates originate from sources other than α_1 acid glycoprotein (not shown). However, in the presence of α_1-acid glycoprotein, the fluorescence intensity of calcofluor increases while in the presence of free carbohydrates, it decreases. In order to clarify this problem, we compared the fluorescence emission spectra of calcofluor in the presence of free carbohydrates and of α_1 - cellulose (Fig. 8.13). We notice that with α_1 - cellulose, the fluorescence intensity of calcofluor increases, a phenomenon similar to that observed when the fluorophore is bound to α_1 - acid glycoprotein.

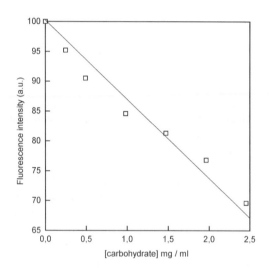

Figure 8.12. Titration of 10 µM of calcofluor with a glycopeptide (one to two amino acids) from α_1-acid glycoprotein . λ_{ex}, 300 nm and $\lambda_{em,}$ 435 nm. Source: Albani, J. R. and Plancke, Y. 1999. Carbohydrate Research. 318, 193-200.

α_1 - Cellulose possesses a well-defined secondary structure, while smaller free carbohydrates in solution do not. However, when bound to the protein moiety of α_1-acid glycoprotein, the carbohydrates have a defined structure in space. This secondary structure will interact with the protein differently than in its absence, i.e., when the carbohydrates are free in solution. The interaction between calcofluor and the highly hydrophilic α_1-acid glycoprotein carbohydrates will induce the important red shift (15 nm) in the emission of the fluorophore (Fig. 3.32).

Calcofluor White is a symmetric molecule with two triazol rings and two primary alcohol functions on both sides of an ethylene bridge. At low concentration of fluorophore the binding to the carbohydrates moiety of α_1-acid glycoprotein occurs by one end of the fluorophore, the shift observed being dependent on the ratio of calcofluor to α_1-acid glycoprotein. Whereas, for calcofluor in excess compared with α_1-acid glycoprotein, the two equivalent parts of calcofluor are involved in the binding, thus affording bridges between distant alcohol functions of the carbohydrate residues. In this case, we observe a red shift of 15 nm. All N-linked sugars such as those of α_1-acid glycoprotein, contain a hexasaccharide Man$_3$ Glc-Nac$_2$ core. The heterogeneity of N-linked carbohydrates originates primarily from the presence, absence, type or length of the sugar side chains attached.

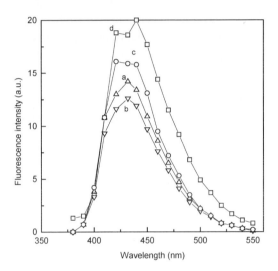

Figure 8.13. Fluorescence emission spectra of 10 µM of calcoftuor in water (a), in the presence of a glycan (b), in the presence of α_1-cellulose (c) and in the presence of α_1-acid glycoprotein (d). λ_{ex} 300 nm. α_1-cellulose was added to the cuvette containing calcoftuor with the glycan, and α_1-acid glycoprotein was then added to the cuvette containing the ftuorophore, the glycans and α_1 cellulose. Source: Albani, J. R. and Plancke, Y. 1999. Carbohydrate Research. 318, 193-200.

Identical binding parameters and the same fluorescence variation were obtained for the calcofluor when bound to α_1-acid glycoprotein molecules prepared from different human sources and by different methods. Therefore, the heterogeneity due either to the protein backbone or to the carbohydrate antennae does not play any fundamental role in the interaction between the calcofluor and α_1 - acid glycoprotein. This conclusion is in good agreement with that we found with TNS : the microheterogeneity of α_1 - acid glycoprotein does not play any role in the interaction of the protein with TNS and is not responsible of the heterogeneous fluorescence of the fluorophore.

Calcofluor interacts with different carbohydrate structures, mainly with structures containing galactose, glucose and mannose. Since α_1-acid glycoprotein is rich in these residues, the interaction with calcofluor should be expected.

Moreover, these results suggest that the spatial organization of the carbohydrates play an important role in the interaction between calcofluor and α_1-acid glycoprotein.

4d. Titration of carbohydrate residues with calcofluor

In the previous study, the binding parameters of calcofluor-α_1-acid glycoprotein complex were determined by following the decrease in the fluorescence intensity of the Trp residues of the protein in the presence of increasing concentrations of calcofluor. However, since the fluorophore binds preferentially to the carbohydrate residues, it is interesting to know whether the data obtained when the fluorescence parameters of calcofluor itself are followed, are identical to those obtained with the fluorescence intensity decrease of the Trp residues. Therefore, we studied the interaction between calcofluor and α_1- acid glycoprotein with calcofluor in presence of high concentration of calcofluor compared to that of the protein. Also, one should indicate that *in vivo*, experiments are usually carried out at excess concentration of calcofluor. Variation of the emission maximum of 220 µM of calcofluor in presence of increasing amounts of sialylated (a) and asialylated α_1-acid glycoprotein (b) is displayed in Fig. 8.14. Only binding of calcofluor to the sialylated protein induces a shift in the emission maximum of the fluorophore. In the absence of protein, the fluorescence maximum is at 438 nm and it shifts progressively to 450 nm with increasing concentrations of sialylated α_1-acid glycoprotein (a). The shift stops when the stoichiometry of 1 calcofluor for 1 sialic acid residue is reached. Addition of asialylated α_1-acid glycoprotein (b) or of free sialic acids (not shown) does not modify the position of the emission maximum of calcofluor.

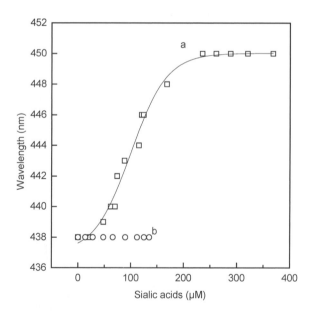

Fig. 8.14. Variation of the emission maximum of 220 µM of Calcofluor with the concentration of sialylated (a) and asialylated α_1- acid glycoprotein (b). λ_{ex}, 300 nm and temperature = 20°C. The proteins concentrations are expressed in sialic acid residues. 1 mol of sialylated α_1- acid glycoprotein contains 16 sialic acid residues. Source : Albani, J. R., Sillen A., Plancke, Y. D., Coddeville, B. and Engelborghs, Y. 2000. Carbohydr. Res. 327, 333-340.

Also, in addition to the shift in the emission maximum, we observe an increase in the fluorescence intensity (Fig. 8.15).

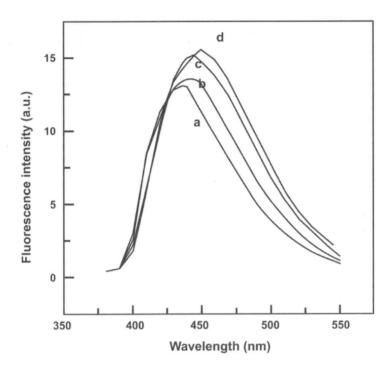

Figure 8.15. Observed fluorescence emission spectra of 194 µM of Calcofluor free in solution (a) and in presence of 3 (b) 11 (c) and 18 (d) µM of α_1-acid glycoprotein. The concentrations expressed in sialic acid are 48, 176 and 288 µM, respectively. λ_{ex} 300 nm, t = 20°C. Source : Albani, J. R., Sillen A., Plancke, Y. D., Coddeville, B. and Engelborghs, Y. 2000. Carbohydr. Res. 327, 333-340.

Figure 8.16 displays the fluorescence intensity increase of calcofluor (194 µM) in the presence of increasing concentrations of sialylated α_1-acid glycoprotein observed at 440 nm. Between protein concentrations 1.2 and 10 µM the fluorescence intensity increase linearly and reaches a plateau in presence of 12.7 ± 1.2 µM of protein (200 µM of sialic acids) and thus at a stoichiometry close to 1:1. The deviation of the first point from the line could be due to a change in specific fluorescence upon increasing the number of bound calcofluor molecules per protein. The linear increase of the fluorescence upon protein addition (between 1.2 and 12 µM) indicates that the dissociation constant should be less than 12 µM.

Addition of increasing amounts of free sialic acids to 212 µM of calcofluor induces a limited decrease in the fluorescence intensity of calcofluor (Figures 8.17a and b.). This decrease indicates that interactions occur in solution between free sialic acid and calcofluor. However, the absence of curvature within the range of concentrations used indicates that we are far to have a complex between calcofluor and free sialic acid.

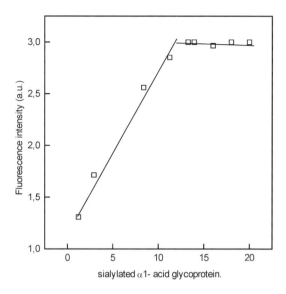

sialylated α1- acid glycoprotein.

Figure 8.16. Titration curve of 194 μM of Calcofluor with sialylated α_1- acid glycoprotein. Saturation occurs for 12.7 μM of protein (200 μM of sialic acids), thus indicative of a stoichiometry of one calcofluor for one sialic acid residue. λ_{ex} = 300 nm and λ_{em} = 440 nm. The fluorescence intensities are corrected for the dilution and for the inner filter effect.

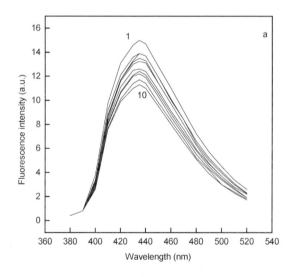

Figure 8.17a. Titration of 212 μM of Calcofluor with sialic acid residues. Aliquots of 31 μM of sialic acid residues solution were added to the Calcofluor. (a) Observed fluorescence spectra of Calcofluor in solution (1) and in the presence of the increased concentration of sialic acid residues. λ_{ex} = 300 nm.

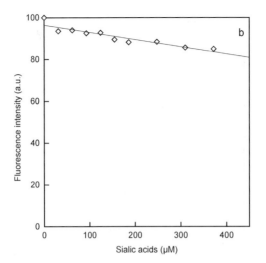

Fig. 8.17b. Normalized fluorescence intensities of 212 μM calcofluor at λ_{em} = 435 nm as a function of the sialic acids concentration. The intensities were corrected for the dilution and for the inner filter effect. Sources of figures 8.16 and 8.17: Albani, J. R., Sillen A., Plancke, Y. D., Coddeville, B. and Engelborghs, Y. 2000. Carbohydr. Res. 327, 333-340.

Figures 8.18a and b exhibit the fluorescence intensity titration of 57 μM of calcofluor with asialylated α_1-acid glycoprotein.

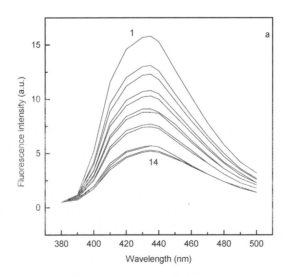

Figure 8.18a. Titration curve of Calcofluor (57 μM) with asialylated α_1- acid glycoprotein. Observed fluorescence spectra of free Calcofluor in solution (spectrum 1) and in the presence of increased concentrations of protein (spectra 2-14). λ_{ex} = 300 nm.

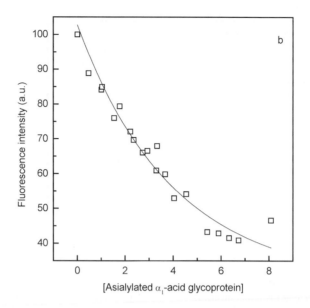

Figure 8.18b. Normalized fluorescence intensities observed at 435 nm as a function of asialylated protein. The number n of binding sites is equal to 12.7 ± 1.2 mol of Calcofluor for 1 mol of protein. Since α_1-acid glycoprotein contains five heteropolysaccharide groups, we would have about two calcofluor molecules for each group. The fluorescence intensities were normalized for the dilution and for the inner filter effect. Source: Albani, J. R., Sillen A., Plancke, Y. D., Coddeville, B. and Engelborghs, Y. 2000. Carbohydr. Res. 327, 333-340.

The intensity decrease is clearly hyperbolic and therefore a mathematical binding analysis can be performed using the following quadratic equation obtained from the definition of the equilibrium constant (assuming all binding sites have the same dissociation constant K_d):

$$L_b = 0.5\,[(nP_0 + L_0 + K_d) - \{(nP_0 + L_0 + Kd)^2 - 4\,n\,P_0\,L_0\}^{1/2}\,] \qquad (8.3)$$

where n is the number of binding sites (assumed to be identical), and P_0 is the protein concentration. The parameter Flu_1 was obtained by extrapolation from a reciprocal plot and found to be 32.5. Using this parameter, non linear least squares fitting gives the continuous line as the best fit with $n = 12.7 \pm 1.2$ and $Kd = 6.8 \pm 2.6$ μM. This should be considered as an average dissociation constant per site.

Titration experiments are usually carried out by following the intensity variation. In the above-described experiments, we show that it is possible to perform titration experiments by following not only the fluorescence intensity variation but also the shift in the position of the emission maximum of calcofluor. This shift is sensitive to sialic acid residues only when they are bound to α_1-acid glycoprotein (Fig. 8.14). Therefore, sialic acids of α_1-acid glycoprotein possess a spatial conformation that can be detected by calcofluor.

Binding of calcofluor to sialylated α_1-acid glycoprotein induces also an increase in the fluorescence intensity of the fluorophore (Fig. 8.15 and 8.16). The titration curve gives a stoichiometry of 1 calcofluor for 1 sialic acid residue. This result is in good agreement with that obtained when the shift of the emission maximum of calcofluor was followed (Fig. 8.14). The dissociation constant could not be determined from the titration curve but must be smaller than the protein concentration at stoichiometry (< 12 μM) which is in agreement with that calculated when binding experiments were performed by following fluorescence intensity decrease of Trp residues.

Addition of free sialic acid to calcofluor decreases the fluorescence intensity of the fluorophore (Fig. 8.17). Thus, the type of interaction between free sialic acid residues and calcofluor is different from that observed when sialic acids are bound to α_1-acid glycoprotein.

Interaction of calcofluor with the other carbohydrate residues occurs also, since we observe a decrease in its fluorescence intensity upon addition of asialylated α_1-acid glycoprotein (Fig. 8.18). The dissociation constant of the calcofluor-asialylated α_1-acid glycoprotein complex is 6.8 μM, a value equal to that measured for the calcofluor-sialylated protein (Fig. 8.11). Thus, we suggest that this binding constant may characterize the interaction between the carbohydrate residues and calcofluor. However, an interaction even weak with the protein matrix cannot be excluded. 12 binding sites were found for the calcofluor-asialylated α_1-acid glycoprotein complex. Since the protein contains five heteropolysaccharide groups, we would have about two calcofluor molecules for each group In the presence of sialic acids the type of interaction of calcofluor with the carbohydrate residues is modified. The dissociation constants measured in absence and in presence of sialic acids do not show that the affinity of calcofluor to sialic acid residues is higher than that for the other carbohydrate residues. However, the titration curves clearly indicate that calcofluor fluorescence is sensitive to the presence or to the absence of sialic acids of α_1-acid glycoprotein.

The fact that the interactions between calcofluor and sialic acids differ whether the latter are on the protein or free in solution, indicates that the sialic acids on α_1-acid glycoprotein possesses a defined spatial conformation. Also, since titration curves with calcofluor are not the same whether the sialic acid residues are present or not, one may conclude that the presence of the sialic acid confers to the carbohydrate residues backbone a spatial conformation that changes upon desialylation of the protein.

The spatial structure of the carbohydrate residues of α_1-acid glycoprotein induces the binding of the calcofluor. The absence of this spatial structure (free carbohydrates in solution) modifies the type of their interactions with the fluorophore.

4e. Fluorescence lifetime of calcofluor bound to α_1-acid glycoprotein

Fluorescence intensity of calcofluor, whether free in water or bound to α_1-acid glycoprotein decays as a sum of four exponentials. When the fluorophore is free in solution, the intensity average lifetime is 0.85 ns. It increases to 4.8 and 3.9 ns when the fluorophore is bound to the sialylated and to the asialylated protein, respectively.

Table 8.1 summarizes the lifetimes obtained for calcofluor free in solution, in the presence of sialylated and asialylated α_1-acid glycoprotein.

Table 8.1. Fluorescence lifetimes (ns) and corresponding fractional intensities of calcofluor free in solution and bound to sialylated and asialylated α_1-acid glycoprotein.

Calcofluor free in solution (3.2 µM).

τ_1	τ_2	τ_3	τ_4	f_1	f_2	f_3	f_4
0,3	0,6	2,28	14,05	0,34	0,6	0,03	0,02
± 0,03	± 0,3	± 0,5	± 4	± 0,1	± 0,1	± 0,01	

Calcofluor (6.6 µM) in presence of 4.7 µM sialylated α_1-acid glycoprotein.

τ_1	τ_2	τ_3	τ_4	f_1	f_2	f_3	f_4
0,41	1,07	4,01	12,13	0,27	0,25	0,15	0,32
± 0,01	± 0,03	± 0,2	± 0,14	± 0,008	± 0,005	± 0,04	

Calcofluor (4.55 µM) in presence of 5.5 µM asialylatyed α_1-acid glycoprotein.

τ_1	τ_2	τ_3	τ_4	f_1	f_2	f_3	f_4
0,35	0,8	3,3	11,42	0,22	0,35	0,16	0,26
± 0,007	± 0,021	± 0,12	± 0,12	± 0,02	± 0,01	± 0,03	

Source: Albani, J. R., Sillen A., Coddeville, B., Plancke, Y. D., and Engelborghs, Y. 1999, Carbohydr. Res. 322, 87-94.

Binding of calcofluor white to α_1-acid glycoprotein induces an increase in the fluorescence lifetime. The four fluorescence lifetimes originate from the structure of the fluorophore itself since four lifetimes do exist when the fluorophore is free in the solvent. When bound to α_1-acid glycoprotein, the values of the four lifetimes do not change significantly, they are almost the same. However, the fractional intensities of the two longer lifetimes increases significantly, 5 times for the lifetime equal to 3 ns and 10 times for the lifetime equal to 12 ns. These results are a clear indication that the origin of the fluorescence lifetime of calcofluor is the structure of the molecule itself. The four lifetimes observed are intrinsic to the calcofluor and probably due to different possible conformations of calcofluor. Its binding to α_1- acid glycoprotein increases the average lifetime as the consequence of the increase in the fractional intensities of the two longest lifetimes. This increase is the result of the type of interaction between the fluorophore and the carbohydrates of the glycoprotein.

The results obtained with Calcofluor White are in good agreement with those found with TNS: the protein microheterogeneity does not play any role in the interaction of the protein with TNS and is not responsible of the heterogeneous fluorescence of the fluorophore. Binding to α_1-acid glycoprotein occurs with a specific interaction with its environment modifying the fractional intensities of the two most important lifetimes. This will lead to the increase of the mean fluorescence lifetime. Simple comparison with the fluorescence lifetime of TNS allows understanding the importance of the structure of a fluorophore in its global fluorescence. TNS shows a very weak fluorescence when free in water (intensity, lifetime and quantum yield) while calcofluor white displays a significant fluorescence in the same conditions. Upon their interaction with a protein, the Brownian motions are less important, increasing by that their mean fluorescence lifetime. However, while two new lifetimes characterizes the fluorescence of bound TNS, the main change in the fluorescence of calcofluor lifetimes is the fractional intensities. This simple comparison allows us to conclude that if one can find common features to most of the fluorophores, each fluorophore has its own specificity and one should avoid trying always to generalize when studies are done on a specific fluorophore within a specific environment.

4f. Binding parameters of calcofluor white - $α_1$-acid glycoprotein by following the fluorescence lifetime of calcofluor

Titration experiment of 3.5 μM of calcofluor with sialylated $α_1$-acid glycoprotein is shown in Figure 8.19. We notice that the curvature begins at 4.5 μM of protein and the lifetime reaches a plateau equal to around 5 ns.

Plotting the data with Eq. 8.2

$$x = n\,P\,/\,(L + K_d)\,, \tag{8.2}$$

where $x = τ\,/\,τ_{max}$, n is the number of sites (=1) and K_d the dissociation constant of the complex. The value of K_d was found equal to 4.5 ± 0.5 μM. This value is in the same range of that (6.6 μM) calculated when the fluorescence intensity of Trp residues of $α_1$-acid glycoprotein was followed and to that (6.6 μM) obtained when titration of $α_1$-acid glycoprotein with calcofluor was performed following fluorescence intensity modification of calcofluor.

In these three different studies, we are monitoring the same type of interaction: calcofluor white – carbohydrate residues.

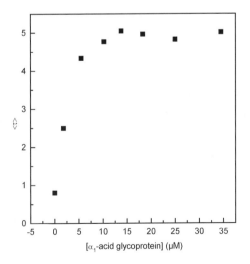

Figure 8.19. Titration of 3.2 μM of calcofluor with sialylated $α_1$-acid glycoprotein following fluorescence lifetime increase of calcofluor.

4g. Dynamics of calcofluor bound to α_1-acid glycoprotein at equimolar concentrations

4g1. Dynamics of calcofluor bound to sialylated α_1-acid glycoprotein

Figure 8.20 displays the normalized fluorescence emission spectra of 10 µM of calcofluor in the presence of 15 µM of α_1-acid glycoprotein obtained at three excitation wavelengths. The maximum (440 nm) of the Calcofluor fluorescence does not change with the excitation wavelength (λ_{ex} 385, 395 and 405 nm). The results obtained clearly indicate that the microenvironment of calcofluor has residual motions independent of the global rotation of the protein, and this may induce the local motions of calcofluor.

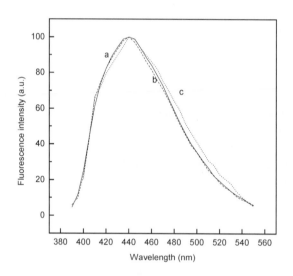

Figure 8.20. Normalized steady-state fluorescence spectra of 10 µM Calcofluor in the presence of 15 µM sialylated α_1-acid glycoprotein recorded at three excitation wavelengths, 385 nm (a), 395 nm (b) and 405 nm (c). The fluorescence maximum is equal to 440 nm at the three excitation wavelengths. Source: Albani, J. R., Sillen A., Coddeville, B., Plancke, Y. D., and Engelborghs, Y. 1999, Carbohydr. Res. 322, 87-94.

Steady-state fluorescence anisotropy of 10 µM of Calcofluor in the presence of 5 µM of α_1-acid glycoprotein ($\lambda_{em} = 435$ nm and $\lambda_{ex} = 300$ nm) was performed at different temperatures. A Perrin plot representation (Fig. 8.21a.) yields a rotational correlation time equal to 7.5 ns at 20 °C. This value is lower than that (16 ns) expected for α_1-acid glycoprotein and thus indicates that calcofluor displays segmental motions independent of the global rotation of the protein. Thus, two motions contribute to the depolarization process, the local motion of the carbohydrate residues and the global rotation of the protein, i.e., a fraction of the total depolarization is lost due to the segmental motion, and the remaining polarization decays as a result of the rotational diffusion of the protein.

The rotational correlation time (= 7.5 ns) obtained from the Perrin plot is the result of the global rotation of the protein and the local motion of Calcofluor White.

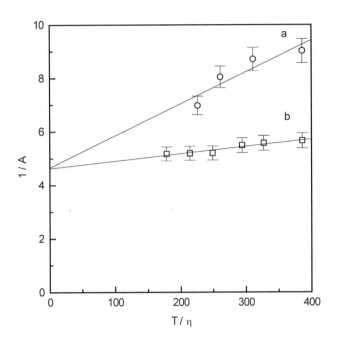

Figure 8.21. Steady-state fluorescence anisotropy vs. temperature over viscosity for 5 µM Calcofluor in the presence of 10 µM sialylated α_1-acid glycoprotein (λ_{ex}, 300 nm, λ_{em}, 435 nm) (plot a), and for 8.5 µM Calcofluor in the presence of 5.5 µM asialylated α_1- acid glycoprotein ((λ_{ex}, 300 nm, λ_{em}, 445 nm (plot b). The data shown are the mean values of two measurements, and they are obtained by thermal variation in the range 15-35°C. The ratio T/η is expressed in Kelvins/centipoise. Source: Albani, J. R., Sillen A., Coddeville, B., Plancke, Y. D., and Engelborghs, Y. 1999, Carbohydr. Res. 322, 87-94.

4g2. Dynamics of Calcofluor bound to asialylated α_1- acid glycoprotein

Figure 8.22 exhibits the normalized fluorescence emission spectra of 7.5 µM of Calcofluor bound to 5 µM of asialylated α_1 -acid glycoprotein obtained at three excitation wavelengths. The maximum of the fluorescence emission spectra of Calcofluor is a function of the excitation wavelength. At 385 nm, the emission maximum is located at 437 nm. It shifts to higher wavelengths (440 and 445 nm) when the excitation wavelengths are 395 and 405 nm, respectively. This is taken as direct evidence that carbohydrate residues in the microenvironment of Calcofluor exhibit restricted motions.

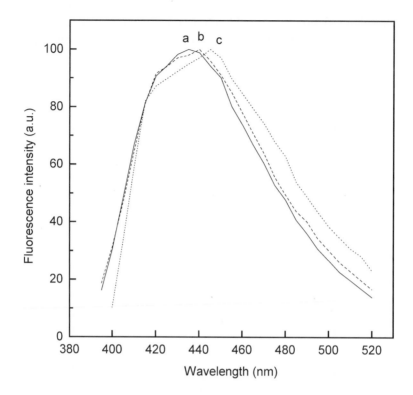

Figure 8.22. Normalized steady-state fluorescence spectra of 7.5 μM Calcofluor in the presence of 5 μM asialylated α_1-acid glycoprotein recorded at three excitation wavelengths, 385 nm (a), 395 nm (b) and 405 nm (c). The fluorescence maximum is at 437 nm (a), 440 nm (b) and 445 nm (c). Source: Albani, J. R., Sillen A., Coddeville, B., Plancke, Y. D., and Engelborghs, Y. 1999, Carbohydr. Res. 322, 87-94.

α_1 - Acid glycoprotein possesses 16 sialic acid residues of molecular mass 314. Thus, the molecular mass of asialylated α_1 - acid glycoprotein is near 36,000. Eq. 5.15 yields a rotational correlation time of 14 ns for the asialylated protein.

Steady-state fluorescence anisotropy of 8.5 μM of calcofluor in the presence of 5.5 μM α_1 -acid glycoprotein (λ_{em} = 445 nm and λ_{ex} = 300 nm) was performed at different temperatures. The Perrin plot representation (Fig. 8.21b) yields a rotational correlation time equal to 19 ns at 20 °C, revealing the absence of a segmental motion. Anisotropy results are in good agreement with those obtained by red-edge excitation spectra experiments, i.e., Calcofluor is bound tightly to the carbohydrate residues of the protein.

Dynamics of carbohydrate residues, whether free in solution or bound to a protein, are usually studied by ^{1}H NMR. The previous described data allow one for the first time to follow the dynamics of a carbohydrate moiety using a fluorescence approach.

4 h. Dynamics of calcofluor bound to α_1-acid glycoprotein at excess concentration of calcofluor

Figure 8.23a displays the variation of the emission maximum of 300 μM of calcofluor in the presence of 5 μM of sialylated α_1-acid glycoprotein as a function of the excitation wavelength. At 385 nm, the emission maximum is located at 447 nm. It shifts to higher wavelengths (452 and 458 nm) when the excitation wavelengths are 400 and 415 nm, respectively. This is taken as direct evidence that the carbohydrate residues in the microenvironment of calcofluor exhibit restricted motions, even if a substantial fraction of the calcofluor remains unbound. For comparison, the emission maximum of 9.8 μM of calcofluor in presence of 10 μM of α_1-acid glycoprotein recorded as a function of the excitation wavelength is equal to 441 nm at all wavelengths (Fig. 8.23b). The results obtained in presence of low and high calcofluor concentrations are in opposition. Experiments performed at equimolar concentrations of calcofluor and α_1-acid glycoprotein (sialylated and asialylated forms) showed that the sialic acid residues of the protein are highly mobile while the other carbohydrate residues exhibit restricted motions. At high concentrations of calcofluor, a kind of stacking occurs inducing a decrease of the motions of the sialic acid residues. Thus, sialic acid residues dynamics are dependent on the calcofluor concentrations.

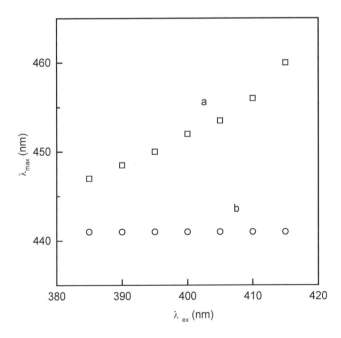

Fig 8.23. The dependence of the fluorescence emission maximum of 300 μM of Calcofluor in the presence of 5 μM sialylated α_1- acid glycoprotein on the excitation wavelengths (a). Plot b shows this dependence for 10 μM of Calcofluor in presence of an equivalent amount of protein. Source: Albani, J. R., Sillen A., Plancke, Y. D., Coddeville, B. and Engelborghs, Y. 2000. Carbohydr. Res. 327, 333-340.

Steady state fluorescence anisotropy of 250 μM of calcofluor in presence of 5 μM of α₁-acid glycoprotein (λ_{em} = 445 nm and λ_{ex} = 300 nm) was performed at different temperatures. A Perrin plot representation (Fig. 8.24) should enable us to obtain information concerning the motion of the fluorophore. The rotational correlation time of the fluororophore Φ_F, using 4.8 as average lifetime, is found to be 16 ± 1 ns at 20°C. This value is equal to the theoretical value (16 ns) expected for α₁-acid glycoprotein revealing the absence of a segmental motion (despite the presence of a substantial fraction of unbound calcofluor). Anisotropy results are in good agreement with those obtained by red-edge excitation spectra experiments, i.e., calcofluor is bound tightly to the carbohydrate residues and follows the global motion of the protein.

The red-edge excitation spectra and anisotropy measurements yield information on the dynamics of the fluorophore and of its environment. At high concentration of calcofluor compared to that of α₁-acid glycoprotein, a red-edge excitation shift is observed (Fig. 8.23a). This shift indicates that the microenvironment of the fluorophore exhibits restricted motions. Therefore, the carbohydrate residues in the vicinity of calcofluor are rigid and do not show any segmental motions.

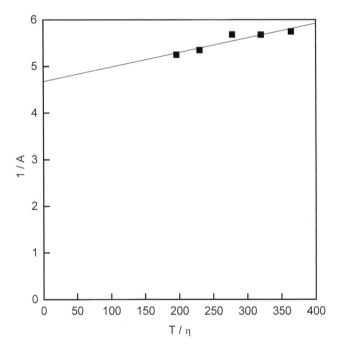

Figure 8.24. Steady-state fluorescence anisotropy vs. temperature over viscosity for 250 μM of Calcofluor in presence of 5 μM of sialylated α₁- acid glycoprotein (λ_{ex}, 300 nm and λ_{em}, 445 nm). The data are obtained by thermal variation in the range 5-35 °C. The ratio T / η is expressed in Kelvins over centipoise. Source: Albani, J. R., Sillen A., Plancke, Y. D., Coddeville, B. and Engelborghs, Y. 2000. Carbohydr. Res. 327, 333-340.

5. Fluorescence properties of the Trp residues in α₁-acid glycoprotein

5a. Fluorescence spectral properties

α₁-Acid glycoprotein contains three Trp residues. One residue, Trp-160, is at the surface of the protein, and two are located in the protein matrix. One of these two Trp residues, Trp-25, lies in a hydrophobic environment at the N-terminal side chain of the protein.

Excitation at 295 nm yields a fluorescence spectrum with a peak equal to 331 nm (Fig. 3.3). The position of the peak is typical for proteins containing tryptophan residues in a hydrophobic environment and results from the contribution of both polar and hydrophobic Trp residues.

Figure 8.25 shows the excitation anisotropy spectrum of α₁-acid glycoprotein at -45°C. The spectrum is typical of Trp residues.

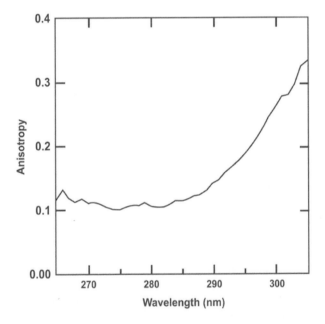

Figure 8.25. Excitation anisotropy spectrum of α₁-acid glycoprotein recorded at -45°C in glycerol/phosphate buffer.

In order to put into evidence the Trp residues that participate in the emission of the protein, quenching resolved emission with cesium were first performed. When a protein contains two classes of intrinsic fluorophore, one at the surface of the protein and the second embedded in the protein matrix, fluorescence intensity quenching with cesium allows obtaining the spectra of these two classes. A selective quenching implies that addition of quencher induces a decrease in the fluorescence observables (intensity, anisotropy and lifetime) of the accessible class. At high quencher concentration the remaining observables measured will reflect essentially those of the embedded fluorophore residues.

Dynamic fluorescence quenching is analyzed by the Stern-Volmer equation:

$$I_o / I = 1 + k_q \tau_o [Q] \qquad (4.6)$$

where I_o and I are the fluorescence intensities in absence and presence of quencher, respectively; k_q the bimolecular diffusion constant; τ_o the mean fluorescence lifetime and $[Q]$ the concentration of cesium added. In presence of a selective quenching, the Stern-Volmer plot shows a downward curvature. The fraction of the fluorescence intensity that is accessible (f_a) is obtained from the modified Stern-Volmer equation

$$I_o / \Delta I = 1 / f_a + 1 / f_a \, K_{SV} [Q] \qquad (4.15)$$

where K_{SV} is the Stern-Volmer for the fluorescence quenching of the surface fluorophore residue by cesium ion and

$$\Delta I = I_o - I \qquad (4.11)$$

Knowing f_a and f_b along the emission spectrum will allow us to draw the spectrum of each class of fluorophore.

Upon binding of a ligand on the protein and if this binding induces structural changes inside the protein, the intensities of the spectrum of the classes can be modified and in the case of the Trp residues, the emission maximum can shift to the higher or lower wavelengths. Figure 8.26 shows the fluorescence spectra of α_1-acid glycoprotein (a) (λ_{max} = 332 nm), of the inaccessible Trp residues (b) (λ_{max} = 324 nm) obtained by extrapolating to infinite concentration of cesium, and of the quenched Trp residue (c) obtained by subtracting spectrum (b) from (a). The emission maximum of the accessible Trp residue is located at 350 nm, a characteristic of an emission from Trp residue at the surface of the protein.

Quenching resolved emission methods indicate that apparently in α_1-acid glycoprotein we have two classes of Trp residues that participate to the emission of the protein. However, this does not mean that there is only two Trp residues that contribute to the global emission of α_1-acid glycoprotein. In fact, two Trp residues could have very close fluorescence so that the quenching method is not adequate to separate them. Thus, other methods should be used to see whether the three Trp residues contribute to the global fluorescence of the protein or not.

A very common method to put into evidence the fluorescence spectrum of each component is the decay associated spectra. This method allows combining the dynamic time-resolved fluorescence data with the steady-state emission spectrum. The fluorescence decays are globally or individually fitted to the n-exponential function (n being the number of species, tryptophan or tyrosine residues for example), and the decay associated spectra are constructed (Krishna and Periasamy, 1997). Figure 8.27 displays the decay associated spectra of Trp residues of α_1-acid glycoprotein performed by Hof et al (1996). The authors found four fluorescence lifetimes instead of the three we obtained. They attributed the peak of 337 nm to Trp-122 with a location between the surface of the protein and its core. The determination of the degree of hydrophobicity around the Trp residues showed that Trp-25 residue is located in a hydrophobic environment and Trp-160 is the most exposed to the solvent.

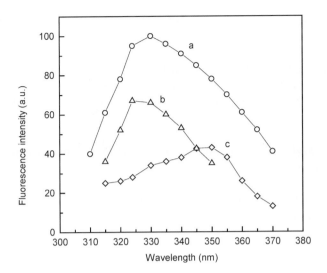

Figure 8.26. Fluorescence spectra of α_1-acid glycoprotein (a) (λ_{max} = 332 nm), of the hydrophobic Trp residues (b) (λ_{max} = 324 nm) and of the surface Trp residue (c). Source: Albani, J. R. 1997, Biochim. Biphys. Acta, 1336, 349-359.

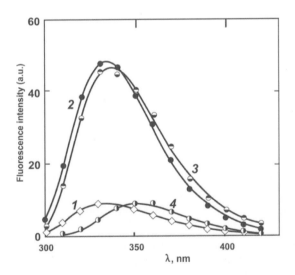

Figure 8.27. Decay-associated emission spectra of α_1-acid glycoprotein. The fluorescence decay time values are 0.22 :t 0.02 ns (component A, spectrum 1), 1.0:t 0.2 ns, (B, 2), 2.5 :t 0.2 ns *(C,* 3), and 8.4 :t 0.5 ns (D, 4). The emission maxima of the log-normal distribution fits are 334 nm (A, 1), 334 nm (B, 2), 337 nm *(C,* 3), and 352 nm (D, 4). Shown are the experimental points and the corresponding log-normal distribution fits. Source: Hof, M., Vajda, S., Fidler, V. and Karpenko, V, 1996, Collect. Czech. Chem. Commun. 61, 808-818. Authorization of reprint accorded by The Institute of Organic Chemistry and biochemistry, Academy of Sciences of the Czech Republic.

In another study, we used a different method that allowed us to deconvolute the fluorescence emission spectra of the Trp residues of α_1-acid glycoprotein. Since our fluorescence data with calcofluor showed that the interaction between the extrinsic fluorophore and α_1-acid glycoprotein differs whether we are working at equimolar concentrations of protein and probe or in presence of excess concentration of calcofluor, we resolved the fluorescence spectra of the Trp residues of α_1-acid glycoprotein in presence of high concentrations of calcofluor.

5b. Effect of high calcofluor concentration on the local structure of α_1-acid glycoprotein

Figure 8.28 displays the fluorescence emission spectra of 14 µM α_1-acid glycoprotein in presence of 10 (a) and 160 µM of calcofluor (b), both obtained at λ_{ex} of 295 nm.

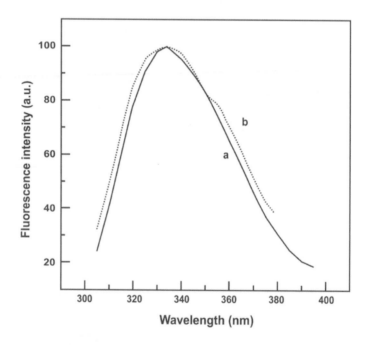

Figure 8.28. Fluorescence emission spectra of 15 µM of α_1-acid glycoprotein in the presence of 10 µM of calcofluor (a) and of 14 µM of α_1-acid glycoprotein in presence of 160 µM of calcofluor (b). Both spectra are obtained at excitation of 295 nm. Source: Albani, J. R., 2001, Carbohydr. Res. 334, 141-151.

We notice that in presence of high concentrations of calcofluor, the bandwidth of the spectrum increases by 6 nm and a shoulder at 350 nm appears. These spectrum disruptions indicate that a possible structural modification of the protein has occurred in presence of high concentrations of calcofluor, i.e., addition of calcofluor at high concentrations would affect the structure of the surrounding microenvironments of the Trp residues.

The disruptions of the emission spectrum of the Trp residues in presence of high calcofluor concentrations would originate from modification of the local structure of the protein and/or from an energy transfer from the Trp residues to the calcofluor. Although a very weak energy transfer occurs, it is improbable that this energy transfer induces the important disruptions of the emission spectra. Recording the fluorescence excitation spectrum of the Trp residues in absence and in presence of low and high calcofluor concentrations can monitor a structural modification within α_1-acid glycoprotein. In fact, the fluorescence excitation spectrum characterizes the electron distribution of the molecule in the ground state. At λ_{em} = 330 nm, calcofluor does not emit and thus, only the excitation spectrum of the Trp residues would be recorded. Therefore, any modification of the fluorescence excitation spectrum in presence of calcofluor would be the result of a structural modification of the protein in the ground state.

Figure 2.11 displays the fluorescence excitation spectrum of the Trp residues of α_1-acid glycoprotein in absence (a) and presence of 10 µM (b) and 120 µM (c) of calcofluor. We notice that in presence of high concentrations of calcofluor, the peak is located at 280 nm instead of 278 nm and a shoulder is observed at 295 nm. Thus, the global shape of the spectrum is modified in presence of high calcofluor concentrations, i.e., binding of calcofluor at high concentrations has modified the local structure of the Trp residues of α_1-acid glycoprotein. The fact that binding of low concentrations of calcofluor (10 µM) does not affect this local structure indicates that a minimum concentration of fluorophore is necessary to induce any structural modification of the protein.

5c. Deconvolution of the emission spectra obtained at low and high concentrations of calcofluor into different components

In order to find out which of the Trp residues that is the most affected by the presence of calcofluor, two deconvolution methods were applied. In the first method, we fitted the spectra obtained as a sum of Gaussian bands. In the second one, we used the method described by Siano and Metzler (1969), and that was then applied by Burstein and Emelyanenko (1996) using a four parametric log-normal function (a skewed Gauss equation). The analysis were done at the three excitation wavelengths (295, 300 and 305 nm), although the results shown are those obtained at λ_{ex} = 295 nm.

5c1. Analysis as a sum of Gaussian bands

In this analysis, we have proposed for each spectrum, one, two or three maxima and we asked the program to draw the best Gaussian curves we can obtain from our spectra.

The best fit was obtained with the sum of two bands. At low concentration of calcofluor, one band is obtained at a maximum of 326 nm and a bandwidth of 23 nm. The second band has a maximum at 347 nm and a bandwidth of 40 nm (Fig. 8.29). Although the positions of the maxima are significant and correspond to those found for the hydrophobic and surface Trp residues, respectively (Fig. 8.26), the values of the bandwidths do not correspond to those known for Trp residues (48 and 59 nm for the hydrophobic and surface residues, respectively). This means that it is difficult to separate the real spectra one from each other, i.e., in presence of low calcofluor concentration, α_1-acid glycoprotein has kept its folded structure.

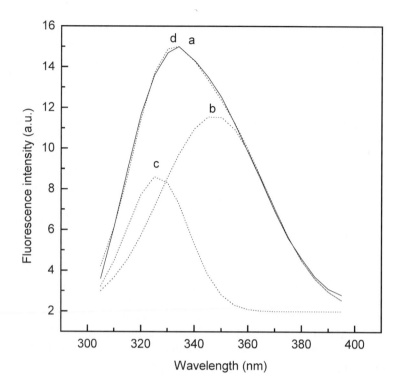

Figure 8.29. Fluorescence emission spectrum of 15 µM of α₁-acid glycoprotein in the presence of 10 µM of calcofluor recorded at excitation wavelength of 295 nm (a). The figure displays also (dotted lines) the two Gaussian components of the fluorescence spectrum. The maximum of the bands are 347 nm (b) and 326 nm (c). These positions correspond to the emission from surface and hydrophobic Trp residues, respectively. However, the bandwidths of the two bands (40 and 23 nm, respectively), are lower than those known for Trp residues, i.e. α₁-acid glycoprotein has kept its folded structure. Spectrum d shows the sum of spectra (a) and (b). Source: Albani, J. R., 2001, Carbohydr. Res. 334, 141-151.

Spectrum (d) shows the sum of spectra (c) and (b). This spectrum overlaps the original spectrum (a) indicating how well the two components fit the original spectrum.

Performing the fits for the spectrum obtained at high concentrations of calcofluor, yields also two bands (Fig. 8.30). One has a maximum at 324 nm and a bandwidth of 28 nm. The second shows a maximum at 349 nm and a bandwidth of 53 nm. We notice that only the bandwidth of the second band is close to that of fluorescence spectrum of Trp residue near the protein surface. Thus, binding of calcofluor at a concentration much higher than that of the protein, induces a modification in the fluorescence spectrum of the surface Trp residue. This modification seems not to be observed for the hydrophobic Trp residues.

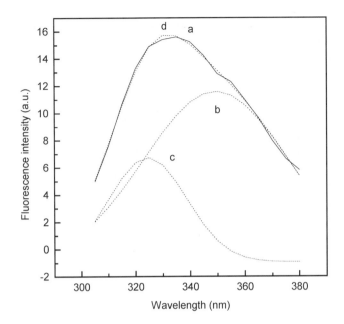

Figure 8.30. Fluorescence emission spectrum of 14 µM of α_1-acid glycoprotein in the presence of 160 µM of calcofluor recorded at excitation wavelength of 295 nm (a). The figure displays also (dotted lines) the two Gaussian components of the fluorescence spectrum. The maximum of the bands are 349 nm (b) and 324 nm (c). These positions correspond to the emission from surface and hydrophobic Trp residues, respectively. The bandwidth of band (b) is equal to 53 nm, indicating that the structure of the protein surface is disrupted. Spectrum d shows the sum of spectra (a) and (b). Source: Albani, J. R., 2001, Carbohydr. Res. 334, 141-151.

5c2. The Ln-normal analysis

In this analysis, the following equation was used :

$$I_{(vm)} = I_m \exp \{ - (Ln\ 2\ /\ Ln^2\ \rho) - Ln^2\ [(a - v)\ /\ (a - v_m)] \} \qquad (8.4)$$

where $I_m = I_{(vm)}$ is the maximal fluorescence intensity. v_m is the wavenumber of the band maximum, $\rho = (v_m - v-)\ /\ (v+ - v_m)$ is the band asymmetry parameter. $v+$ and $v-$ are the wavenumber positions of left and right half maximal amplitudes. $a = v_m + H\rho\ /\ (\rho^2 - 1)$ and H is the bandwidth : $v+ - v-$

Equation 8.4 allows us in principle to draw a spectrum that matches the entire one obtained technically. This is correct only if the recorded spectrum originates from one fluorophore (i.e. Trp in solution) or from compact protein within a folded structure. However, when the experimental spectrum originates from two fluorophores (i.e., mixtures of tyrosine and tryptophan in solution) or from a disrupted protein that has two classes of Trp residues, the calculated spectrum using equation 8.4 does not match with the recorded one. In the present work, all the spectra are drawn as the intensity as a function of the wavelength.

Figure 8.31 shows the experimental fluorescence spectrum of pure L- Trp in water. One can notice that analysis with Eq. 8.4 yields a spectrum (squares) that matches the experimental one (line).

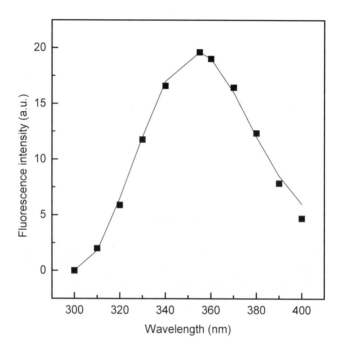

Figure 8.31. Experimental fluorescence emission spectrum of pure L-Trp in water (line) and that obtained by Eq. 8.4. Source: Albani, J. R., 2001, Carbohydr. Res. 334, 141-151.

Figure 8.32 displays the experimental fluorescence spectrum of a mixture of L-tyrosine and L-tryptophan in water (line, spectrum a). The presence of a small shoulder around 303 nm indicates that the L-tyrosine contributes to the fluorescence emission. Analysis of the data using equation 3 yields a fluorescence spectrum (squares, spectrum b) that does not match with the experimental one, especially within the region where the tyrosine emits. The calculated spectrum (b) corresponds to that of the tryptophan alone. Subtracting spectrum (b) from (a) yields a spectrum with a maximum at 303 nm corresponding to that of the tyrosine residues.

Figure 8.33 shows the experimental fluorescence spectrum of Trp residues of 15 μM of α_1-acid glycoprotein ($\lambda_{ex} = 295$ nm) in presence of 10 μM of calcofluor (line) and the calculated spectrum using Eq. 8.4 (square). We notice that the calculated spectrum matches with the experimental one. Therefore, binding of low concentrations of calcofluor to α_1-acid glycoprotein does not perturb the structure of the protein.

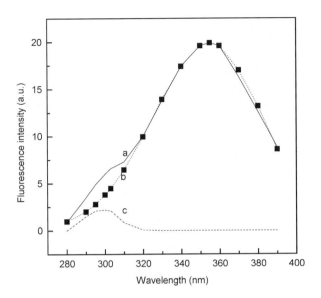

Figure 8.32. Experimental fluorescence emission spectrum of a mixture of L-tyrosine and L-tryptophan in water (line, spectrum a) and that approximated by Eq. 8.10 (squares, spectrum b). Subtracting spectrum (b) from (a) yields a fluorescence spectrum (c) characteristic of tyrosine. λ_{ex} = 260 nm. Source: Albani, J. R., 2001, Carbohydr. Res. 334, 141-151.

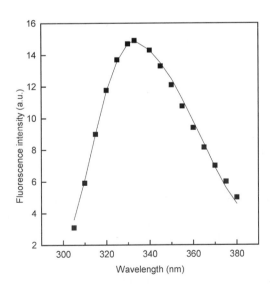

Figure 8.33. Experimental fluorescence emission spectrum of the Trp residues of α_1– acid glycoprotein (15 µM) in the presence of 10 µM Calcofluor (line) and that approximated by Eq.8.4 (squares). λ_{ex} = 295 nm. Source: Albani, J. R., 2001, Carbohydr. Res. 334, 141-151.

Figure 8.34 displays the fluorescence emission spectrum of 14 μM of α_1-acid glycoprotein in presence of 160 μM of calcofluor (a). Equation 8.4 yields a fluorescence spectrum (square, spectrum b) that does not fit the experimental one. Subtracting spectrum (b) from (a) yields two spectra (c and d) with maxima located at 330 and 355 nm, respectively. We notice that the three spectra (b, c and d) participate significantly to the total emission of α_1-acid glycoprotein. The result obtained in presence of high concentration of calcofluor with comparison to that obtained at low concentration of the extrinsic fluorophore, indicates clearly that binding of calcofluor at high concentration on α_1-acid glycoprotein disrupts the structure of the protein.

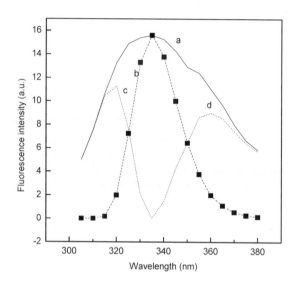

Figure 8.34 Experimental fluorescence emission spectrum of the Trp residues of 14 μM α_1– acid glycoprotein in the presence of 160 μM Calcofluor, spectrum (a) and that approximated by Eq. 8.4 , spectrum (b). Subtracting spectrum (b) from (a) yields two spectra (c) and (d) with maxima located at 330 and 355 nm, respectively. $\lambda_{ex} = 295$ nm. Source: Albani, J. R., 2001, Carbohydr. Res. 334, 141-151

The results obtained with the Ln-normal analysis shows that this method is more suitable than the one that consists of deconvoluting the recorded emission spectrum into the sum of two spectra. The presence of an emission spectrum with a maximum located at 337 nm indicates that three components can best describe the emission of the Trp residues of α_1-acid glycoprotein. An emission at 337 nm corresponds to that of a Trp residue located in an area between the core of the protein and its surface.

The positions of the two maxima (347 and 326 nm) found in our analysis are identical to those (350 and 324 nm) we have already obtained by performing fluorescence intensity quenching with cesium (Fig. 8.26). However, the Ln-normal analysis performed on α_1-acid glycoprotein in presence of high calcofluor concentrations is much more accurate than the simple analysis with the sum of different spectra or even with the quenching experiments, since it allows to reveal the presence of a third component with a maximum located at 337 nm.

Analysis of the data with a sum of two gaussian curves or by the Log-normal analysis indicates clearly that binding of calcofluor at high concentrations induces a disruption in the structure of α_1-acid glycoprotein. Apparently, calcofluor induces a pressure at the surface of the protein that denatures it partially. This effect is identical to that observed when hydrostatic pressure is applied on proteins (Ruan and Weber, 1993; Ruan and Balny, 2002).

The Log-normal analysis method shows also that binding of calcofluor at high concentrations disrupts the protein structure. This method is very sensitive to the modifications occurring around the fluorophore. In fact, in presence of folded protein structure, the calculated emission spectrum fits perfectly the experimental one (Fig. 8.33). However, when the structure of the protein is altered, we obtain a calculated spectrum that is different from the experimental one (Fig. 8.34, spectrum b, $\lambda_{max} = 337$ nm). Also, in this case, we can derive two other emission spectra that correspond to the hydrophobic ($\lambda_{max} = 330$ nm) and surface Trp residues ($\lambda_{max} = 355$ nm). This deconvolution is not possible in the presence of a compact protein (Fig. 8.33). Therefore, the Log-normal analysis allows comparing disrupted and compact proteins with two or three classes of Trp residues, since in the case of a compact protein the analysis gives a result identical to that obtained for a free tryptophan in solution.

Since the carbohydrate residues cover almost 90% of the surface of α_1-acid glycoprotein, it is normal that binding of calcofluor to the carbohydrate residues affects mainly the surface of the protein. However, this does not mean that the environments of the two other Trp residues could not be affected by the presence of calcofluor, although the effect would be less important than the one observed at the surface of the protein. Finally, one should indicate here that disruption does not mean necessarily denaturation.

Dynamics of tryptophan residues of α_1-acid glycoprotein in presence of low and high concentrations of calcofluor were performed by the red-edge excitation spectra method and Perrin plot. Let us remind that in the red-edge excitation spectra studies, it is the tryptophan environment which is relaxing around the excited fluorophore, while in the fluorescence anisotropy experiments, the displacement of the emission dipole moment of the tryptophan which is monitored. Figure 8.35 displays the normalized fluorescence spectra of 15 µM of α_1-acid glycoprotein in presence of 10 µM of calcofluor obtained at three excitation wavelengths (295, 300 and 305 nm). Spectrum c obtained at λ_{ex} equal to 305 nm has a shape globally similar to spectra a and b, obtained at λ_{ex} equal to 295 and 300 nm, but with a shoulder at 352 nm that is more evident. Therefore, it appears in the first instance that the Trp residues of α_1-acid glycoprotein are behaving differently to different excitation wavelengths. However, the bandwidth of the spectrum increases only by 2 nm when excitation occurred at 305 instead of 295 nm, indicating that the emission observed at λ_{ex} of 305 nm still occurs from hydrophobic and surface Trp residues. The shoulder observed is not typical to the presence of Trp residues present at the surface of the protein, since the emission spectrum of the hydrophobic Trp residues of crystals of α_1-acid glycoprotein obtained with λ_{ex} of 305 nm possesses a shoulder at 352 nm that is more evident than when excitation occurs at 295 nm. Also, the bandwidth of the spectrum does not change with the excitation wavelength. Therefore, the results obtained in figure 8.35 do not demonstrate that the Trp residues are behaving differently with the excitation wavelength.

Analyzing the spectra obtained at low concentration of calcofluor by a sum of two gaussians indicate that at the three excitation wavelengths, the positions of the maxima of the two spectra are 347 and 326 nm (data not shown for λ_{ex} equal to 300 and 305 nm). The absence of a shift in the position of the maximum with the excitation wavelength indicates that the Trp residues are not behaving differently with the excitation wavelength but means that Trp residues of α_1-acid glycoprotein displays continuous local motions independently of the global rotation of the protein. Red-edge excitation spectra experiments are confirmed by the Perrin plot (Fig. 8.36). In fact, in presence of low and high calcofluor concentrations, the rotational correlation time of the tryptophan residues in α_1-acid glycoprotein is found equal to 8.5 and 5.4 ns, respectively. These values are lower than the rotational correlation time of α_1-acid glycoprotein (16 ns) and indicate that Trp residues display free motions.

The red-edge excitation spectra and the Perrin plots showed that binding of calcofluor at low or / and high concentration to α_1-acid glycoprotein does not modify the internal dynamics of the Trp residues. At high calcofluor concentrations, the dynamics of the protein is maintained (Fig. 8.36) but its structure is altered (Fig. 8.28). In fact, the fluorescence excitation and emission spectra of the Trp residues (Fig. 2.11 and 8.28, respectively) are disrupted at high calcofluor concentrations, which is not the case when the experiments were performed at low calcofluor concentrations.

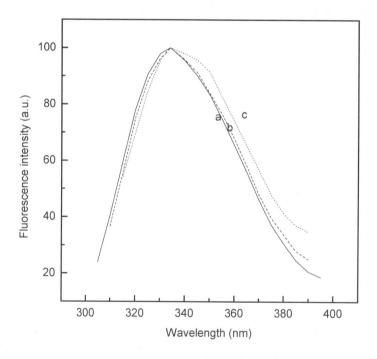

Figure 8.35. Normalized Steady-state fluorescence spectra of 15 μM of α_1-acid glycoprotein in the presence of 10 μM calcofluor recorded at three excitation wavelengths, 295 (a), 300 (b) and 305 nm (c). The fluorescence maximum is equal to 334 nm at the three excitation wavelengths. Source: Albani, J. R., 2001, Carbohydr. Res. 334, 141-151.

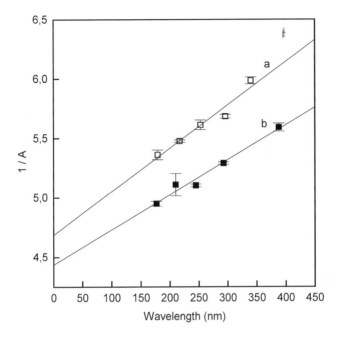

Fig. 8.36. Perrin plots for 15 μM α₁-acid glycoprotein in the presence of 10 μM Calcofluor (plot a) and in the presence of 150 μM Calcofluor (plot b) (λ_{ex} = 300 nm and λ_{em} = 330 nm). The data are obtained by thermal variation in the range of 5-35°C. Source: Albani, J. R., 2001, Carbohydr. Res. 334, 141-151

6. Förster energy transfer experiments from Trp residues to calcofluor white

In PBS buffer, free calcofluor white displays a fluorescence that is modified in presence of α₁-acid glycoprotein (λ_{ex} = 295 nm). Figure 8.37 shows emission spectra of a fixed amount of α₁-acid glycoprotein before and after addition of variable amounts of calcofluor white, the excitation wavelength being 295 nm. As the calcofluor concentration increases, its emission around 450 nm increases, while emission of the Trp residues (at 333 nm) decreases. An isoemissive point is observed at 395 nm. The analysis of the decrease of the fluorescence intensity of the Trp residues yields a stoichiometry for the protein-calcofluor complex equal to 1 (see figure 8.11).

The decrease of the fluorescence intensity of the Trp residues in presence of calcofluor white is the result of an energy transfer Förster type from the Trp residues to the extrinsic probe. The efficiency of this energy transfer depends on three parameters, the distance R between the donor (Trp residues) and the acceptor (calcofluor white), the spectral overlap between the fluorescence spectrum of the donor and the absorption spectrum of the acceptor and the orientation factor κ^2 which gives an indication on how the dipoles of acceptor in the fundamental state and donor in the excited state are aligned.

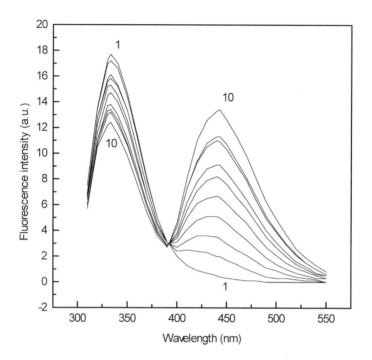

Figure 8.37. Fluorescence spectra of α_1-acid glycoprotein, before (spectrum 1) and after addition of calcofluor white. λ_{ex}, 295 nm. Aliquots of 2.1 μM of calcofluor white were added to a solution of 15 μM of α_1-acid glycoprotein. At spectrum 10, [calcofluor white] = 19 μM. Source: Albani, J. R., 2003, Carbohydr. Res. 338, 2233-2236.

Figure 8.38 displays the normalized fluorescence emission spectrum of Trp residues in α_1-acid glycoprotein (a) and absorption spectrum of calcofluor (b). From the overlap of the two spectra we have calculated the overlap integral J

$$J = \frac{F_{d(\lambda)} \varepsilon_{a(\lambda)} \lambda^4 d\lambda}{F_{d(\lambda)} d\lambda} \qquad (6.10)$$

J was found equal to $0.447 \times 10^{-4} M^{-1} cm^3$.

The Forster distance R_o (in cm) at which the efficiency of energy transfer is 50% was calculated with equation 12:

$$R_o^6 = 8.8 \times 10^{-25} (\kappa^2 n^{-4} Q_d J) \qquad (8.5)$$

where κ^2 is the orientation factor (= 2/3), n the refractive index (= 1.33) and Q_d the average quantum yield (= 0.064).

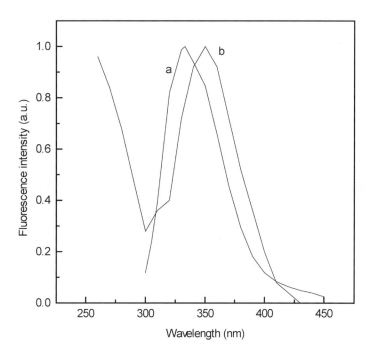

Figure 8.38. Fluorescence emission spectrum of Trp residues of 15 μM of α₁-acid glycoprotein (a) and absorption spectrum of 5 μM of calcofluor (b). Buffer, PBS, pH 7. Source: Albani, J. R., 2003, Carbohydr. Res. 338, 2233-2236.

R_o is found equal to 17.3 Å. This small distance suggests the presence of a weak energy transfer between Trp residues and calcofluor white.

Quenching efficiency (E) is equal to

$$E = 1 - \frac{I}{I_o} = 1 - \frac{\tau}{\tau_o} \qquad (6.7)$$

where τ and I are the mean fluorescence lifetime and intensity in the absence (τ_o I_o) and in the presence of calcofluor white (τ and I).

The value of E calculated at infinite concentrations of calcofluor white was obtained by plotting 1/E as a function of 1/[calcofluor white] (Fig. 8.39). E was found equal to 0.45.

The distance that separates the donor from the acceptor was calculated using equation 6.8:

$$R = (R_o^6 \ x \ (1-E)/E \ )^{1/6} \qquad (6.8)$$

R is equal to 17.95 Å.

The value of 0.45 found for E suggests that energy transfer mechanism from Trp residues to calcofluor occurs within a time close to the fluorescence lifetime. The constant rate of the energy transfer (k_t) can be calculated from equations 8.6 and / or 8.7:

$$k_t = (1 / \tau_o) (R_o / R)^6 \qquad (8.6)$$

and

$$k_t = 1 / \tau_o - 1 / \tau_r \qquad (8.7)$$

where τ_r is the radiative lifetime. $1 / \tau_r$ and $1 / \tau_o$ are equal to 0.4376×10^9 and $0.028 \times 10^9 \, s^{-1}$, respectively. Equations 15 and 16 yield very close values for k_t, $0.357 \times 10^9 \, s^{-1}$ and $0.4 \times 10^9 \, s^{-1}$, respectively. We notice that k_t is 41 to 46% of the sum of the three rate constants of equation 16 confirming the value (45%) found for E (Fig. 8.39).

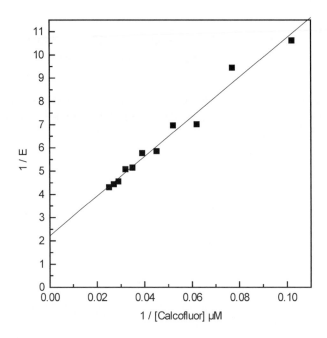

Figure 8.39. Determination of the efficiency (E) of energy transfer from Trp residues of α_1-acid glycoprotein to calcofluor white. The value of the efficiency determined at the extrapolation is equal to 0.45. Source: Albani, J. R., 2003, Carbohydr. Res. 338, 2233-2236.

Different mechanisms could account for the quenching of Trp residues fluorescence of α_1-acid glycoprotein upon binding of calcofluor white. One such mechanism is energy transfer between the tryptophans to Calcofluor white. Another possibility is a conformational change induced by calcofluor white binding, altering by that the environment around the tryptophans. Also, both mechanisms could occur simultaneously. At the concentrations of calcofluor white and α_1-acid glycoprotein used in the present work, important conformational changes around Trp residues can be ruled out. In fact, the position of the maximum (333 nm) of the Trp residues does not change in presence of calcofluor white (Fig. 8.37). The presence of an isoemissive wavelength means that the stoichiometry of the α_1-acid glycoprotein – calcofluor white complex is 1:1. This result is in good agreement with our other results (Fig. 8.28).

Calcofluor white interacts preferentially with the carbohydrate residues of α_1-acid glycoprotein and thus binds to the glycosylation site of the protein and does not bind directly to the protein matrix. This binding property of the probe and the presence of the carbohydrate residues around the extrinsic fluorophore decrease the quenching efficiency of the fluorescence of Trp residues. Although energy transfer does exist between the Trp residues and calcofluor, it is weak as it is revealed by the values of E (0.45) and R_o (17.3 Å). The calculated distance R (17.95 Å) between donor and acceptor has no physical meaning because of the presence of three potential donor tryptophan residues located in different areas of the protein. The presence of three Trp residues renders the analysis of the data very complex. In fact, we are observing an energy transfer mechanism that is occurring from the three Trp residues to the acceptor molecules. In this case, the distance measured does not allow establishing a structural map. However, it can be used to follow important structural modifications within the protein.

7. Relation between the secondary structure of carbohydrate residues of α_1-acid glycoprotein and the fluorescence of the Trp residues of the protein

α_1-Acid glycoprotein consists of two moieties, the protein and the carbohydrate residues. Both moieties are physiologically interrelated and one cannot dissociate the structure or/and the dynamics of one moiety from the other. The two moieties are important for the physiological functions of α_1-acid glycoprotein. We have found that asialylation of α_1-acid glycoprotein modifies the secondary structure of the carbohydrate residues and thus the fluorescence properties of the Trp residues should differ whether α_1-acid glycoprotein is sialylated or asialylated. In order to study the possible relation between the secondary structure of the carbohydrate residues of α_1-acid glycoprotein and the fluorescence of the Trp residues of the protein, we calculated the efficiency of quenching and the radiative and non-radiative rate constants.

Quantum yields of the sialylated and asialylated α_1-acid glycoprotein are 0.0645 and 0.0385, respectively. Since emission of α_1-acid glycoprotein can be interpreted in terms of two contributions (surface and hydrophobic domains), it would be possible to estimate the quantum yield of each class of Trp residues.

α_1-Acid glycoprotein contains 3 Trp residues that contribute to the fluorescence of the protein. If we consider that the two hydrophobic Trp residues participate equally to the "blue" emission, thus one can write:

$$Q = \frac{2\,Q_{blue} + Q_{red}}{3} \qquad (8.8)$$

Table 8.2 shows the values of Q, Q_{blue} and Q_{red} for the sialylated and asialylated α_1-acid glycoproteins.

Table 8.2. Fluorescence quantum yield of the global emission (Q), the red Trp residues (Q_{red}) and the hydrophobic Trp residues (Q_{blue}) for the sialylated and asialylated α_1-acid glycoproteins. The values are from 6 experiments.

	sialylated α_1-acid glycoproteins	asialylated α_1-acid glycoproteins
Q	0.0645 ± 0.005	0.0385 ± 0.0036
Q_{red}	0.0780 ± 0.007	0.0462 ± 0.0085
Q_{blue}	0.0585 ± 0.006	0.0346 ± 0.0022

Source: Albani, J.R. 2003, Carbohydr Res. 338, 1097-1101.

We can notice that the quantum yields of asialylated α_1-acid glycoprotein are lower than those obtained for the sialylated protein. Therefore, asialylation of the protein induces a quenching of the fluorescence quantum yield of α_1-acid glycoprotein of both classes of Trp residues.

Table 8.3 shows the values of the fluorescence lifetimes of the Trp residues of the sialylated and asialylated α_1-acid glycoproteins. We notice that the asialylation of α_1-acid glycoprotein induces a quenching of the fluorescence lifetimes. The lifetimes found for Trp residues in α_1-acid glycoprotein are the result of the combined emission of the three Trp residues since the three residues contribute to the global emission of the protein. These values are within the same range of those found for the most Trp residues in proteins. However, we are going to see in paragraph 8.b of this chapter, that it is possible to attribute the longest lifetime to the surface residue (Trp-160 residue) and the 1.4 ns lifetime to the blue or hydrophobic Trp-residues.

Table 8.3. Fluorescence lifetimes (in ns) of the Trp residues of the sialylated and asialylated α_1-acid glycoproteins.

Sialylated α_1-acid glycoprotein

τ_1	f_1	τ_2	f_2	τ_3	f_3	τ_o
0.354	0.101	1.664	0.66	4.638	0.238	2.285
± 0.034	± 0.05	± 0.072	± 0.03	± 0.342	± 0.01	

asialylated α_1-acid glycoprotein

τ_1	f_1	τ_2	f_2	τ_3	f_3	τ_o
0.197	0.07	1.42	0.65	3.61	0.28	1.948
± 0.037	± 0.01	± 0.05	± 0.02	± 0.15	± 0.03	

Source : Albani, J.R. 2003, Carbohydr Res. 338, 1097-1101.

The mean fluorescence lifetime is the second order mean (Lakowicz, 1999):

$$\tau_o = \Sigma \, f_i \, \tau_l \tag{2.35}$$

and

$$f_i = \beta_i \tau i \; / \; \Sigma \; \beta_i \tau i \tag{2.36}$$

where β_i are the preexponential terms, τ_i are the fluorescence lifetime and f_i the fractional intensities. $\lambda_{em} = 330$ nm.

The efficiency of quenching is equal to

$$E = 1 - \frac{\tau_t}{\tau_o} = 1 - \frac{Q_t}{Q_o} \tag{6.7}$$

where τ and Q are the mean fluorescence lifetime and quantum yield in the absence (τ_t Q_t) and in the presence of sialic acid (τ_o and Q_o).

Table 8.4 shows the values of E calculated from the mean lifetime and quantum yield and from the lifetime and quantum yield of the surface and hydrophobic Trp residues.

Table 8.4.Values of E calculated from the mean lifetime and quantum yield and from the lifetimes and quantum yields of the surface (longest lifetime) and hydrophobic Trp residues ($\tau =$ 1.8 or 1.4 ns).

$E_{(\tau_o)}$	$E_{(\tau)\,(blue)}$	$E_{(\tau)\,(red)}$	$E_{(Q)}$	$E_{(Q)\,(blue)}$	$E_{(Q)\,(red)}$
0.148	0.147	0.220	0.403	0.408	0.408

Source : Albani, J.R. 2003, Carbohydr Res. 338, 1097-1101.

The value of E calculated from the quantum yield is higher than that measured from the lifetime. Therefore, fluorescence lifetimes of Trp residues of α_1-acid glycoprotein are the result of an energy transfer Förster type or by electron transfer to neighboring amino acids and of the molecular collisions of the Trp residues with their environments.

This local motion is indicated by the fact that the value of E calculated from the quantum yields of each class of Trp residues is equal to that calculated from the mean quantum yield and is higher than the value of E obtained from the lifetimes. This means that the two classes of Trp residues display local motions, a result that confirms the one obtained with the anisotropy measurements.

The fact that the fluorescence parameters of α_1-acid glycoprotein decrease in the asialylated form indicates that the carbohydratess are closer to the protein matrix in the absence of the sialic acids than in their presence. Removal of the sialyl groups causes the carbohydrate residues to contract closer to the core of the protein. This is likely to be the case, since the sialyl group is charged. Thus, the Trp residues microenvironment of α_1-acid glycoprotein should be more polar in the asialylated form.

In this case, the radiative rate constant k_r

$$k_r = 1 / \tau_r = Q / \tau_o \qquad (8.9)$$

and the non- radiative rate constant k_{nr}

$$k_{nr} = (1 / Q) / \tau_o \qquad (8.10)$$

of the sialylated and asialylated α_1-acid glycoprotein are not the same. The values of k_r and k_{nr} for the sialylated and asialylated α_1-acid glycoprotein calculated from the mean values of Q and τ and of each class of Trp residues are shown in table 8.5.

Table 8.5. Values of the radiative (k_r) and non radiative (k_{nr}) rate constants calculated from the mean values of the quantum yield (Q) and lifetime (τ_o) and from the values of Q and τ of the red Trp residues and the hydrophobic Trp residues for the sialylated and asialylated α_1-acid glycoproteins.

Sialylated α_1-acid glycoprotein

$<k_r>$	$<k_{nr}>$	$k_{r\,(red)}$	$k_{r\,(blue)}$	$k_{nr\,(red)}$	$k_{nr\,(blue)}$
0.028	0.409	0.0168	0.035	0.198	0.566

Asialylated α_1-acid glycoprotein

0.0197	0.494	0.0128	0.0244	0.2642	0.679

Source : Albani, J.R. 2003, Carbohydr Res. 338, 1097-1101.

    The radiative rate constant of indole increases when the polarity of the environment decreases (Privat et al. 1979). We notice from table 8.5 that the k_r of the sialylated α_1-acid glycoprotein is higher than that of the asialylated protein. Thus, presence of sialic acids decreases the polarity of the environment of the Trp residues. This suggests that the carbohydrate residues are closer to the Trp residues in absence of sialic acids.

    This is also observed for the surface (Trp-160) and hydrophobic residues, since the k_r calculated for each class of Trp residue are higher for the sialylated α_1-acid glycoprotein than for the asialylated protein. This means that asialylation of α_1-acid glycoprotein affects the secondary structure of the whole carbohydrates of the protein, i.e., asialylation induces a motion of the carbohydrates toward the protein matrix. Since the polarity of the microenvironments of hydrophobic Trp residues are also affected, then these residues are close to the carbohydrate residues and are not buried in the protein core. These results are in good agreement with the model we have already described: the N-terminal side chain adopts a spatial conformation so that a pocket in contact with the buffer is induced and to which Trp-25 residue belongs. The five carbohydrate units are linked to the pocket, i.e., the glycosylation site belongs to the pocket.

Comparing the k_{nr} values obtained for the sialylated and asialylated α_1-acid glycoprotein, we notice that the k_{nr} of the sialylated protein is lower than that of the asialylated one (see table 8.5). Thus, the molecular interaction in α_1-acid glycoprotein is more important in the asialylated protein, i.e., in the absence of sialic acid residues the carbohydrate residues are closer to the protein matrix and thus to the Trp-25 and Trp-160 residues.

All our data, lifetime, quantum yield, k_r and k_{nr} indicate that in the asialylated α_1-acid glycoprotein, compared to the sialylated protein, a fluorescence quenching of the Trp-residues is occurring as the result of a strong interaction with the carbohydrate residues. In the asialylated form, the carbohydrate residues are closer to the protein surface and matrix than in the sialylated form.

The results obtained and the conclusions drawn from the data on the Trp residues are confirmed by the fluorescence lifetimes measurements obtained on calcofluor. In fact, the mean fluorescence lifetime of calcofluor bound to the carbohydrate residues of sialylated α_1-acid glycoprotein decreases from 4.8 ns to 3.9 ns in the asialylated one. The efficiency of quenching is equal to 0.1875, a value close to that (0.148) found for the Trp residues. Therefore, in presence of sialic acids, the spatial conformation of the carbohydrate residues is different from that observed in their absence. Asialylated α_1-acid glycoprotein possesses the carbohydrates structure and backbone closer to the protein matrix than the sialylated form, inducing by that a decrease in the fluorescence lifetime of calcofluor as the result of the molecular interaction between the carbohydrate residues and the protein matrix.

8. Effect of the secondary structure of carbohydrate residues of α_1-acid glycoprotein on the local dynamics of the protein

8a. Fluorescence emission intensity of α_1-acid glycoprotein as a function of temperature

Figure 8.40 displays the fluorescence intensities of sialylated (8 μM) (a) and asialylated (10 μM) (b) α_1-acid glycoprotein as a function of temperature. We notice that the fluorescence intensities of the sialylated protein are higher than those observed for the asialylated form, confirming that quantum yield of sialylated α_1-acid glycoprotein is much higher than that of the asialylated protein. The global profile observed for the two proteins is not at all a monotonous function of temperature and some differences in these two profiles do exist. For asialylated α_1-acid glycoprotein, from –46 to –16°C, the fluorescence intensity is quenched by increasing temperature. Between –16 and –7°C, the intensity increases with temperature. Above –7°C, increasing temperature once again results in intensity quenching. This profile was also observed for ATCase, although the temperature at which the change of the fluorescence intensity occurs differs from that of α_1-acid glycoprotein (Royer et al. 1987). We can interpret this anomalous temperature behavior as follows. Between –46 and –16°C, increasing temperature causes an increase in motion that brings the tryptophan residues in contact with a nearby quencher. Therefore, one may assume ground-state complex formation. At temperature between –16 and –7°C, thus upon increasing the temperature within this range, tryptophan motions increase, and the probability of ground-state complex formation decreases. Therefore, we observe an increase in the fluorescence intensity. Finally, at temperatures higher than 7°C, motions continue to increase, collisional quenching becomes more important and fluorescence intensity decreases. One should indicate that quenching induced by ground-state complex decreases the fluorescence intensity but lifetimes. However, collisional quenching will induce a decrease in both intensities and lifetimes.

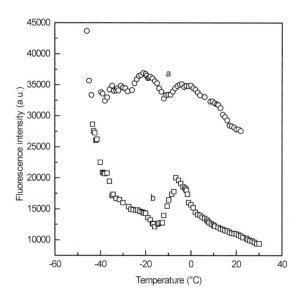

Figure 8.40. Fluorescence intensity of Trp residues of sialylated (a) and asialylated (b) α_1-acid glycoprotein vs temperature. Source: Albani, J. R. 2004, Chemistry and Biodiversity, 1, 152-160. Authorization of reprint accorded by Wiley Interscience.

For sialylated α_1-acid glycoprotein, although the intensity modifications with temperature can be explained by the presence of different types of quenching, the global profile is more complex than in the asialylated protein as the result of the difference in the local structures, around mainly the surface Trp residue.

The global profile of the fluorescence intensity with temperature appears to be common to different proteins such as ATCase and α_1-acid glycoprotein. In the present work, the profile was obtained from the fluorescence of the three Trp residues. This does not mean that the three Trp residues follow the same intensity variation with temperature. In fact, for example, in ATCase each of the two Trp residues possess its own profile (Royer et al. 1987). The fact that the global profile is not a monotonous function of temperature indicates clearly that the amino acids of a protein can undertake different types of interactions with the neighboring amino acids and can possess different types of local motions. This conclusion is in good agreement with our results showing that fluorescence lifetimes of Trp residues of α_1-acid glycoprotein are the result of an energy transfer Förster type or by electron transfer to neighboring amino acids and of molecular collisions of the Trp residues with their environments (Table 8.5).

Comparison of the profiles of sialylated and asialylated α_1-acid glycoprotein shows that the profile of asialylated protein is more structured than that observed for the sialylated protein. The peak is sharper and well-defined while the profile of the sialylated protein is larger and less well structured. This means that in the sialylated α_1-acid glycoprotein, the two classes of Trp residues or at least one of them (in this case it would be the Trp residue of the protein surface), is less hindered by the surrounding amino acid residues.

In the asialylated α_1-acid glycoprotein, the carbohydrate residues are closer to the protein surface than in the sialylated protein. Therefore, the amino acids of the protein surface will be closer one to each other than in the sialylated form. This will induce an intra-protein dynamics energy that is more coherent and better defined than in the case of the sialylated protein. This leads us to conclude that the sialylated α_1-acid glycoprotein has a conformation that is looser than the asialylated protein, at least at the surface of the protein.

8b. Fluorescence emission anisotropy of α_1-acid glycoprotein as a function of temperature

Perrin plot and red-edge excitation spectra experiments performed on Trp residues of sialylated and asialylated α_1-acid glycoprotein have shown that in both proteins the intrinsic fluorophore displays local motions and are surrounded by a flexible environment. However, the above two mentioned methods yield information on the mean residual motion and can in no way give an indication on the dynamics of each class of Trp residues. In fact, the exposed tryptophan residue should be expected to rotate much more freely than the hydrophobic residues. In order to study the dynamics behavior of each class of Trp residues, steady-state measurements of emission anisotropy at different temperatures (-45 to + 30°C) can be carried out. This method (the Weber's method) known also as the Y-plot, allows deriving parameters characteristic of the environment of the rotating unit, such as the thermal coefficient of the frictional resistance to the rotation of the fluorophore.

In order to study the dynamics behavior of each class of Trp residues, we measured the anisotropy between –46 and 28°C. The classical Perrin plot was modified by expanding the viscosity term with temperature around 0°C :

$$\eta = \eta_o \, e^{\, b(To-T)} \tag{8.11}$$

where η_o is the viscosity at T_o (0°C) and b is the thermal coefficient of the viscosity. Replacing this expression for η in the Perrin equation and taking logarithms yield (Weber et al. 1984)

$$Y = Ln\left(\, (A_o/A) - 1 \, \right) - Ln\left(RT\tau_o/V \right) = -Ln\,\eta_o + b\,(T - T_o) \tag{8.12}$$

where A_o is the limiting anisotropy; A is the value of the anisotropy at each Kelvin temperature T ; R is the gas constant; τ_o is the fluorescence lifetime at each temperature and V is the rotational volume (= 36.5 ml / mol). V was calculated from the classical Perrin plot . A plot of Y vs T gives a slope of b, the thermal coefficient to resistance to a rotation. For fluorophores in pure solvents, Y vs T is a straight line and the slope b_s is dependent upon the viscosity of the solvent. For fluorophores in peptides and proteins, Y vs T gives two slopes: the first one is obtained at low temperatures and is equal to b_s, the second is obtained above a critical temperature T_c, specific to each protein or peptide. At this critical temperature, protein-solvent interactions cease to be important and intra-protein interactions dominate. For proteins with two intrinsic fluorophore molecules, three slopes are observed, one solvent slope b_s, and two protein slopes b_1 and b_2. This feature is obtained only if the critical temperatures T_{c1} and T_{c2} differ by more than 10°C and if the values of b_1 and b_2 are sufficiently different.

Figure 8.41 shows the Y-plot for Trp residues in sialylated (b) and asialylated (a) α_1-acid glycoprotein.

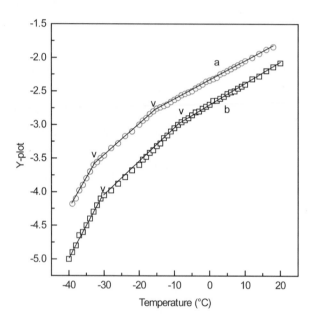

Figure 8.41. Y vs. temperature for sialylated (b) and asialylated (a) α_1 -acid glycoprotein. $\lambda_{ex} =$ 300 nm. A bandpass filter, with the maximum at 334 nm, was used for the emission. Protein concentrations were equal to 8 μM. The arrowheads indicate the positions of the critical temperatures. Sources : plot a: Albani, J. R. 1996, Biochim. Biophys. Acta., 1291, 215-220. Plot b: Albani, J. R. 2004, Chemistry and Biodiversity, 1, 152-160.

The solvent slope b_s is equal to 10% for both proteins. This value in the same range of that (8.5-9.2%) found for Trp residue in pure solvent (Weber et al. 1984; Scarlata et al. 1984; Rholam et al.1984.) The first critical temperature observed was found at −30°C and −33°C for the sialylated and asialylated forms, respectively. At this temperature, the amplitude (α) and the angle (θ) of the rotation of the buried Trp residues calculated from Eqns 5.19 and 5.20 :

$$A(0) / A_o = 1 - \alpha \qquad\qquad (5.19)$$

where α is the residual motion of the fluorophore, $A(0)$ is the anisotropy measured at the critical temperature ($A(0) = A_2 = 0.273$ and 0.252 for the sialylated and asialylated proteins, respectively). The average angular displacement θ of the fluorophore inside the protein is:

$$\text{Cos}^2\,\theta = 1 - 2\alpha/3 \tag{5.20}$$

For the sialylated α_1-acid glycoprotein, the values of α and θ are 0.032 and $8°$ of arc, respectively. These values are close to those (0.045 and $10°$ of arc) found for the hydrophobic Trp residues in the asialylated α_1-acid glycoprotein.

A second break in the Y-plot occurs at a critical temperature equal to $-9°C$, instead of $-16°C$ observed for the asialylated form. At $-9°C$ (in the sialylated protein), the values of α and θ corresponding to the amplitude and the angle of rotation of the Trp residue of the protein surface of the sialylated protein were 0.077 and $13°$ of arc. These values are close to those (0.084 and $14°$ of arc) obtained for the asialylated protein.

The region before the first temperature break is characteristic of the buried Trp residues, the second region going from the first to the second temperature breaks is characteristic of the surface Trp residue and the one that begins at the second break is characteristic of the solvent. The value of b_1 is a linear-weighted combination of each temperature of the protein slope of the first tryptophan b_{u1} and the solvent slope of the second :

$$b_1 = f_1\,b_{u1} + f_2\,b_s \tag{8.13}$$

Similarly, the second slope b_2 is a linear-weighted combination of the two protein slopes, b_{u1} and b_{u2}:

$$b_2 = f_1\,b_{u1} + f_2\,b_{u2} \tag{8.14}$$

where f_1 and f_2 are the fractional contributions of each tryptophan class of the total fluorescence intensity at each temperature. b_u depends on the viscosity of the protein medium surrounding the tryptophan and on the specific interactions of the tryptophan with its neighboring amino acid residues.

Assuming that f_1 and f_2 do not change significantly with temperature, one can calculate from Fig. 8.41 the values of b_{u1} and b_{u2}. As the two components with lifetimes equal to 1.66 and 4.64 ns represent 90% of the fluorescence intensity decay, we normalized their fractional contributions (0.7 and 0.3 instead of 0.66 and 0.24, respectively) and we neglected the third short-lifetime component. In solving for b_{u1}, the lifetime of 1.66 ns ($f_1 = 0.7$) was assumed to correspond to the buried Trp residues. The inverse assumption yields negative value for b_{u1}. For the sialylated α_1-acid glycoprotein, our analysis gives values of 0.0286 and 0.0666 for b_{u1} and b_{u2}, respectively, instead of 0.023 and 0.039 obtained for the asialylated α_1-acid glycoprotein (Albani, 1996).

Y-plot experiment clearly indicates that the dynamics properties of the surface Trp residue in the sialylated protein differ from those observed in the asialylated form. In fact, while the first critical temperatures are very close for both proteins (-33 and $-30°C$), the second critical temperatures characteristic of the surface Trp residue differ.

This temperature is –16°C and –9°C in the asialylated and sialylated proteins, respectively. This means that intra-protein interactions in the vicinity of the surface Trp residue are much more important in the sialylated form of the protein. This could be explained by the fact that, if the surface Trp residue has an important freedom of rotation in the sialylated protein, its interaction with the neighboring amino acids occurs easily, a fact that will induce the presence of a critical temperature higher than if the same Trp residue has less freedom of rotation such as in the asialylated protein.

The presence of sialic acids increases b_{u2} from 0.039 to 0.0666. Thus, in presence of sialic acids, the resistance to rotation of the surface Trp residue is less important than in their absence. Therefore, in the asialylated α_1-acid glycoprotein, the surface Trp residue has much higher degree of freedom than in the asialylated protein. This is explained by the fact that in the asialylated α_1-acid glycoprotein, the carbohydrate residues are closer to the protein surface and thus closer to the surface Trp residue. This will increase its contact with the neighboring amino acids decreasing the quantum yields and the fluorescence lifetimes (Tables 8.2 and 8.3). The radiative k_r and non-radiative k_{nr} constants of both surface and hydrophobic Trp residues are dependent upon the sialylation of the protein. In fact, k_r values of the surface and of hydrophobic Trp residues in the sialylated protein are higher than those in the asialylated protein (Table 8.5). The presence of sialic acid decreases the polarity of the environment of the three Trp residues. This suggests that the carbohydrate residues are closer to the three Trp residues in the absence of sialic acids.

Comparing the k_{nr} values obtained for the sialylated and asialylated α_1-acid glycoprotein, we notice that the k_{nr} of the sialylated protein is lower than that of the asialylated one (Table 8.5). Thus, the molecular interactions in α_1-acid glycoprotein are more important in the asialylated protein, i.e., in the absence of sialic acid residues the carbohydrate residues are closer to the protein matrix and thus to the Trp-residues.

Since in the Y-plot, we did not observe any modification in the value of b_{u1}, we may conclude that the interaction between the carbohydrate residues and the protein matrix does not necessarily induce the compactness of the hydrophobic domain of the protein moiety. This compactness is however observed for the protein surface.

In the Y-plot experiments, we measured the anisotropy every 1°C. Since the temperature is known to be $\pm$ 0.1°C, T_c is then determined to at least $\pm$ 1°C. Also, the large number of temperatures studied permits a much greater certainty of the values of the anisotropy at each temperature. The error on the anisotropy lies between $\pm$ 0.00025 and $\pm$ 0.0015. Repeating the experiments at least 3 to 5 times allows also minimizing the error. Thus, the error obtained on b_u does not exceed $\pm$ 0.6%.

It is important to mention that the Weber's method allows the measurement of an important parameter, which is the thermal coefficient of the frictional resistance b_u indicating that the immediate environment opposes to the rotation of the fluorophore (Weber et al. 1984). This parameter cannot be measured by other methods such as the classical Perrin plot, the anisotropy decay with time or with the red-edge excitation spectra. b_u is determined in the proteins by the interactions of the fluorophore with the amino acids in the immediate vicinity. The other methods do not give an exact idea on the definition of the microenvironment of the fluorophore. Is it one, two, few amino acids around the Trp residue or an important region in contact with the fluorophore?

9. Tertiary structure of α₁-acid glycoprotein : first model describing the presence of a pocket

The absence of crystallographic data on α₁-acid glycoprotein is the result of the difficulty in obtaining stable protein crystals. Therefore, we performed different types of fluorescence experiments in order to be able to describe a model for the tertiary structure of α₁-acid glycoprotein.

We show from the experiments we described with calcofluor that the carbohydrate residues possess a secondary structure that is modified whether the protein is asialylated or not. Calcofluor binds to the carbohydrate residues and thus to a hydrophilic binding site.

Also, we found that TNS binds to a hydrophobic site on α₁-acid glycoprotein and thus to a site different from that of calcofluor.

α₁-Acid glycoprotein possesses three Trp residues, one at the surface (Trp-160) and two in hydrophobic areas of the protein (Trp-25 and Trp-122). The three Trp residues contribute to the global fluorescence of the protein. The emission maxima was found equal to 347 nm (Trp-160), 337 nm (Trp-122) and 324 nm (Trp-25).

Red-edge excitation spectra and anisotropy experiments indicate clearly that the Trp residues and the calcofluor displays free local motions, while TNS is tightly bound to the protein. Also, experiments with calcofluor showed that the sialic acid residue displays free motion while the carbohydrate residues that are close to the glycosylation site are tightly rigid.

These results are not sufficient by themselves to give a description of the tertiary structure of α₁-acid glycoprotein. Other complementary experiments are necessary. The main problem is the presence of the Trp residues in different regions of the protein. Therefore, we decided first to perform quenching resolved emission anisotropy experiments with cesium at 20°C and anisotropy measurements from –45° to + 20°C. When a protein contains two classes of intrinsic fluorophore, one at the surface of the protein and the second embedded in the protein matrix, fluorescence intensity quenching with cesium allows obtaining the spectra of these two classes. A selective quenching implies that addition of quencher induces a decrease in the fluorescence observables (intensity, anisotropy and lifetime) of the accessible class. At high quencher concentration the remaining observables measured will reflect essentially those of the embedded fluorophore residues. In this case, one can determine the fraction of fluorescence intensity that is accessible (f_a). Knowing f_a along the emission spectrum will allow us to draw the spectrum of each class of fluorophore.

Upon binding of a ligand on the protein and if this binding induces structural modifications inside the protein, the intensities of the spectrum of the classes can be modified and in the case of the Trp residues, the emission maximum can shift to the higher or lower wavelengths.

When a protein contains fluorophore residues located at the surface and in the core as it is the case for α₁-acid glycoprotein, the results obtained from the classical Perrin plot contain contributions from all residues. In order to obtain information on the motion of each class of fluorophore residues, one may follow anisotropy and intensity variations as a function of quencher concentration (Quenching Resolved Emission Anisotropy) or / and anisotropy and lifetime variations with temperature (-50 to +35°C) (Weber method).

QREA method allows determining the fraction f_b and the intrinsic anisotropy of the non accessible residues and knowing the value of the anisotropy in absence of quencher, one can determine the limiting anisotropy of the accessible residues.

The Weber method provides parameters characteristic of the environment of the rotating unit such as the limiting anisotropy and the thermal coefficient to resistance to a rotation. Also, the method permits to separate the motion of different classes of fluorophore.

The experiments were done in absence and presence of progesterone. The results obtained are shown in Tables 8.6 to 8.8.

Table 8.6. Anisotropies (A) and the fractional resistance to a rotation (b_u) of Trp residues of α_1-acid glycoprotein measured with the Weber method in absence and presence of progesterone

	Hydrophobic Trp residues	Trp-160 residue
Progesterone (-)	$A_2 = 0.252$	$A_1 = 0.242$
Progesterone (+)	$A_2 = 0.233$	$A_1 = 0.201$
Progesterone (-)	$b_{u2} = 0.039$	$b_{u1} = 0.023$
Progesterone (+)	$b_{u2} = 0.0127$	$b_{u1} = 0.0308$

Sources: Albani J. R., 1996 Biochimica Biophysica Acta 1291, 215-220 and Albani J. R., 1997, Biochimica Biophysica Acta 1336, 349-359.

Table 8.6 shows that binding of progesterone to α_1-acid glycoprotein decreases the values of the anisotropies of both Trp residues. Thus, the motions of the internal fluorophores increase. Also, in presence of progesterone, the resistance to rotation of hydrophobic Trp-residue is less important than that of the Trp-106 residue. Since binding of progesterone occurs near hydrophobic Trp-residues, changes observed near Trp-160 residue are indicative of long-distance effects between the binding site and the protein surface. The values of the anisotropies of the two classes of Trp residues, whether in the absence or the presence of progesterone, are very close. This means that the two classes display very close motions.

The anisotropies of both Trp-residues of α_1-acid glycoprotein measured at 20°C are very close (Table 8.7), (a result identical to that obtained with the Weber method, see Table 8.6) indicating that the two Trp residues are highly mobile. This result is confirmed by the values of the rotational correlation times (3 and 5 ns) lower than the global rotational correlation time of α_1-acid glycoprotein. When the Trp residue is embedded in the protein core and does not show any residual motions such as in the *Lens culinaris* agglutinin, its rotational correlation time will be close to that of the protein and its anisotropy will be higher than that of the surface Trp residue.

We notice however, that the anisotropies are more important in the Weber method. This can be explained by the fact that in the Weber method, anisotropies were calculated at low temperatures (-33 and -16°C for the buried and surface Trp residues, respectively), while in the QREA method, the anisotropies were measured at 20°C. At low temperatures, the global rotation of the protein and the local motions around the Trp residues are decreased.

Table 8.7. Comparison of the anisotropies of the two classes of Trp residues of α_1-acid glycoprotein and Lens culinaris agglutinin. Measurements were performed at 20°C with the Quenching Resolved Emission Anisotropy method.

α_1-acid glycoprotein (Φ_P = 17 ns)

	hydrophobic Trp residue	Trp-160 residue
Anisotropy	0.178	0.155
Φ_F	3 ns	5 ns

Lens culinaris agglutinin (Φ_P = 21 ns)

	Embedded Trp-residues	Surface Trp-residues
A	0.257	0.112
Φ_F	17 ns	2.715 ns

Sources : Albani, J. R., 1996, Journal of Fluorescence, 6, 199-208 and Albani, J. R. 1999, Spectrochimica Acta , Part A, 55, 2353-2360.

Table 8.8 shows that binding of progesterone on α_1-acid glycoprotein induces a shift in the emission maximum of both Trp-residues. Therefore, a modification of the local structure occurs near the Trp-residues upon binding of progesterone to the protein.

Table 8.8. Position of the emission maximum of Trp residues of α_1-acid glycoprotein in absence and presence of progesterone.

	Hydrophobic Trp- residue	Trp-160 residue
Progesterone (-)	324 nm	350 nm
Progesterone (+)	330 nm	345 nm

Source: Albani J. R., 1997, Biochimica Biophysica Acta 1336, 349-359.

The fact that binding of progesterone on α_1-acid glycoprotein induces a modification of the local structure and the dynamics of the protein indicates that these modifications are important for the binding of the ligand.

Although we have shown that the three Trp-residues contribute to the global fluorescence of α_1-acid glycoprotein (paragraph 5c of this chapter) quenching resolved emission anisotropy and the Weber method allowed giving a description on the mean local dynamic of the Trp-residues. The dynamic of the surface Trp residue is well separated from the two other Trp residues.

Results obtained reveal that both classes of Trp residues exhibit very close local motions. In fact, both QREA experiments and the Weber's method show that the difference between the anisotropy of the surface and that of the buried Trp residues is not important. Although the values indicates that free motions of the surface Trp residue are slightly more important than those of the buried Trp residues, the global dynamics of both classes are close. Therefore, there are no significant constraint domains in α_1-acid glycoprotein within the microenvironment of the Trp residues.

This observation is different from what we know on the dynamics of proteins in general, where the protein matrix has in general more restricted motions than its surface. For examples, (1) the protein matrix of lysozyme shows restricted motions while the protein surface displays a high degree of freedom (the values of the anisotropies of the surface Trp residue and of the buried one are 0.16 and 0.26, respectively) (Eftink, M, 1983).

(2) We measured recently the anisotropies of the Trp residues buried in the protein core of the lectin *Lens culinaris* agglutinin (LCA) and of those at its surface. We found that the protein core has restricted motions compared to the protein surface (the values of the anisotropies of the surface and of the buried Trp residues are 0.112 and 0.257, respectively) (Albani, 1996).

Progesterone binds to the protein core of α_1-acid glycoprotein near Trp 25 residue that belongs to the N-terminal side chain. The five carbohydrate units are linked to this portion of the polypeptide moiety rendering this section of the protein hydrophilic. Reading these last three lines indicates that two major problems arise: 1) How a hydrophobic molecule such as progesterone can bind to hydrophobic core of the protein without going through the hydrophilic protein surface? 2) The N-terminal fragment is hydrophobic and in the same time it is hydrophilic since the glycosylation site belongs to this fragment. In other term, the N-terminal fragment does not constitute the protein core, otherwise the hydrophilic carbohydrate residues will be linked within the protein core and then go up to the surface of the protein. This feature belongs to the science fiction than to the exact science. Thus, we have proposed the following model : the N-terminal fragment should be in contact with the solvent, and would adopt a spatial conformation so that a pocket in contact with the buffer is induced, and to which Trp 25 and possibly Trp-122 residues belongs. The microenvironment of the two hydrophobic Trp residues is not compact or rigid (low anisotropy and thus high mobility). The five carbohydrate units are linked to the pocket. Thus, progesterone can bind directly to this pocket since it diffuses from the buffer immediately to its binding site within or at the surface of the pocket. This binding site will be mainly hydrophobic since progesterone is hydrophobic.

This model is in good agreement with the fact that binding site of progesterone on α_1-acid glycoprotein contains both a hydrophobic and a polar surface of the protein. Also, the possible presence of this pocket has been evoked from binding experiments as a function of the pH (Urien et al.1991).

Binding site of TNS on α_1-acid glycoprotein is within the pocket in a hydrophobic and a rigid region. Since, the five carbohydrate units are linked to the pocket rendering this section of the protein hydrophilic, it seems that the pocket possesses two different domains: one hydrophobic, the other hydrophilic. TNS and progesterone will bind specifically to the hydrophobic domain of the pocket.

The secondary structure of the carbohydrate residues, absent in many other glycoproteins or when the carbohydrates are free in solution, would help to explain the physiological role of these residues. This secondary structure is probably related to the fact that the carbohydrates are linked to the pocket inducing by that a specific spatial conformation. It is difficult to predict from fluorescence studies the spatial structure of the carbohydrate residues.

10. Are there any other alternative fluorescence methods other than the QREA or the Weber's method to put into evidence the presence of a pocket within α₁-acid glycoprotein ?

In theory, QREA seems to be very simple. However, the measurements are difficult when using a selective quencher and where there is a small change in the steady-state anisotropy values. The reason for this small variation is that only a small fraction of the total fluorescence intensity is quenched, thus limiting the observed change in the steady-state anisotropy. This is not the case when a non-selective quencher such as oxygen is used. For example, fluorescence quenching of Trp residues in LADH by oxygen induces an increase in the steady-state anisotropy from 0.077 to 0.125 at 24°C (Lakowicz et al. 1983). Also, it was found that oxygen efficiently quenches the fluorescence of porphyrin-myoglobin. This quenching induces an increase in the steady-state anisotropy from 0.037 to 0.218 mM oxygen concentration to 0.0875 in the presence of 88 mM oxygen (Albani and Alpert, 1986). Also, using cesium ion as a quencher rather than iodide ion renders the experiments more difficult. In fact, efficiency of quenching with cesium is lower than that with iodide (Burstein et al. 1973). However, since α₁-acid glycoprotein is negatively charged, it is impossible to perform the quenching experiments with iodide. Quenching experiments can also be carried out with acrylamide. However, the problem with acrylamide is that it is a large molecule so that the quenching reaction involves physical contact between the quencher and the excited indole ring. Thus, the interaction between the two molecules can be kinetically described in terms of a collisional and a static component (Eftink and Ghiron, 1976). Therefore, to be sure that only dynamics quenching occurs, we performed the quenching experiments with cesium ion.

Quenching resolved emission anisotropy experiments could be performed at emission wavelengths in the blue (< 330 nm) and red (> 330 nm) portions of the spectrum to yield a more consistent data surface. However, this could be possible if at each edge of the fluorescence spectrum the emission occurs mainly from the buried or the surface Trp residues. Unfortunately, this is not the case since for example at 315 nm, the fractional contribution to the total fluorescence of the surface Trp residue is 42%.

Time-resolved anisotropy decay may also be used to obtain information on the dynamics of Trp residues. However, the technique does not give detailed information on the motion of each class of Trp residues. When the fluorophore has segmental motions, the anisotropy will be analyzed as a sum of exponential decays:

$$A(t) = \Sigma \, A_i \exp(-t / \Phi_i) \tag{8.15}$$

The Φ_i term contains the global rotation of the protein, the residual motion of the Trp residues and the motion of the microenvironment around the Trp residues. In a system with one Trp residue or with two Trp residues located in the same region, fluorescence anisotropy decay would give information on the dynamics of the area where the Trp residue(s) is present. For example, fluorescence anisotropy decay experiments were performed on the myosin rod from rabbit skeletal where two Trp residues per heavy chains are located in the same hydrophobic site (Chang and Ludescher, 1994), and on the Trp residue of the porcine pancreatic phospholipase A_2 located in the active site of the protein (Kuipers et al. 1991). When a protein contains Trp residues located at the surface and in the core, the results obtained from the fluorescence anisotropy decay measurements contain contributions from all the residues. In order to obtain information on the motion of each class of Trp residues with the anisotropy decay method, we may carry out the experiment in the presence of a very high concentration of a selective quencher of the surface Trp residue only, a condition that is very difficult to obtain, or we may combine site-directed mutagenesis and time-resolved fluorescence measurements.

Also, it will be possible to measure the fluorescence anisotropy decay at certain wavelengths that are specific to one class of fluorophore but not the other. The decay measurements performed on a protein where the surface Trp residue is oxidized, would give information on the dynamics of the buried Trp residues. This is true if oxidation of the Trp residue surface did not induce any modification in the local or/and global structure and dynamics of the protein.

α_1-Acid glycoprotein contains three Trp residues, one at the surface and two within a hydrophobic domain, thus a fluorescence anisotropy decay applied to such a system, would give information on the mean motion of the two classes of Trp residues. Also, it is difficult to separate the two motions by performing measurements at different wavelengths, since the fluorescence spectra of both classes overlap at all wavelengths. Site-directed mutagenesis on the protein is not possible for the moment. Moreover, it is not possible to work at infinite concentration of cesium ion, therefore, the best methods to study the dynamic behavior of each class of Trp residues of α_1-acid glycoprotein are for the moment the Weber's method and the quenching resolved emission anisotropy with cesium ion. Still, these experiments are limited since they do not allow separating the spectral or the dynamical properties of the two hydrophobic Trp residues.

11. Experiments giving proofs of the presence of a pocket within α_1-acid glycoprotein

11a. Binding of progesterone to α_1-acid glycoprotein

Titration of α_1-acid glycoprotein with progesterone was performed following fluorescence intensity modifications of calcofluor bound to the protein.

Binding of progesterone to sialylated α_1-acid glycoprotein induces a decrease in the fluorescence intensity of calcofluor, accompanied by a shift to shorter wavelengths of the emission maximum. The maximum is located at 438 nm in absence of progesterone and shifts to 431 nm in presence of 62 μM of progesterone (Fig. 8.42). The intensity decrease is the result of the binding of progesterone to the protein. Plotting the fluorescence intensity at a fixed wavelength (440 nm) as a function of progesterone concentration yields a monophasic curve, suggesting one binding site (Fig. 8.43a).

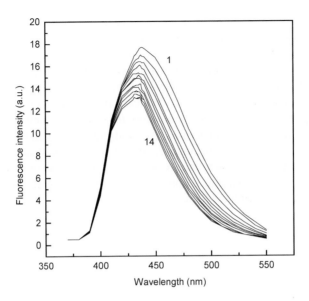

Figure 8.42. Titration of 42 μM calcofluor bound to 8 μM α_1-acid glycoprotein with progesterone. λ_{ex} = 310 nm. Spectrum 1 is obtained in absence of progesterone while spectrum 14 is obtained in presence of 62 μM progesterone.

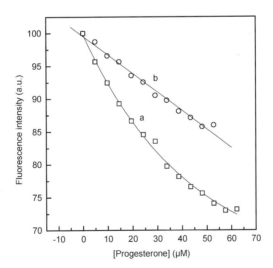

Figure 8.43. (a) Fluorescence intensity variation of 42 μM of calcofluor bound to 8 μM of α_1-acid glycoprotein as a function of progesterone. λ_{ex} = 310 nm and λ_{em} = 440 nm. (b) Fluorescence intensity titration of free calcofluor in presence of increasing concentrations of progesterone. Source: De Ceukeleire, M. and Albani, J. R. 2002, Carbohydrate Research 337, 1405-1410.

The intensity decrease yields a dissociation constant K_d for the α_1-acid glycoprotein-progesterone complex equal to 60 ± 1 µM.

Addition of progesterone to a solution of calcofluor induces a decrease in the fluorescence intensity of the fluorophore without any shift in the emission maximum (Fig. 8.44). The decrease of the fluorescence intensity is linear (Fig. 8.43b), indicating the difficulty in obtaining a complex between free calcofluor and progesterone. Therefore, the shift observed in the emission maximum of calcofluor white upon binding of progesterone, indicates that the hydrophobic ligand binds to α_1-acid glycoprotein in proximity of calocofluor, i.e. near the glycosysation site.

Calcofluor binds to the carbohydrate residues and the variation in the fluorescence parameters observed upon binding of progesterone to the protein is the result of the interaction between the ligand and the carbohydrate residues.

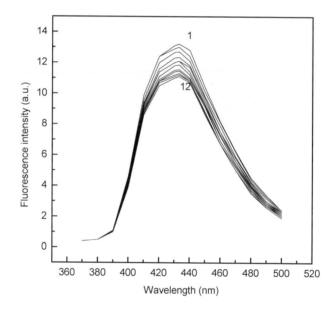

Figure 8.44. Titration of 28 µM of calcofluor with progesterone. Spectrum 1 is obtained in absence of progesterone while spectrum 12 is obtained in the presence of 52 µM of ligand. $\lambda_{ex} = 310$ nm.

Fluorescence parameters such as the intensity and the position of the emission maximum are sensitive to the modifications occurring in the microenvironment of the fluorophore. In fact, emission energy is important when the environment of the fluorophore shows a low polarity, inducing by that a spectrum with a maximum located at low wavelength. Since the fluorescence intensity is proportional to the number of emitted photons, less the surrounding environment is polar, more the number of the emitted photons is important and much higher will be the intensity.

Binding of progesterone to α_1-acid glycoprotein induces a decrease in the emission intensity and a shift in the position of the maximum, a shift absent when progesterone was added to calcofluor free in solution. Therefore, the result obtained in presence of α_1-acid glycoprotein clearly indicates that progesterone and calcofluor are very close, i.e., progesterone interacts with the carbohydrate residues of α_1-acid glycoprotein. This interaction is occurring in a defined region of the protein, where the presence of progesterone renders the region more and more hydrophobic, affecting by that the position of the maximum of the fluorescence spectrum of calcofluor. Since progesterone is a hydrophobic molecule, its interaction with the carbohydrate residues does not take place at the surface of the protein (a hydrophilic zone), but in a well defined area of the protein that is the pocket formed by the N-terminal segment.

However, the changes in the fluorescence (blue shift and intensity quenching) do not necessarily indicate a direct (physical) interaction between progesterone and calcofluor. Progesterone could potentially bind at a separate site within the pocket inducing a protein conformational change within the pocket altering by that the fluorescence properties of the probe. Although this indirect mechanism cannot be ruled out, since the progesterone binds to the pocket where the five carbohydrate units are linked, a direct (physical) interaction between them cannot be excluded.

The fact that the glycosylation site of α_1-acid glycoprotein is within the pocket would explain why the carbohydrate residues adopt a spatial conformation that could be the key of the different physiological roles of the protein.

11b. Binding of hemin to α_1- acid glycoprotein

In chapter 6 paragraph 3, we have described experiments showing that hemin binds to α_1- acid glycoprotein in a hydrophobic pocket. These experiments are a direct proof of the presence of a pocket in α_1- acid glycoprotein. In the absence of crystallographic data on the protein, the results described in paragraphs 11a and 6.3 are the only direct experiments performed up to now that put into evidence the presence of a pocket in α_1- acid glycoprotein.

12. Homology modeling of α_1- acid glycoprotein

α_1- Acid glycoprotein belongs to a family of ligand binding proteins which seem to display a similarity of function and a similarity of both primary and tertiary structure (Pervais & Brew, 1985; Flower, 1996). The term lipocalin is derived from the greek words 'lipos', meaning fat, and 'calyx', meaning cup (Pervais and Brew 1985.) The crystal structures of several lipocalins have been solved and show an 8-stranded anti-parallel beta-barrel fold well conserved within the family. These proteins possess a main cavity delineated by the β-barrel and which is not shielded from the solvent although its wall is formed by hydrophobic residues (see for example, Spinelli et al. 2002, for a description of the three dimensional X-ray structure of the boar salivary lipocalin). The protein, a monomer of molecular mass near 20 kDa, is a pheromone-binding protein expressed in the submaxillary glands of the boar (Loebel et al. 2000).

Therefore, the model we described (see paragrapgh 9 of this chapter) using fluorescence spectroscopy showing the presence of a pocket in α_1- acid glycoprotein is in good agreement with that already described to other lipocalin family proteins.

Kopecky et al (2003) applied a combination of infra-red and Raman spectroscopy with molecular modeling to study α_1- acid glycoprotein. The authors reach the conclusion that a pocket does exist in α_1- acid glycoprotein.

The authors coupled methods that can compensate their particular disadvantages and add their advantages.

Figure 8.45 depicts the logical scheme of the proposed approach.

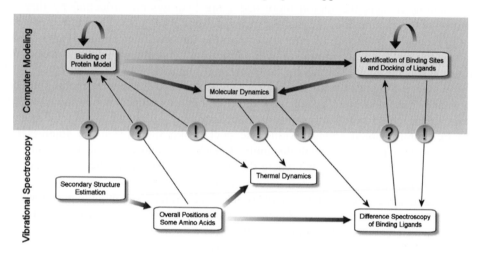

Figure 8.45. Scheme of the new strategy for proper determination of protein structure combines vibrational spectroscopy with computer modeling. Interrogation and exclamation marks indicate collation of the models with experimental data and detail explanation of experiments, respectively. Curtosey from: Vladimír Kopecký Jr., Rüdiger Ettrich, Kateřina Hofbauerová and Vladimír Baumruk.

The authors explained their scheme as follows: "As the first step, vibrational spectroscopy enables to measure protein samples without any special requirements on preparation or conditions. Thus, estimation of the secondary structure content and in some cases even a number of segments with regular secondary structure in the studied protein can be easily obtained. Despite the fact that this information is loaded with error, a homology model structure of the protein should not be in steep discrepancy with it. The model should reflect these experimental data and, if not, a set of restrains leading to a different protein fold must be used for remodeling. Subsequently, more detail information concerning about some particular amino acid residues can be obtained from Raman spectra, e.g. local environments and torsion angles of aromatic residues, presence and conformation of S-S bridges etc. This type of information particularly points to alignment errors or other shortcomings of the computer model and can be used as a starting geometry for its optimization, enlarging the set of constraints used for building of the protein model. After this step the 3D model of the protein consistent with vibrational spectral data is finally gained. Whether in energetically minimized structure only or after a molecular dynamics run, the model can explain many aspects of the behavior observed in thermal dynamics experiments, e.g. identification of amino acids residues surfacing up from the protein core upon heating.

One of the most important features is the possible identification of binding sites and docking of ligands into the protein structure, which can lead to a better drug design.

Vibrational difference spectroscopy can identify changes in the secondary structure and amino acids influenced by ligand binding. Nevertheless, the situation can be more complicated due to the amino acid composition in the binding site. Only the amino acids with aromatic side chains can be identified by Raman difference spectroscopy upon ligand binding. All structural aspects observed in difference spectra must be present in the binding site identified in the model structure. If not, another docking process must be used. When all experimentally demanded conditions are met, the computer model can identify the composition of the binding site in detail and thus predicts appropriate steps in mutagenesis or modification of ligands. We can conclude that this approach containing several self-corrective steps brings always a valuable contribution to the knowledge of the protein 3D structure, despite of corrections or shortcomings that have to be made, even if X-ray or NMR analysis succeeds in future."

The second derivative of the FTIR spectrum of α₁- acid glycoprotein (Fig. 8.46) that allows resolution of overlapping spectral components, validate the high content of extended β-sheets (presence of a strong band at 1637 cm⁻¹). Presence of a band at 1692 cm⁻¹ suggests that β-sheets are antiparallel, in agreement with the repeated +1 topology β-barrel presented in the model of the protein moiety.

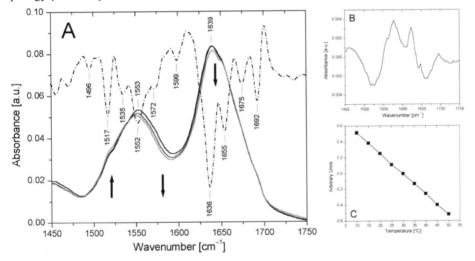

Figure 8.46. FTIR spectra of native AGP in the 10–50 °C temperature range. (A) Solid lines represent FTIR spectra of AGP recorded at 10, 30 and 50° C while arrows label band changes due to increasing temperature. The dash-dot line is associated with the second derivative (15 pts) of the spectrum at 10° C. (B) The second subspectrum of the PCA of the set of temperature dependent FTIR spectra of AGP. (C) The corresponding PCA coefficient associated with the second subspectrum depicted at (B). Source: Kopecky, V. Jr., Ettrich, R., Hofbauerova, K. and Baumruk. V, 2003, Biochemical Biophysical Research Communications. 300, 41-46.

At this point, however, the model can only provide a rough estimation of the protein fold, and cannot serve as a model on the atomic level, e.g. as a basis for docking experiments.

Therefore the model of the protein moiety of α₁- acid glycoprotein was verified experimentally in as many details as possible following the scheme depicted on Figure 8.45. The secondary structure estimation by least-squares analysis of FTIR amide I and II bands, at 1636 cm⁻¹ and 1552 cm⁻¹ respectively, and of Raman amide I band, at 1662 cm⁻¹ leads to 15% of α-helix, 41% of β-sheet, 12% of β-turn, 8% of bend and 24% of unordered structure. Also, the authors found unusual *trans-gauche-trans* conformation of S–S bridges, which corresponds to the presence of Raman marker band at 541 cm⁻¹. Raman difference spectroscopy revealed proximity of Trp to the binding site by the shift of W17 mode at 884 cm⁻¹ (in the difference spectrum) indicating changes in Trp NH-bond donation. The authors found only one – Trp[122] – close to the hydrophobic pocket, thus it very precisely identified the binding site. Both FTIR and Raman spectroscopy showed increase of β-sheet – ca. 4%, and decrease of α-helix – ca. 3%, upon progesterone binding. It is in an excellent agreement with behavior of α-helix in the first loop above the β-barrel in the model, which is transformed into antiparallel β-sheet upon progesterone binding to the hydrophobic pocket located inside the β-barrel.

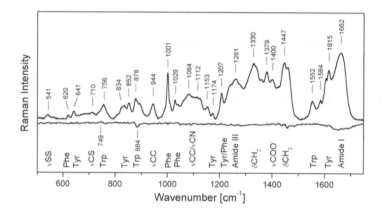

Figure 8.47. Raman spectrum of native form α₁- acid glycoprotein. (Bottom) Raman difference spectrum of native α₁- acid glycoprotein minus α₁- acid glycoprotein with bound progesterone. Positions of amino acids with aromatic side chains, with respect to their environment or surface of the protein, have been also determined. Curtosey from : Vladimír Kopecký Jr., Rüdiger Ettrich, Kateřina Hofbauerová and Vladimír Baumruk.

All molecular dynamics studies of the complete protein should be carried out in the presence of the carbohydrate moiety, which strongly stabilize the structure. Unfortunately, molecular dynamics modeling in the presence of such a huge amount of sugar compound, 42%, does not lead to reliable results at the present state of the art. FTIR experiments with thermal dynamics are in agreement with this observation and support the important role of the sugar compound for the protein stability. Although no simply observable changes of FTIR spectra as a function of temperature can be seen on Fig. 8.46A, the principle component analysis (PCA) can resolve those changes. As a multivariate mathematical technique, PCA reduce spectra to their lowest dimension, thus each spectrum can be expressed as a linear combination of loading coefficients and orthogonal subspectra.

The second subspectrum, which reflects changes in the spectral set, depicted on Fig. 8.46B corresponds to a) rearrangement in β-structures – mainly increasing content of turn structures – by bands at 1641 cm⁻¹ and 1661 cm⁻¹, and b) changes of the band at 1518 cm⁻¹ – probably surfacing of Tyr residues – with increasing temperature. The loading coefficients, depicted at Fig. 8.46C, reflect extreme thermal stability of α_1- acid glycoprotein and, in general, correspond to the "breathing" of β-barrel. Finally, according to the scheme in Fig. 8.45, ligand binding to α_1- acid glycoprotein was explored. Raman difference spectroscopy (Fig. 8.47) revealed proximity of Trp to the binding site by the shift of W17 mode at 884 cm⁻¹ (in the difference spectrum) indicating changes in Trp NH-bond donation. Only one – Trp[122] – is found close to the hydrophobic pocket, thus it very precisely identified the binding site. Both FTIR and Raman spectroscopy showed increase of β-sheet – ca. 4%, and decrease of α-helix – ca. 3%, upon progesterone binding. It is in an excellent agreement with behavior of α-helix in the first loop above the β-barrel in the model, which is transformed into antiparallel β-sheet upon progesterone binding to the hydrophobic pocket located inside the β-barrel. Thus, if the model should bring new valuable results then it must also explain previous experimental results, which were not used for the construction of the model.

Combining computer modeling of proteins with vibrational spectroscopy is a good approach to study the structure of α_1- acid glycoprotein, despite the fact that this approach could be criticized for non 100% reliability of the 3D structure. Figures 8.48 and 8.49 display the tertiary structure of α_1- acid glycoprotein in absence and presence of progesterone as proposed by Kopecky et al (2003).

Figure 8.48. Three-dimensional stereo picture of α_1- acid glycoprotein. The picture shows the +1 topology β-barrel that the protein has common with other members of the lipocalin family. Source: Kopecky, V. Jr., Ettrich, R., Hofbauerova, K. and Baumruk. V, 2003, Biochemical Biophysical Research Communications. 300, 41-46.

Figure 8.49. Three-dimensional stereo picture of α_1- acid glycoprotein. The hydrophobic pocket in the middle of the barrel holds a progesterone molecule. Source: Kopecky, V. Jr., Ettrich, R., Hofbauerova, K. and Baumruk. V, 2003, Biochemical Biophysical Research Communications. 300, 41-46.

13. Dynamics of Trp residues in crystals of human α_1- acid glycoprotein

13a. Introduction

Crystallization of α_1- acid glycoprotein proved to be somewhat difficult, probably because of the high solubility and the large carbohydrate moiety of the protein. By using lead ion (0.007 M), the solubility of α_1- acid glycoprotein is reduced significantly, approximately equal to the effect of 40% ethanol. Also, the addition of a mixture of equal amounts of acetone and methanol decreases the solubility much further, to yield crystals under appropriate conditions.

However, because of the difficulty in obtaining the crystals, since 1953, crystallization of α_1- acid glycoprotein has not been performed. Therefore, there are no structural or dynamical studies (such as X-ray diffraction, NMR, or fluorescence) on crystals of α_1- acid glycoprotein.

We report here the development of a method that allowed us to obtain crystals of α_1-acid glycoprotein complexed to progesterone. Furthermore, we investigated the dynamics of the microenvironments of the two buried Trp residues in the crystals of the protein-progesterone complex, by the red-edge excitation spectra method.

13b. Protein preparation

In 1953, F. Schmid described a method of crystallization where the solution is heated in order to dissolve α_1- acid glycoprotein and to evaporate the methanol-acetone mixture. We attempted to crystallize the protein by following the same procedure without any success. We reached 70°C and failed to obtain a limpid solution. We repeated the experiment five times without any success.

In our procedure, we did not heat the solution, but we performed a slow evaporation at 5°C.

Our crystals were obtained as follows: 32 mg of the lyophilized protein was dissolved in 460 μl of twice-distilled water, 30 μl of the stock solution of progesterone was then added to the solution of α_1- acid glycoprotein, and finally, 12 μl of a 1 M lead acetate solution was mixed with the α_1- acid glycoprotein – progesterone complex. The final solution was kept at 5°C for 2 h. Three hundred microliters of a precooled (5°C) mixture of methanol-acetone (1:1) was then added drop by drop, with careful stirring. The final solution was kept at 5°C for 1 year. During this time important crystals of α_1-acid glycoprotein-progesterone complex were formed (Fig. 8.50). One should indicate that we obtained different forms of crystals, something that allows us to perform our diffraction studies.

The important growth of the crystals could be the result of the slow growth at low temperatures. The presence of a high concentration of carbohydrate in α_1- acid glycoprotein plays an important role in the global crystallization and induces the spatial conformation and orientation of the protein in the crystal (Wyss and Wagner, 1996).

Crystals were mounted on a micropipette cone and were placed in the center of the cuvette holder of the fluorometer. Observed fluorescence intensities were corrected for the background intensities of a crystallized buffer solution, obtained at saturated concentrations of sodium phosphate buffer.

X 56.8   0.2 mm

Figure 8.50. Photograph (x 57) of crystals of α_1- acid glycoprotein-progesterone complex, obtained in presence of lead acetate. Source : Albani, J. R. 1998, Journal of Fluorescence, 8, 213-224. Authorization of reprint accorded by Kluwer Academic Publishers.

In crystals of α_1- acid glycoprotein-progesterone complex, the proteins are linked together with the carbohydrates. This protein-protein interaction is strong, inducing a decrease in the global rotation of the whole system and thus allowing observation of the electron density of the system. Progesterone, a hydrophobic ligand, was added to the solution of α_1- acid glycoprotein in order to decrease the hydrophilicity of the protein and to facilitate its crystallization. Since we did not attempt to prepare the crystals without adding progesterone, we are not able, for the moment, to define the role of the ligand in the procedure of crystallization. The organization of the protein in the crystals was identified by small- and wide-angle x-ray diffraction studies and will be described in paragraph 14.

13c. Fluorescence Excitation and Emission Spectra

Figure 8.51 displays the fluorescence excitation spectrum of the crystallized α_1- acid glycoprotein. The value of the bandwidth (38 ± 1 nm) indicates that the Trp residues are responsible for the fluorescence of the protein in the crystal.

The steady-state emission spectrum of the crystals obtained with unpolarized excitation light (Fig. 8.52A) does not overlap the emission spectrum of α_1- acid glycoprotein - progesterone complex obtained in solution (Fig. 8.52B), but overlaps the spectrum of hydrophobic Trp residues of the protein (Fig. 8. 53). Thus, the fluorescence observed for the crystal is characteristic of Trp residues embedded in the protein matrix. Therefore, the two Trp residues surrounded by a hydrophobic environment have the same microenvironments in crystal and in solution.

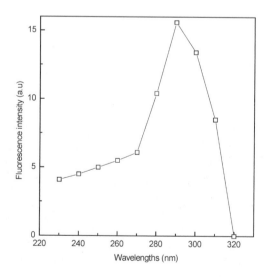

Figure 8.51. Fluorescence excitation spectrum of a crystal of α_1- acid glycoprotein - progesterone complex. λ_{em} = 335 nm. The spectrum was recorded by excitation with unpolarized light. Source : Albani, J. R. 1998, Journal of Fluorescence, 8, 213-224. Authorization of reprint accorded by Kluwer Academic Publishers.

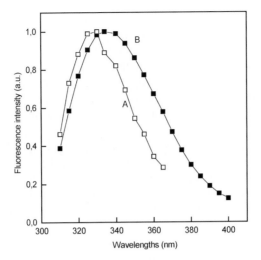

Figure 8.52. Fluorescence emission spectra of crystals of α_1- acid glycoprotein - progesterone complex (A) and of a solution of α_1- acid glycoprotein - progesterone complex (B) both obtained at an excitation wavelength of 295 nm with unpolarized light. Source : Albani, J. R. 1998, Journal of Fluorescence, 8, 213-224. Authorization of reprint accorded by Plenum Com.

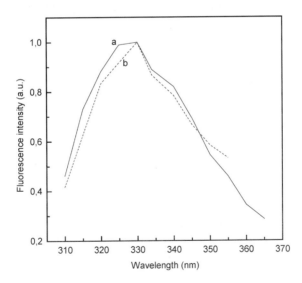

Figure 8.53. Fluorescence emission spectra of crystals of α_1- acid glycoprotein - progesterone complex (a) and of hydrophobic Trp residues in solution in presence of progesterone (b). λ_{ex} = 295 nm. Spectrum b was obtained from fluorescence intensity quenching by cesium of a solution of α_1- acid glycoprotein - progesterone complex, after extrapolating to $[Cs^+]$ = ∞. Source : Albani, J. R. 1998, Journal of Fluorescence, 8, 213-224. Authorization of reprint accorded by Kluwer Academic Publishers.

Since the fluorescence emission spectrum of the crystals is characteristic of Trp residues present within a hydrophobic environment, this means that the fluorescence of the surface Trp residue is completely quenched. This result may be explained by the fact that the protein-protein interaction in the crystal occurs via the surface of the protein. The Trp residue at the surface will be in contact with two proteins, facilitating high energy transfer to the neighboring amino acids. This result is in good agreement with that found by x-ray diffraction studies, i.e., the proteins are linked together at their surface by the carbohydrates.

When excitation was performed with a vertical polarized light, whether on solution of α_1- acid glycoprotein or on the crystal, the bandwidths of the spectra are 10 nm smaller than those observed when the spectra were recorded with unpolarized light. In the crystal, the bandwidth is equal to 35 ± 1 nm instead of 43 ±1 nm, and in solution, it is equal to 41 ± 1 nm instead of 54 ± 1 nm. We do not know whether this phenomenon is due to the polarizer used or not.

Excitation at 300 nm gives an emission maximum in the same range as that obtained with excitation at 295 nm. Therefore, there is no evidence of tyrosine fluorescence contribution at excitation wavelengths equal to or higher than 295 nm. This contribution may account for a blue-shifted spectrum.

13d. Dynamics of the microenvironments of the hydrophobic Trp residues

Red-edge excitation spectra method is very sensitive to the changes that occur in the microenvironment of the Trp residues. For example, Trp residues of intact lens protein from rat exhibit an emission shift of 14 nm upon varying the excitation from 290 to 308 nm. Photodamaging the lens induces an emission shift of 24 nm (Rao et al. 1989).

Figure 8.54 displays the fluorescence spectra of Trp residues of crystals of α_1- acid glycoprotein -progesterone complex obtained by exciting with vertically polarized light at three excitation wavelengths, 295, 300, and 305 nm. The maximum (331 nm) of the tryptophan fluorescence of the crystals does not change with the excitation wavelength. The same result was obtained in the absence and presence of progesterone for α_1- acid glycoprotein in solution. The values of the centers of gravity are 3.0208×10^4 cm^{-1} (331 nm), 3.015×10^4 cm^{-1} (331.7 nm), and 3.0303×10^4 cm^{-1} (330 nm) at λ_{ex}, 295, 300, and 305 nm, respectively. Since the center of gravity of the fluorescence spectrum does not change with the excitation wavelength, emission occurs after relaxation process. In this case, the emission energy and thus the emission maximum and the center of gravity of the spectrum are independent of the excitation wavelength. Our results indicate that the microenvironment of the Trp residues is not rigid. This is in good agreement with those found in solution for α_1- acid glycoprotein prepared by chromatographic methods. The fast dipolar relaxation of the microenvironment of the two hydrophobic Trp residues could be explained by the water content of the protein crystal (35%). Therefore, crystallization of α_1- acid glycoprotein does not modify the structure and the dynamics of the protein core. Sine we are observing the emission from the two hydrophobic Trp residues, we are monitoring the dynamics around both Trp residues. The red-edge excitation spectra method does not allow studying the dynamic behavior of each tryptophanyl residue. We do not know from our experiments whether or not the two Trp residues have the same degree of freedom.

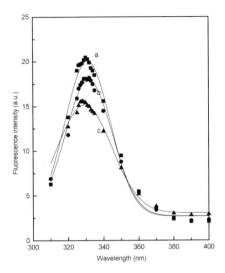

Figure 8.54. Steady state fluorescence emission spectra of a crystal of α_1- acid glycoprotein - progesterone complex recorded at three excitation wavelengths, 295 nm (a), 300 nm (b) and 305 nm (c). The dotted lines show the spectra obtained by considering the maximum equal to the centers of gravity, 331, 331.7 and 330 nm, at λ_{ex} 295, 300 and 305 nm, respectively. The spectra were obtained with a vertically polarized excitation light. Source : Albani, J. R. 1998, Journal of Fluorescence, 8, 213-224. Authorization of reprint accorded by Kluwer Academic Publishers.

Rotation of the crystal around the excitation beam by steps of 45 ± 5° does not induce any modification in the red-edge excitation spectra results. At all angles (0, 45, 90, 135, and 180°), no shift was observed either in the fluorescence maximum or in the center of gravity of the spectrum (data not shown). This result indicates clearly that at all the angles we are monitoring the dynamics of the same microenvironments of the Trp residues in the crystal of protein. At all angles, we are exciting the dipoles of the two buried Trp residues.

The fact that the emission maximum does not change upon rotation of the crystal is, in our case, another proof that the emission occurs from the buried Trp residues. Otherwise, we would observe a shift to at least 345 nm and an increase in the bandwidth, if at certain angles we are observing an emission from the Trp residue of the surface.

Since the fluorescence intensity of a fluorophore is proportional to the square of the absorption transition probability, then it should vary with the angular dependence of the fluorophore's dipole with respect to the direction of the excitation beam. If the orientation of the dipoles of the Trp residues is modified as the result of the motion of the Trp residues, then the fluorescence intensity will change with the angle of the rotation of the crystal. In the absence of motions, the fluorescence intensity will remain constant at all positions of the crystal, since we are monitoring a definite orientation of the dipoles. One also should note that we are monitoring the fluorescence of two Trp residues. Therefore, the fluorescence intensity variation, if any, will be the result of the relative mobility of each Trp residues.

The fluorescence intensity variation as a function of angles of rotation obtained at three excitation wavelengths, 295, 300, and 305 nm, is shown in Fig. 8.55. The dynamics of the microenvironment of the Trp residues will induce the motion of the fluorophore. This motion, within the nanosecond range, will modify the orientation of its corresponding dipole, thus affecting its absorption and its emission. Since we have two Trp residues, the results obtained in Fig. 8.55 clearly indicate that at least one dipole is changing its orientation with time as the result of the dynamics of the corresponding Trp residue. This motion will modify the orientation of the corresponding dipole and thus will affect its absorption and emission. Thus, our data shows clearly that the hydrophobic tryptophan residues display constant motions in the crystals such as we observed in solution and the crystals obtained have conserved the native structure of the protein in solution. This condition has allowed us to go further and to study the three dimensional structure of α_1- acid glycoprotein.

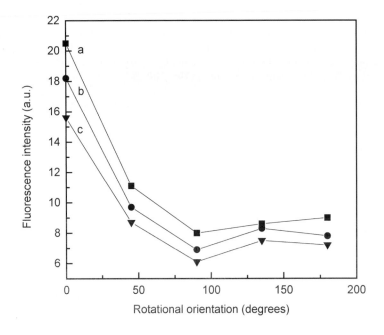

Figure 8.55. Dependence of the fluorescence intensity on the crystal orientation with respect to the vertical direction of the excitation light. Fluorescence intensities are taken at the peaks of the spectra. Excitation wavelengths are are 295 nm (a), 300 nm (b) and 305 nm (c). Source : Albani, J. R. 1998, Journal of Fluorescence, 8, 213-224. Authorization of reprint accorded by Kluwer Academic Publishers.

14. Structural studies of human α_1-acid glycoprotein followed by X-rays scattering and transmission electron microscopy.

14a. Small angle diffraction studies (SAXS)

SAXS performed on a single needle crystal show the presence of reinforcements in the meridional direction when the needle is aligned parallel to the equator (Fig. 8.56a) and in the equatorial direction when it is aligned perpendicular to the equator (Fig. 8.56b). Thus, the preferred direction of the planes of the crystallites is parallel to the axis of growth of the needle. The mean interreticular distance d can be calculated from the position of both the external and internal circumferences of the corona, d is found to be between 30 and 50 Å. These values are characteristic of those found for macromolecules such as proteins.

A central reinforcement (arrowhead) in the same equatorial direction to those obtained in the corona is also observed. This means that another periodicity does exist in the structure of the complex, i.e., the proteins are packed together in crystallite forms and the different packages are separated by a distance that lies between 100 and 200 Å. However, this is the minimal distance that separates the packages of protein since the higher intensity of the central reinforcement is within the well and thus it cannot be observed.

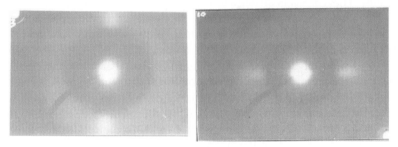

Figure 8.56. S.A.X.S. pattern of one needle crystal, obtained with the axis of the needlle parallel (left panel) and perpendicular (right panel) to the equator.

Figure 8.57 displays the SAXS pattern of the crystal, obtained on the adjacent side of the needle to that used in Fig. 8.56. Two interreticular distances can be calculated: 30 Å < d_1 < 50 Å (arrow) and 60 Å < d2 < 70 Å (arrowhead). d_1 is the same distance defined and determined in Fig. 8.56 while d_2 characterizes the interreticular distance between the proteins, perpendicular to the axis of growth of the needle. Also, data of figure 8.57 suggest that α_1-acid glycoprotein is little more elongated than a globular protein, although fluorescence studies have indicated that considering α_1-acid glycoprotein in solution as spherical is correct. Therefore, we can consider the protein α_1-acid glycoprotein as if it occupies the volume of a cylinder of a height H equal to 65 Å and a diameter of 40 Å. The volume of one protein within the crystal would be equal to 81680 Å^3. This volume is close to that (72453 Å^3) of a spherical protein of molecular weight equal to 41000. Thus, the crystal volume per unit weight, V_m, was calculated to be 1.99 Å^3 / Da.

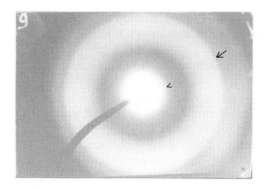

Figure 8.57. S.A.X.S. pattern of one needle crystal, obtained with the X-ray beam perpendicular to the main axis of the elongated proteins.

This value is within the normal range for protein crystals (Matthews, 1968; Miki et al. 1990; Kita et al. 1991; Harada et al. 1991).

SAXS show a central corona with reinforcements indicating preferential directions of the crystallites. This corona is diffused as the result of the macromolecular lattice and of the internal dynamics of the protein. In fact, fluorescence studies have indicated that the protein displays constant and important motions.

We notice also that to the difference of the corona of Fig. 8.56, the one displayed in Fig. 8.57 does not have any reinforcements. This result is consistent with an arrangement of the proteins planes with interreticular distances d_1 and d_2, respectively perpendicular and parallel to the direction (z) of growth of the needle crystal, where the proteins are in juxtaposition one to each other. Therefore, the protein-protein interaction in the crystal occurs via the surface of the protein. This result is in good agreement with that obtained on the fluorescence of the crystals. In fact, fluorescence emission spectrum of the crystals occurs from Trp residues buried in the protein core and that the fluorescence of the surface Trp residue is completely quenched. The Trp residue at the surface is in contact with two proteins, facilitating high energy transfer to the neighboring amino acids.

Also, we notice that in all the SAXS patterns (Fig. 8.56 and 8.57), both the external and internal diameters of the corona are well cut-shaped. This means that the lattitude of the displacement of the lattice (thermal motions, displacement of the macromolecular lattice) is limited by a physical and well identified energy barrier. These physical barriers are the result of the compact configuration of the juxtaposed proteins. In figure 8.56, the X-ray beam is parallel to the principal axis of the proteins (c) and thus allows determining d_1. In figure 8.57, the X-ray beam is perpendicular to (c) and thus allows determining the values of both d_1 and d_2. Also, since in Fig. 8.57, the X-ray beam is parallel to the preferred direction of the planes of the crystallites, i.e., perpendicular to these planes, the reinforcements observed in figure 8.56 will not be observed, and only a homogeneous distribution of the proteins is detected.

14b. Wide angle diffraction studies (WAXS)

14b1. Carbohydrate residues studies

When the film was placed 5 cm far from the needle, and when the X-ray beam reaches it with an inclination of 20° of arc to the principal axis of the needle, the diffraction pattern obtained with the Luzzati camera constitutes a cross (Fig. 8.58).

The periodicity (I) of the helix was calculated from the distance S that separates the equatorial and the first stratum, with the formula of Polanyi (Tadokoro, 1979) (Fig. 8.59)

$$I \sin \alpha = m \lambda \tag{8.16}$$

where

$$tg \, \alpha = S / r \tag{8.17}$$

where r is the distance between the sample and the film (r = 50 mm), λ is the wavelength of the X-ray beam, $\lambda = 1.54$ Å and S = 2.5 mm. Equation 8.17 yields a value of I of 29 ± 2 Å.

Figure 8.58. W.A.X.S. pattern of the needle crystal, obtained with the Luzzati setup. The distance film- to- speciman is 50 mm. The spots constitute crosses indicating the presence of a helical structure.

The distance that separates two successive strands is obtained from the position of the most intense spot in the vertical azymuth. Using the formula of Bragg

$$\lambda = 2 \, d \sin \alpha \tag{8.18}$$

where $\alpha = 7°$ of arc and $\lambda = 1.54$ Å. d is found equal to 6 Å. This value characterizes the diffraction from the carbohydrate residues.

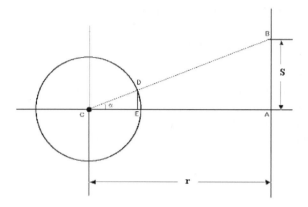

Figure 8.59. Geometrical determination of helical structure. S is the distance that separates the equatorial passing through the center of the crystal and the first stratum.

Thus, since the periodicity of the helix is equal to 30 Å and the distance between two successive carbohydrates is equal to 6 Å, we are observing the diffraction from 5 successive carbohydrates in the helix. Since each carbohydrate chain is composed of 7 residues, the X-ray diffraction studies allowed detecting five residues.

14c. Electron microscopy studies

Transmission electron microscopy observations were carried out with a Jeol 200 CX operated at 200 kV. In order to avoid fusion of the samples, the experiments were carried out on areas that make between 50 to 100 μm of thickness. Electron microscopy was applied to crystals that have the form of a leaf (Fig. 8.60).

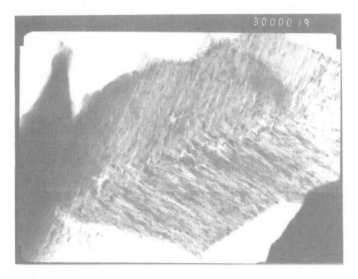

Figure 8.60. Electron microscopy photograph of Crystals of α_1–acid glycoprotein-progesterone complex.

Analysis of figure 8.60 allowed us to observe stacking of molecules of α_1-acid glycoprotein. In fact, different types of constraints, that correspond to objects of different nature (protein matrix, carbohydrate residues and lead acetate), are observed (not shown).

Diffraction studies obtained on a leaf yield diffused spots (Fig. 8.61) characteristic of a fiber pattern of a helical structure of macromolecules. This result is in good agreement with that obtained with the X-ray diffraction studies and is associated with the structure of the carbohydrate residues (Fig. 8.58).

The periodicity (J) of the observed helix was calculated from Eq. 8.19

$$J\,(\text{Å}) = 35.5 \,/\, D(\text{mm}) \qquad\qquad (8.19)$$

where D is the distance between two consecutive layers of spots in the diffraction pattern obtained with a camera length of 137 cm. The value of J was found equal to 12 Å. This value is lower than that found for the whole carbohydrate residues and thus represents the distance of two consecutive carbohydrate residues. In fact, the length corresponding to five successive carbohydrate residues is equal to 30 Å, as it has been obtained from the X-rays diffraction studies (Fig. 8.58).

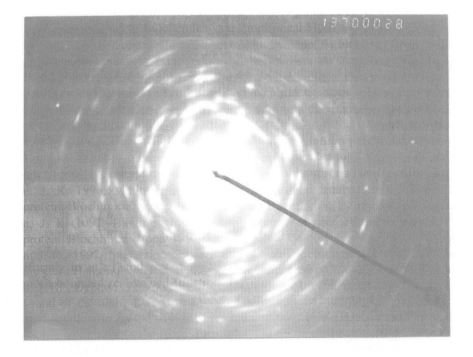

Figure 8.61. Electron microscopy diffraction pattern of crystals of carbohydrate residues in α_1-acid glycoprotein-progesterone complex. The spots constitute crosses indicating the presence of a helical structure.

The distance h between two successive atoms within the carbohydrate skeleton was obtained from the position of the higher spot located at the 12th layer of the vertical azimuth. h was found equal to about 1 Å.

The diameter D of the helix can be calculated with the equation :

$$tg \; \alpha = (1 \; / \; R) \; / \; (N \; / \; J) \qquad\qquad (8.20)$$

where R = D / 2, N is the number of strata (= 5), J the periodicity (= 12 Å) and α the angle between the direction of the spot and the principal axis of the helix. Equation 8.20 yields for R a value of 1.8 Å, i.e., a diameter D equal to 3.6 Å.

Therefore, while X-ray diffraction studies allowed the to detect the presence of 5 carbohydrates, electron microscopy diffraction studies allowed us to obtain information concerning the distance between the neighboring atoms within one carbohydrate residue.

The pattern shown in figure 8.61 indicates the presence of two helices that form an angle of 104° which is the result of the repulsion forces that may exist between the different antenna of the carbohydrates that adopt an extended conformation.

The surface covered by one carbohydrate is found equal to 1492 Å^2. Since the protein possesses 5 carbohydrate residues, the total surface of the protein covered by the carbohydrates is 7460 Å^2.

If we consider α_1-acid glycoprotein as a cylinder of diameter equal to 40 Å and a length of 65 Å (Fig. 8.57), its surface will be equal to 8164 Å^2. Thus, the area covered with the carbohydrate structures is equal to 91%.

The WAXS and the transmission electron microscopy patterns allowed us to find out that the carbohydrate residues of α_1-acid glycoprotein are organized within a helical structure. Fluorescence spectroscopy allowed us to find out that carbohydrate residues of α_1-acid glycoprotein possess a secondary structure that is present only when the carbohydrates are linked to the protein. Therefore, the carbohydrates and the protein matrix constitute two complementary entities that give the lipoprotein its specific physiological properties.

X-ray diffraction studies allowed us to detect 5 carbohydrates of 7. These 5 residues are the most external to the glycosylation sites. In fact, the five carbohydrate units are linked to the N-terminal fragment of the polypeptide moiety rendering this section of the protein hydrophilic. The N-terminal fragment is in contact with the solvent, and adopts a spatial conformation so that a pocket in contact with the buffer is induced. Thus, the five carbohydrate units are linked to the pocket. Therefore, the two carbohydrate residues which are the most proximal to the glycosylation sites belong to the pocket. The other five residues are developed on the external surface of the protein and would be more detected by diffraction techniques.

Also, the helix structure of the carbohydrate residues would be coherent with the presence of a pocket from where the first wheels of the helix emerge.

Observing the helix from X-rays and electron microscopy was very difficult, because we had to find out the right crystals orientation relatively to the incident beams. This sensitivity was most revealed in the electron microscopy. In fact, modifying for one degree or even less the tilt of the goniometer was sufficient to loose the observation of the helix pattern.

Chapter 9

STRUCTURE AND DYNAMICS OF HEMOGLOBIN SUBUNITS AND OF MYOGLOBIN

1. Introduction

The principal functions of hemoglobin and myoglobin are oxygen transport (case of hemoglobin) and its storage in muscles (case of myoglobin). Other hemoproteins play the role of electron transfer molecules (case of cytochromes). First, let us make a brief description of the structures of myoglobin and hemoglobin, then let us see how the structure is modified upon oxygen binding.

Myoglobin molecule, an oblate spheroid, approximately 44 by 44 by 25 Å, is built up from eight connected pieces of α helix (Fig. 9.1). The heme is in a pocket and is surrounded by protein except for the edge that contains the two polar propionic acid side groups, which stick out of the pocket into the surrounding water. A histidine-to-iron coordination links covalently the heme to the globin. The main stabilizing forces for the heme appear to be hydrophobic contacts between the heme plane and the lining of the pocket. The state of ligation of myoglobin (deoxymyoglobin, metmyoglobin, oxymyoglobin and carbonmonoxymyoglobin) modifies the position of the iron relative to the heme plane.

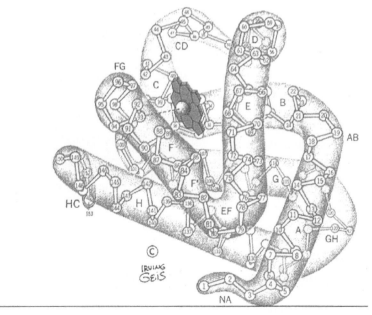

Figure 9.1. Three dimensional structure of Sperm whale myoglobin. Irwing Geis (1908-1997).

Hemoglobin is a tetramer composed from two α and two β chains. The molecule is roughly spherical, 64 by 55 by 50 Å. The four heme pockets are all exposed at the surface of the molecule. The heme groups of chains α_1 and β_2 are particularly close, as are those of α_2 and β_1 (Fig. 9.2). Hemoglobin exists in two different conformations: a T or tense state with low or zero affinity for substrate, and an R or relaxed state with high affinity for substrate. In the T-state, the iron atom lies 0.6 Å out of the mean plane of the porphyrin plane. In the R-state, the iron atom is closer to the heme plane. For a more extensive description of the structure of hemoglobin and myoglobin, the reader is invited to see (Dickerson and Geis, 1983).

However, what about the dynamics of the proteins? Does it play any important role in binding of small ligands such as oxygen? In fact, structural modification observed upon oxygen binding and their importance could be the results of the internal motions of the proteins. Binding of a ligand could simply be dependent on its diffusion within the protein and thus on the structural flexibility of the protein. These dynamics concern the local and global motions of the protein matrix and the dynamics of the porphyrin embedded in the heme pocket.

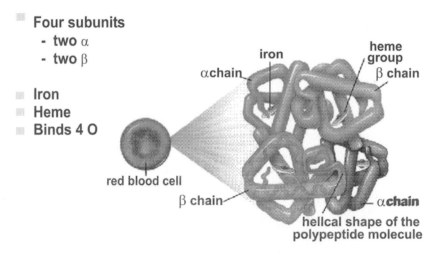

Figure 9.2. Structure of hemoglobin

2. Dynamics of Trp residues in hemoglobin and in its subunits

Fluorescence was used to study the dynamic character of proteins. However, these studies were limited to proteins having high fluorescence quantum yield. Due to the lack of performed instrumentations and the assumption that intrinsic fluorescence from hemoproteins does not exist, attempts to detect fluorescence from these hemoproteins were never realized before 1976. The low hemoproteins fluorescence was explained by the presence of a very strong energy transfer from the tryptophan to the heme.

Binding of ligands such as O_2 or CO on hemoglobin follows a cooperative mechanism. The model commonly used to explain this cooperativity is that hemoglobin is in equilibrium between a low affinity state, tense (T) form and a relaxed (R) high affinity form. Literature is rich with papers describing and discussing the allosteric mechanism of hemoglobin (Perutz, 1970; Weber, 1984; Srinivasan and Rose, 1994; Vasquez et al. 1998; Unzai et al. 1998; Safo et al. 2002; Nagatomo et al. 2002; Riccio et al. 2002; Tsai et al. 2003). These two extreme forms of hemoglobin have not the same tertiary and quaternary structures.

Human hemoglobin has 6 Trp residues, one in each alpha subunit (α_{14}) and two in each beta subunit (β_{15} and β_{37}). The energy transfer, Forster type from the tryptophan to the heme group is very important and more probable than fluorescence. Results on fluorescence lifetime of human hemoglobin and its subunits were published for the first time by (Alpert and Lopez-Delgado, 1976). The authors reported that the fluorescence decays were non exponentials for all investigated hemoproteins, however a detailed analysis of the data was not given due to the fact that excitation pulse was greater than 1 ns, which makes impossible to obtain correct information concerning the values of the picosecond decays. The results obtained are reported in Table 9.1.

Table 9.1. Fluorescence lifetimes of human hemoglobin and of its subunits measured with the synchrotron facility at LURE, Orsay, France.

Derivative	τ
Met Hb	3.5 ns at 3°C.
Met Hb	2.4 ns at 25°C.
HbO_2	1.5 ns at 25°C.
αO_2	1 ns at 3°C.
αO_2	0.5 ns at 25°C.
βO_2	1.4 ns at 25°C.

Precision on the values are 0.25 ns. $\lambda_{ex,}$ 270, 280 or 290 nm. Source: Alpert, B. and Lopez-Delgado, R. 1976, Nature, 1976, 445-446.

At excitation wavelength equal to 290 nm, fluorescence occurs from the Trp residues only. The results indicate that energy transfer Trp → heme is not 100% efficient, thus other processes must occur in the proteins favoring Trp emission.

Later on, other studies were performed showing that fluorescence from hemoglobin occurs. Very often, β_{37} Trp-residue was considered as the origin of the fluorescence in hemoglobin although this assignment is in contradiction with the results shown in Table 9.1.

Also, differences in the fluorescence observables between the oxy and deoxystates were assigned to a change in the microenvironment of the β_{37} Trp due to a conformational change from the deoxy (T) to the oxy (R) states (Hirsch et al. 1980; Itoh et al. 1981; Hirsch and Nagel, 1981).

This assignment cannot be valid for different reasons:

1) Hemoglobin contains 6 Trp residues and not 1 Trp residue, thus we do not see why only one Trp is sensitive to conformational changes?

2) β_{37} Trp residue is the nearest Trp in the hemoglobin to the heme, thus energy transfer could be important.

3) β_{37} is located at the $\alpha_1\beta_2$ contact area. The R-T switch induces a modification in this contact region. Thus fluorescence modifications were considered to represent a structural change. This assumption does not take into consideration the fact that proteins do have internal motions that may change with the ligand and thus induce a difference in the fluorescence spectra or/and lifetimes. These protein motions concern the whole protein areas and thus every amino acid.

4) The subunits do not exhibit a R-T switch. What would be in this case the interpretation of the fluorescence of the isolated subunits if their fluorescence differ in the oxy and deoxystates?

Fluorescence lifetimes of hemoglobin and its subunits were measured in three ligation states, oxy, deoxy and carbomonoxy. The results indicate that hemoglobin and its subunits present similar fluorescence. Their decays depend on the ligation state (Table 9.2).

Table 9.2. Fluorescence decay parameters of human hemoglobin subunits obtained with the time correlated single photon counting method using a sync-pumped picosecond dye laser as the excitation source.

Derivative	$\tau_1(ps)$	$\tau_2(ns)$	$\tau_3(ns)$	F_1	F_2	F_3
βO_2	95	2.65	6.5	50	21	29
βCO	90	2.55	6.45	41	27	32
$\beta(deoxy)$	90	2.30	6.3	50	21	29
αO_2	80	2.2		80	20	
αCO	85	2.3		65	35	
$\alpha(deoxy)$	65	1.9		77	23	
HbO_2	90	1.9	5.4	24	40	37
$Hb(deoxy)$	70	1.8	4.9	30	41	29
$HbCO$	70	1.8	4.9	30	45	25

Sources: Szabo, A.G., Krajcarski, D., Zuker, M. and Alpert, B. 1984. Chem. Phys. Letters. 108, 145-149 and Albani, J., Alpert, B., Krajcarski, D.T. and Szabo, A.G. 1985. FEBS Letters. 182, 302-304.

Three components were found for the hemoglobin (6 Trp residues) and the tetrameric beta subunits (8 Trp residues), while the alpha chain (one Trp residue), its fluorescence decays with two components. What we may conclude from these results?

1) We cannot assign the fluorescence in hemoglobin to the β_{37} Trp residue unless the chains in the hemoglobin are in a conformation that inhibited completely the emission of the other Trp residues.

2) It is very difficult, sometimes impossible, to assign a specific fluorescence lifetime to a specific Trp residue.

3) The tetrameric beta subunit does not present any T-R conformational change, however the fluorescence decay is three exponentials as for the human hemoglobin.

Thus it seems that there is no justification for attributing the fluorescence in hemoglobin to rigid states like T and R.

Then, what could be the origins of these multiexponential decays?

- A ground state heterogeneity due to the presence of an equilibrium between different conformers. The relative orientations Trp-heme are rigid and different from one conformer to another; each conformer presents a specific fluorescence lifetime. The three exponential fluorescence decay from the hemoglobin rules out the presence of the two static conformations T and R supposed to be in equilibrium.

- Emission from two rigid species would imply that the fractional contributions of each fluorescence lifetime should be the same whatever the state of ligation, which is not the case (Table 9.2).

- Different non-relaxing states of the Trp emission. The time dependence may vary from a protein to another or from a ligation state to another. In this case, at least one of the amplitudes (the pre-exponential factors) should be negative which is not the case. Thus, emission from two or different relaxing state is to be ruled out for the Trp emission of hemoglobin and its subunits.

- The Trp-heme orientations change continuously with time. The dynamic of the protein depends on the ligation state. Thus the number and the values of lifetime components will be dependent on the internal mobility of the proteins. The two following examples will illustrate this phenomenon: the hemoglobin alpha chain (one Trp residue) exhibits double exponential fluorescence decay while the sperm whale met-myoglobin (2 Trp residues) exhibits mono-exponential fluorescence decay. This may be explained by the fact that the alpha chain has internal motions different from those in the myoglobin. In one case, two average dynamic conformations are mathematically obtained, while in the second case only one average conformation is observed.

The two fluorescence lifetimes of the Trp residue of alpha chain in hemoglobin could be assigned to two spatial positions of the Trp residue relative to the heme. One of these positions favors energy transfer from the Trp to the heme giving a low fluorescence lifetime in the picosecond range. The second position is not in favor of the energy transfer yielding by that a fluorescence lifetime in the nanosecond scale.

3. Properties of protoporphyrin IX in different solvents and in apomyoglobin

3a. Chemical structure of porphyrins

Porphyrins are composed of four pyrrolic rings linked by four methane bridges (Fig. 9.3) (Falk, 1975). Two nitrogen atoms hold each one a hydrogen atom. Strong bases can remove the two central hydrogens while in presence of acid the two other nitrogens can be protonated.

After loosing its two hydrogen atoms, the pyrrolic cycle can bind a meal forming a metalloporphyrin. Figures 9.4 and 9.5 show respectively the intramolecular distances and the angles of the chemical bonds of the porphyrin (Webb and Fleischer, 1965).

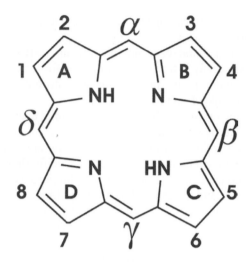

Figure 9.3. Chemical structure of the porphyrin ring. Source: Falk, J. E. 1975. in « Porphyrins and metalloporphyrins". Edited by Kevin M. Smith. Elsevier Scientific Publishing Company.

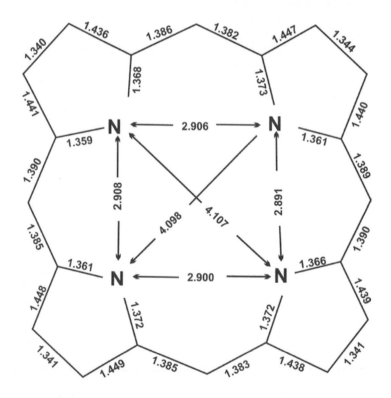

Figure 9.4 Porphyrin intramolecular distances (Å). Source: Webb, L. E. and Fleischer, E. B. 1965. J. Chem. Phys. 43, 3100-3111.

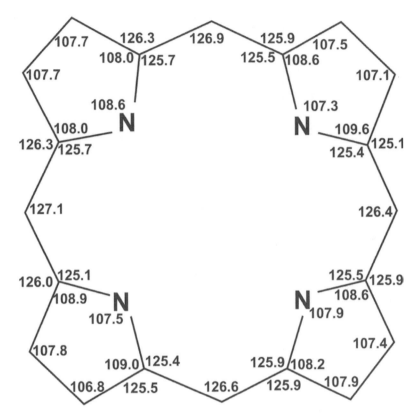

Figure 9.5. Porphyrin bond angles (degrees). Source: Webb, L. E. and Fleischer, E. B. 1965. J. Chem. Phys. 43, 3100-3111.

3b. Spectral properties of protoporphyrin IX in different solvents

Porphyrins can be dissolved in different solvents except in water where it aggregates. Absorption spectrum of porphyrins is very sensitive to their states of aggregation and the fluorescence emission occurs principally from the monomer state. Therefore, it would be possible to obtain the absorption spectrum of the monomers by recording the fluorescence excitation spectrum of a very dilute solution of porphyrin. The Soret band located between 350 and 450 nm (Fig. 9.6) is attributed to the electronic transitions between atoms of carbons of the tetrapyrrolic cycle. These transitions occur from the last occupied orbital of symmetry A_{1u} to the first empty orbital of symmetry E_g (Fig. 9. 7).

It is important to remind that binding of iron to the porphyrin induces a 100% fluorescence quenching of the porphyrin. Binding of zinc or tin induces a loss in the fluorescence of the porphyrin ring, but this loss is not total. Therefore, heme does not fluoresce while zincporphyrin and tinporphyrin display intrinsic fluorescence that can be used to study the structure and dynamics of hemoproteins.

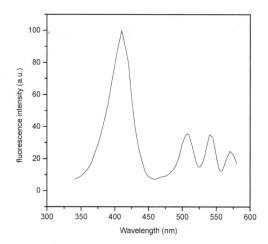

Figure 9.6. Fluorescence excitation spectrum of protoporphyrin IX dissolved in glycerol-water solution and recorded at − 10°C 340 to 575 nm.

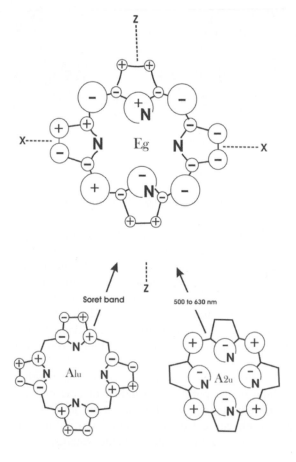

Figure 9.7. Attribution of the electronic transitions of porphyrin.

In the visible between 500 and 650 nm, porphyrins display 4 absorption bands. The intensity ratios between these bands are function of the nature of lateral chains "carried" by the pyrrolic ring. These bands are the results of electronic transitions of nitrogen atoms of the pyrrolic ring. These transitions occur from the full electronic orbitals of A_{2u} configuration to the empty orbitals of the E_g configuration (Fig. 9.7).

Metalloporphyrins show two degenerated bands α and β that result respectively from the association of bands I and III and bands II and IV.

Table 9.3 displays the positions of the absorption peaks of porphyrin IX in solvents of different polarities and dielectric constants. The table shows that water induces a shift to the blue of bands I and II and a shift to the red of bands III and IV of the absorption spectrum compared to the spectrum obtained in other solvents.

Table 9.3. Positions of the absorption peaks of protorphyrin IX obtained at 20°C in solvents of different polarities (Z) and dielectric constants (D).

Solvent	D	Z	Soret	IV	III	II	I
H_2O	80	94.6	400	506	541	568	621
DMSO	49	71.1	405	504	539	575	629
DMF	37	68.5	406	505	539	575	631
Methanol	33	83.6	403	504	539	574	629
Ethanol	24	79.6	401	503	538	576	627
Dioxane / eau 70% / 30%	16	82.8	402	503.5	538	574	630
Dioxane	2	72	405	504	538	575	630.5

The values of Z and D are from: Turner, D.C. and Brand, L. 1968. Biochemistry. 7, 3381-3390. DMF : dimethylformamide. DMSO: Dimethylsulfoxide.

The presence of water disrupts the absorption spectrum of protoporphyrin IX as it can be seen in figure 9.8a and b where absorption spectra are displayed in pure dioxane and in a solvent of 70% dioxane-30% water. The spectrum distortion is the result of porphyrin aggregation by water.

Emission of protoporphyrin IX occurs with two bands, one highly intense band located at around 630 nm and the second located around 690 to 700 nm (Fig. 9.9). Table 9.4 gives the dependence of the first emission peak of protoporphyrin IX on the medium.

Table 9.4. Position of the first emission band of protoporphyrin IX obtained at 20°C in solvents of different polarities (Z) and dielectric constants (D).

Solvent	D	Z	λ_{em} (nm)
H_2O	80	94.6	623
DMSO	49	71.1	631
DMF	37	68.5	632
Methanol	33	83.6	632
Ethanol	24	79.6	632
Dioxane	2	72	635

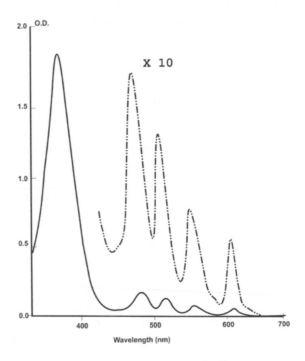

Figure 9.8a. Absorption spectrum of proptoporphyrin IX in dioxane.

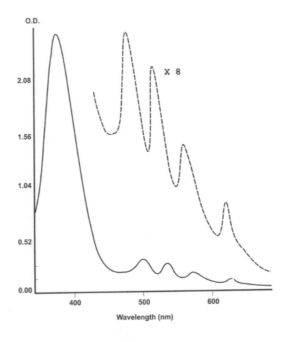

Figure 9.8b. Absorption spectrum of protoporphyrin IX in 70% dioxane-30% water.

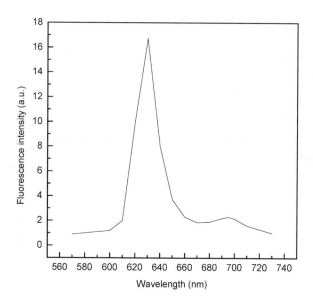

Figure 9.9. Fluorescence spectrum of protoporphyrin IX in DMSO. λ_{ex} = 514 nm.

To the difference of other fluorophores such as TNS or calcofluor, the position of the emission peak of porphyrin does not depend on the medium polarity or dielectric constant (Table 9.4 and Fig. 9.10).

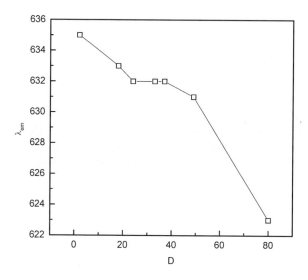

Figure 9.10. Position of emission maximum of porphyrin IX in solvents of different polarities

Also, dependence of the fluorescence emission on the medium viscosity was studied by varying the Ficoll concentration. Addition of sucrose allows maintaining the polarity of the medium constant. Figure 9.11 clearly shows that the viscosity has no effect on the maximum emission of protoporphyrin IX.

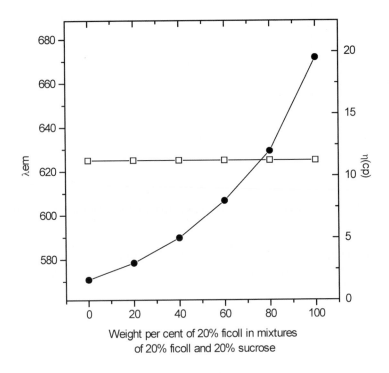

Figure 9.11. Correlation between the emission peak (●) of protoporphyrin IX with viscosity η (□) of sucrose -Ficoll mixtures.

We measured also the fluorescence lifetime of protoporphyrin IX at 20°C in different solvents. The results shown in Table 9.5 indicate clearly that the fluorescence intensity decays with two lifetimes. The major component (94 % of the total fluorescence) corresponds to 17 ns. Water decreases the contribution of the larger component and increases the heterogeneity of the emission. The mean fluorescence lifetime decreases in presence of water indicating that aggregation is occurring between the porphyrin molecules. Fluorescence quenching via energy transfer and/or local dynamics induces a decrease in the mean fluorescence lifetime

Dimerization of protoporphyrin IX in aqueous solution has been studied by fluorimetric techniques. Fluorescence intensity increase linearly with the concentration of protoporphyin in solution up to 0.05 µM. Beyond this concentration, the intensity is no more linear and goes downward as the result of the porphyrin aggregation. In an organic solution, the fluorescence intensity of porphyrin increases linearly with the increase of the porphyrin concentration (Margalit et al. 1983).

Table 9.5. Values of the fluorescence lifetimes (ns), the corresponding fractional intensities of protoporphyrin IX measured in different solvents and the mean fluorescence lifetime τ_0.

Solvent	τ_1	τ_2	τ_0	f_1 (%)	f_2 (%)
DMF	17.270 $\pm$ 0.029	3.947 $\pm$ 0.200	16.34	93	7
DMSO	16.936 $\pm$ 0.067	3.070 $\pm$ 0.212	16.10	94	6
Ethanol	17.747 $\pm$ 0.061	4.325 $\pm$ 0.376	17.07	95	5
Dioxane	17.404 $\pm$ 0.109	4.283 $\pm$ 0.305	16.61	94	6
Dioxane / water 70% / 30%	17.484 $\pm$ 0.086	4.562 $\pm$ 0.131	15.67	86	14
Water	15.412 $\pm$ 0.182	2.148 $\pm$ 0.071	11.70	72	28

3c. Spectral properties of protoporphyrin IX bound to apomyoglobin (Mb^{desFe})

Addition of protoporphyrin IX to a solution of apomyoglobin induces an increase of the optical density of bound protoporphyrin observed at 408 nm. At the inflection point of the titration curve, we obtain a stoichiometry of 1 porphyrin for 1 apomyoglobin (Fig. 9.12).

Absorption spectrum of the complexed porphyrin is characteristic of the absorption of porphyrin in neutral medium (Fig. 9.13).

Fluorescence emission spectrum of Mb^{desFe} shows a peak located at 629 nm. The fluorescence lifetime is dominated by decay equal to 17.8 ns with fractional contribution equal to 98%. The peak of the first emission band is close to that of the emission occurring when protoporphyrin is dissolved in water. This could indicate that the heme pocket is polar, although X-ray diffraction studies (Fraunenfelder et al. 1979) and fluorescence studies using TNS as a probe (Lakowicz and Keating-Nakamoto, 1984) have indicated that heme pocket of myoglobin is hydrophobic. Anyway, this hydrophobic feature of the heme pocket is characteristic to all hemoproteins. Therefore, one may conclude that porphyrin cannot be an indicator of the polarity of the heme pocket.

Fluorescence emission of protoporphyrin IX, to the difference of TNS, is not sensitive to the polarity or the viscosity of the surrounding environment, nevertheless the probe can be used to study the dynamics of the local environment.

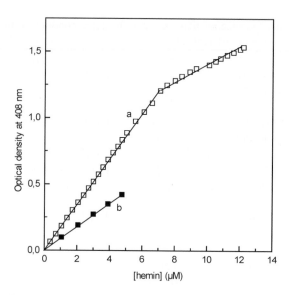

Figure 9.12. Titration curve of 7.5 µM of apomyoglobin with hemin (plot a). After each addition of hemin, the optical density at 408 nm was recorded. The inflection point occurs at 6.5 µM of hemin. Thus, the stoichiometry of the complex is 1:1. Plot b corresponds to the absorption of free hemin.

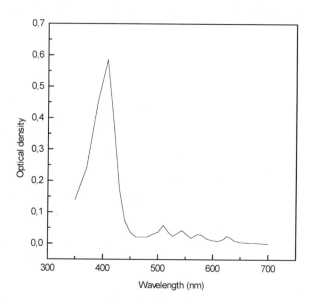

Figure 9.13. Absorption spectrum of porphyrin myoglobin (Mb^{desFe}) at pH 7 and at 20°C.

4. Dynamics of protoporphyrin IX embedded in the heme pocket

In order to investigate the dynamics of the heme pocket of myoglobin, we have used the fluorescence properties of myoglobin labeled with metal-free protoporphyrin (Mb^{desFe}). Steady-state measurements of emission anisotropy as a function of temperature and viscosity, or under conditions of oxygen quenching give information on the dynamics of the fluorescent dye. With these two techniques we have revealed an angular displacement of the porphyrin and obtained the activation energy of its motions. Limiting polarization at λ_{ex} ($P_o = 0.267$) was obtained at an emission wavelength of 630 nm and at $-40°C$. Measurements were performed with a SLM instrument.

Fluorescence lifetimes were obtained by the cross-correlation phase and modulation method. At 20°C we found fluorescence lifetimes of 17.8 and 2.3 ns for the porphyrin and zincporphyrin, respectively.

4a. Protein rotational correlation time

The rotational correlation time (ϕ_P) of myoglobin is equal to 9.2 and 5.5 ns at 15 and 35°C. The fluorescence lifetime (17.8 ns) of the porphyrin inside the myoglobin molecule is about twice the value of ϕ_P. Thus, a Perrin plot representation obtained by varying the temperature should enable us to obtain a correct experimental value of ϕ_P. In fact, during the fluorescence lifetime, the protein has enough time to make a complete rotation around it axis, independently of whether porphyrin displays free motions or not. In fact, the rotational correlation time measured from the Perrin plot at 15 and 35°C are in good agreement with those calculated theoretically (see Table 9.6).

Table 9.6. Rotational correlation time ϕ_P of Mb^{desFe} obtained according to the classical Perrin plot, to the quenching emission anisotropy and to the theoretical equation.

Temperature (°C)	Rotational correlation time (ns)		
	Perrin Method	Quenching experiment	Calculated
15	10	7.8	9.2
35	5.8	4.7	5.5

Source: Albani, J. and Alpert, B. 1986. Chem. Phys. Letters. 131,147-151.

4b. Activation energy of the porphyrin motions in the heme-pocket

Isothermal depolarization experiments were carried out at different temperatures (10 to 35°C) to characterize the thermal dependence of the local porphyrin motions. Fig. 9.14 shows data obtained by addition of sucrose at 15 and 35°C. The extrapolated polarization P(o) of the Perrin plot exhibits a clear trend to increase when the temperature decreases.

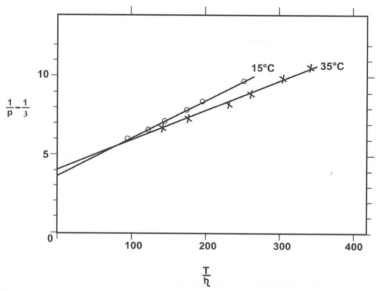

Figure 9.14. Perrin plots of Mb^{desFe} obtained by isothermal variation of the viscosity at 15 and 35°C upon addition of sucrose. $\lambda_{ex} = 514$ nm and $\lambda_{em} = 630$ nm. Source: Albani, J. and Alpert, B. 1986. Chem. Phys. Letters. 131, 147-151.

The energy barrier necessary for free rotation of the fluorescent dye in the myoglobin conjugate can be obtained by the equation given by (Wahl and Weber, 1967):

$$\left(\frac{1}{P} - \frac{1}{3}\right) / \left(\frac{1}{P_{(O)}} - \frac{1}{3}\right) = A \exp(-E_a / RT) \tag{9.1}$$

where P_o is the limiting polarization, E_a the activation energy and A is a constant. This equation can be written in the Arrhenius form:

$$\text{Log}\left(\frac{1}{P_{(O)}} - \frac{1}{3}\right) = cte + \frac{E_a}{2.3 \; RT} \tag{9.2}$$

The slope of the Arrhenius plot gives $E_a = 0.9$ cal/mole. The porphyrin motions require a very low energy to be activated. This means that the dye displays free motions inside the heme-pocket.

4c. Residual internal motions of porphyrin

The fluorescence anisotropy A of a probe ligated to a protein is given by

$$A(\tau) = \frac{\alpha \, A_{(O)}}{1 + (1/\phi_T + 1/\phi_P)\tau} + \frac{(1-\alpha) \, A_{(O)}}{1 + \tau/\phi_P} \tag{5.21}$$

where τ is the fluorescence lifetime in presence of a precise quencher concentration, ϕ_T and ϕ_P are the rotational correlation times for the residual motions of the fluorophore inside the protein and the protein rotation which drives the fluorophore, respectively. α and $1 - \alpha$ are the weighting factors for the respective depolarizing processes. When the fluorescence lifetime τ decreases by collisional quenching, the anisotropy $A(\tau)$ increases and tends to the limiting anisotropy A_o. However, for a complete quenching ($\tau = 0$ ns), the existence of the residual motions will lead to an extrapolated anisotropy $A_{(O)}$ different from A_o. The ratio $A_{(O)} / A_o$ reveals the relative importance of the residual motions:

$$A_{(O)} / A_o = 1 - \alpha \qquad (5.19)$$

and the average angular displacement θ of the fluorophore inside the protein is given by

$$\cos \theta = (1 - 2\,\alpha\,/\,3\,)^{1/2} \qquad (5.20)$$

Fluorescence quenching experiments with oxygen produce emission with non-exponential decays. So, measurements of the emission anisotropy as a function of added collisional quencher were made with the steady fluorescence intensity, which integrates the different weighted fluorescence lifetimes. Typical quenching emission anisotropy plots of $1\,/\,A$ versus F/F_o of Mb^{desFe} are shown in Fig. 9.15. The plots are linear to within experimental error and can be described by the simplified relation:

$$\frac{1}{A} = \frac{1}{A_{(O)}} = \frac{\tau_o\,F\,/\,F_o}{\phi_P\,A_{(O)}} \qquad (9.3)$$

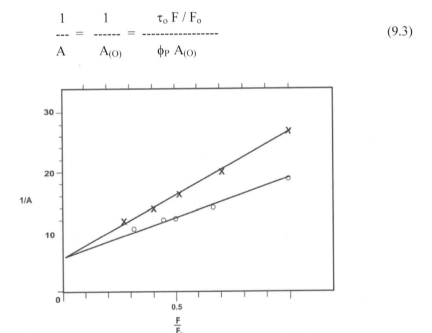

Figure 9.15. Quenching emission anisotropy plots of Mb^{desFe}. Φ_P was obtained from the slope of $1\,/\,A$ versus relative fluorescence intensity $F\,/\,F_o$. Experiments were carried out at 15 and 35°C with $\lambda_{ex} = 514$ nm and $\lambda_{em} = 630$ nm. Source :Albani, J. and Alpert, B. 1986. Chem. Phys. Letters. 131, 147-151.

The rotational correlation times ϕ_P obtained at 15 and 35°C are presented in Table 1.6. These values are slightly lower than the rotational correlation time of the protein. These smaller values are certainly due to the internal porphyrin motions ϕ_T. This rotational diffusion will induce an extrapolated anisotropy $A_{(O)} = 0.174$ different from the limiting one $A_o = 0.195$. As

$$P = 3A / (2+A) \qquad (5.4)$$

the value of $A_{(O)}$ is in good agreement with the extrapolated polarization $P_{(O)} = 0.245$ determined by the Perrin plot obtained by varying the temperature. The values of the weighting factor α, and the rotational angle θ of the residual porphyrin motions are 0.1098 and 15° of arc, respectively. The values of ϕ_T at 15 and 35°C are 1.9 and 0.5 ns, respectively.

The low activation energy found with the Arrhenius equation is very well correlated with the presence of a local motion and explains the fact that even on complete quenching,

$$\tau = \tau_o \, F / F_o \qquad (9.4)$$

the limiting polarization of the porphyrin is not reached.

Removal of iron does not affect the tertiary structure of myoglobin. We do not know if and how the apoprotein fluctuations are affected by the absence of the iron. However, since the activation energy of the porphyrin motions is very small, the rotation of this dye is certainly allowed by the free volume of the heme pocket and not by the fluctuations of the amino acids covering the pocket cavity.

4d. Effect of metal on the porphyrin dynamics

Free porphyrin motions could be considerably altered in the presence of a metal ion, such as the iron, which covalently links the porphyrin to the protein. Linkages between heme and protein involve iron and the porphyrin part of the heme. Iron is linked to an imidazole group of the proximal histidine, His-93. The other heme protein bonds involve, on the heme side, the propionic acid groups, the vinyl groups and the porphyrin as a whole. The two heme propionates help to stabilize the heme by making hydrogen bonds to the side chains of the distal histidine (His-64) and an arginine.

If linkages of the porphyrin part of the heme have a great importance in the iron-protein bond for stabilizing the heme protein complex, the presence of iron which links covalently the porphyrin to the protein could alter considerably the free porphyrin motion.

Heme is not fluorescent, thus, in order to study motion of porphyrin in presence of metal, we have chosen fluorescent zinc porphyrin embedded in the heme pocket of apomyoglobin. In fact, energy transfer from porphyrin to zinc is important but not complete. Fluorescence lifetime of porphyrin decreases from 17.8 in absence of zinc to 2.1 ns in its presence. Although fluorescence lifetime is smaller than the rotational correlation time of the protein, tight interaction between zinc porphyrin and the amino acids of the pocket will yield a rotational correlation time of the fluorophore equal to that of the protein. Figure 9.16 displays the Perrin plot of the $Mb^{Fe \rightarrow Zn}$ obtained at different temperatures ($\lambda_{em} = 600$ nm and $\lambda_{ex} = 517$ nm).

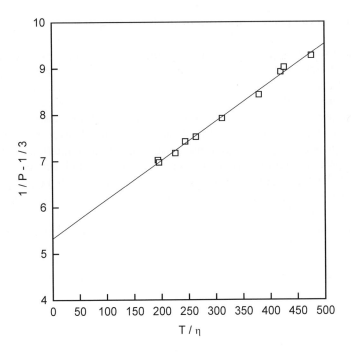

Figure 9.16. Steady-state fluorescence polarization versus temperature / viscosity ratio for $Mb^{Fe \rightarrow Zn}$. Data were obtained by thermal variation of temperature. $\lambda_{ex} = 517$ nm and $\lambda_{em} = 600$ nm.

The rotational correlation times measured at 15 and 35°C are equal to 5.5 and 3.6 ns. These values, within the range of the theoretical values of myoglobin, indicate that the zinc porphyrin is tightly linked to the heme pocket of myoglobin. However, the fact that the experimental rotational correlation time is lower than the theoretical value could be an indication of the presence of very slow motions of the porphyrin ring within its plane. The extrapolated value of P is equal to 0.176. This value, identical to that (0.1744) measured for the polarization Po at - 20°C, does not reveal the presence of motions. Thus, Perrin plot experiments show that zincprotoporphyrin IX is rigid within the heme pocket. This result is in good agreement with that found using Raman spectroscopy. In fact the authors found that the zinc-imidazole bond is anomalously weak and the orientation of the proximal histidine is suboptimal for binding to the zinc and that it is held relatively rigid in the heme pocket (Andres and Atassi, 1970). However, the possibility of having very fast local motions not detectable with our studies should not be excluded. Anyway, since the iron-imidazole bond is stronger than the zinc-imidazole one (Feitelson and Spiro, 1986), one can extrapolate and conclude that the heme presents very weak motions almost non-existent within its pocket in the myoglobin.

5. Dynamics of the protein matrix and the heme pocket

In the above studies with Trp residues and porphyrins our focus was on the dynamics of the fluorophores. These fluorophores belong to the protein matrix (case of Trp residues) or to the heme pocket (case of the porphyrins). Therefore, the dynamics of the fluorophores are correlated with their environment. A method that helps to study dynamic processes is oxygen quenching. Oxygen is a small non-charged molecule that diffuses inside macromolecules without any problem. However, when studying fluorescence quenching with oxygen, the calculated rate of migration of the quencher through the protein interior assumed isotropic diffusion and hence represented an average value. To address the question of anisotropic diffusion in the protein interior, i. e. varying diffusion rates in different protein domains, two or more fluorophores with different fluorescence lifetimes and located at the same site can be used. For example, fluorescence quenching can be performed on protoporphyrin IX and zinc protoporphyrin IX adducts of myoglobin. The rationale for this approach is that the shorter fluorescence lifetime of the zinc adduct (2 ns) compared to the metal-free adduct (17.8 ns) would allow us to probe a region closer to the heme pocket.

Addition of oxygen to the solutions of proteins decreases the emission intensities of the porphyrins. Figs 9.17 and 9.18 show the traditional mode of presentation for such data:

$$I_o / I = 1 + K_{SV} [Q] = 1 + k_q \tau_o [Q] \qquad (4.6)$$

where I_o and I are the fluorescence intensities in the absence and presence of quencher respectively, k_q the bimolecular diffusion constant, τ_o the mean fluorescence lifetime and $[Q]$ the concentration of oxygen in the solution.

For $Mb^{Fe \to Zn}$ the Stern-Volmer plots are linear. At 20 °C the value of the kinetic constant k is 7.6×10^{-9} s^{-1}. For Mb^{desFe} the data show an upward curvature. In this case, the initial slope of the Stern-Volmer plot will give the value of the dynamic quenching constant inside the protein, regardless of whether this upward curvature characterizes a static component, a time-dependent diffusion, or a quencher partition dependent. In fact, as there is no reason for the oxygen molecules to accumulate in the protein interior, the interpretation of the complete quenching curves is complex. For the moment, we are limiting our analysis to the migration process in the low oxygen concentration regime. In addition, the quenching process requires the migration of the oxygen through the protein to the vicinity of the fluorophore. So the Stern-Volmer bimolecular quenching constant represents the average of the details of quencher migration in the highly organized macromolecular structure. Thus, sophisticated theoretical equations used for diffusion studies in a homogeneous medium do not seem appropriate to describe migration processes through the protein matrix. For these reasons, we used the Stern-Volmer rate constant as a simple indicator of the changes in the oxygen random-walk mechanism.

The value obtained for the quenching constant k in Mb^{desFe} is equal to 8.4×10^8 M^{-1} s^1 at 20°C. This value corresponds to an intermediate one between simultaneous contributions of different rates: oxygen diffusion in water (k = 1.3×10^{10} M^{-1} s^{-1}), oxygen penetration into the protein (k = 3.8×10^8 M^{-1} s^{-1}), migration inside the protein (k = 2×10^8 M^{-1} s^{-1}) and the effective quenching rate which is too large to be measured by existing techniques. The quenching rate obtained in this case, $k = 8.4 \times 10^8 M^{-1} s^1$, shows a value close to those of the penetration and migration processes and thus most probably represents these events.

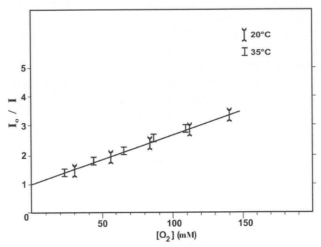

Figure 9.17. Plots of the fluorescence quenching of $Mb^{Fe \to Zn}$ by oxygen. From the standard deviation of the plots, the same slope was obtained at 20 and 35°C. Source: Albani, J. and Alpert, B. 1987. Eur. J. Biochem. 162 : 175-178. Authorization of reprint accorded by Blackwell Publishing.

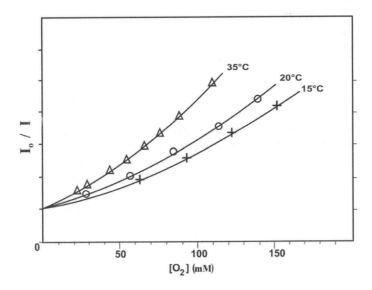

Figure 9.18. Stern-Volmer plots of the fluorescence intensity quenching of Mb^{desFe} by oxygen. The $k\tau_o$ values were obtained from the initial slopes of each plot. Source: Albani, J. and Alpert, B. 1987. Eur. J. Biochem. 162 : 175-178. Authorization of reprint accorded by Blackwell Publishing.

Therefore, this kinetic constant is an indication of oxygen migration through the protein matrix far from the heme region. In the case of $Mb^{Fe \rightarrow Zn}$, as the result of the short fluorescence lifetime, $\tau_o = 2.1$ ns, the dynamic quenching process of zinc-protoporphyrin requires oxygen diffusion in spatial proximity to the fluorophore. Thus, the quenching constant, 7.6×10^9 $M^{-1}s^{-1}$, appears essentially to represent oxygen migration in and near the heme pocket. Oxygen accessibility for these two regions of the protein can be studied since the binding of the two probes is not accompanied by significant changes in the structure of the protein.

Oxygen quenching of $Mb^{Fe \rightarrow Zn}$ seems to be independent of the temperature, while that of Mb^{desFe} is found to be temperature-dependent. As the slopes of the Stern-Volmer plots are equal to $k \tau_o$ and since the fluorescence lifetimes, τ_o of both porphyrins do not change with temperature, the variation of the slope reflects the change that occurs in the diffusion of the oxygen with the temperature. From the slopes at low quencher concentrations, we found values of k equal to $8.4 \pm 0.4 \times 10^8$ $M^{-1} s^{-1}$ and $13 \pm 0.4 \times 10^8$ $M^{-1} s^{-1}$ at 20° and 35°C respectively. To be sure that the protein medium determines the quenching rate and not the temperature dependence of the viscosity of the external water, the diffusion constant of oxygen in water was measured at 20° and 35°C. The oxygen diffusion constant does not exhibit real temperature dependence (1.3×10^{10} $M^{-1} s^{-1}$ and 1.5×10^{10} $M^{-1} s^{-1}$, respectively).

Thus, oxygen displacement is temperature-dependent when these molecules progress through the apoprotein medium and necessitates activation energy of about 29 kJ. When molecules reach the space of the heme cavity domain, collisions with the protein diminish and oxygen diffusion becomes independent of the temperature with practically no activation energy. Thus, microscopic details of the ligand migration follow the rapid fluctuations of the different regions of the protein.

According to our analysis, it is the lifetime of the excited state of the fluorophore, which is the determining factor in revealing the existence of different domains. While the quenching process retains its dynamic character, the details of this quenching in different domains of the protein as investigated by the oxygen migration process are fundamentally different.

The 17.8-ns lifetime of the protoporphyrin IX emission is sufficient to permit oxygen situated in any part of the protein at the instant of fluorophore excitation to reach the porphyrin during the persistence of the excited state; therefore the calculated migration rate *k* for oxygen through the protein represents an average value for all domains of the protein.

For zinc protoporphyrin, on the other hand, the emission lifetime is shorter (2.1 ns) and during the persistence of the excited state, only the few oxygen molecules present in the region of the protein directly surrounding the fluorophore are involved in the quenching process. Since the zincprotoporphyrin is tightly bound to the amino acids of the heme pocket (Fig. 9.16), fluctuations within the heme pocket and at its surface are almost absent. Therefore, oxygen can diffuse much easily when it reaches the heme pocket area without being inhibited by the fluctuations of the amino acids of the heme pocket, as it is the case when oxygen diffuses in the protein matrix.

The fact that oxygen diffusion constant near the heme is significantly higher than that through the protein matrix, indicates a heterogeneity in the accessibility of the oxygen to the heme. Thus the diffusion of the quencher molecules is not restricted to a unique path to reach the heme.

Displacement of oxygen through the protein is affected by its fluctuations. Raising the temperature increases the globin fluctuations and favors the diffusion of the oxygen molecules through the apoprotein. This feature is not observed in the heme-pocket region. Thus ligand diffusion capacity through the different parts of the protein depends critically on the constraints and fluctuation characteristics of each domain in the protein.

6. Significance of the upward curvature

Figure 9.19 shows the analysis of the quenching data of Mb^{desFe} (Figure 9.18) in terms of static and dynamic quenching, at 15 and 35°C. In this case, the Stern-Volmer equation can be written as (Eftink and Ghiron, 1976; Narasimhulu. 1988):

$$I_o / I = (1 + K_{SV} [O_2]) (1 + K_a [O_2]) \qquad (4.27)$$

where K_{SV} and K_a are the dynamic and static (association) constants, respectively. Development of equation yields:

$$\frac{I_o / I - 1}{[O_2]} = (K_{SV} + K_a) + K_{SV} K_a [O_2] \qquad (4.29)$$

Plotting $(I_o / I - 1) / [O_2]$ as a function $[O_2]$ should allow to calculate the values of the two constants. However, one can notice that it is impossible to obtain values for the two constants. In fact, only dynamic quenching constants were obtained, equal to 13.2 and 12.3 M^{-1} at 35 et 15°C, respectively. This is probably due to the fact that one of the two constants is equal to zero. In the absence of iron, oxygen molecules do not bind to myoglobin, which eliminates the static component.

In such a case, the upward curvature can be the result of accumulation of high oxygen concentration within the heme pocket. Oxygen molecules are in direct physical contact within the porphyrin. The probability of oxygen-porphyrin collision increases with the increase of oxygen pressure. Oxygen will accumulate within the heme pocket inducing by that an apparent static quenching.

This phenomenon is referred as "sphere of action" within which the probability of quenching is unity. The sphere of action is characterized by a volume v. Determination of this volume is possible with Eq. 9.5 (Frank and Vavilov, 1931)

$$I_o / I = (1 + K_{SV} [Q]) e^{V[Q]} \qquad (9.5)$$

V is the volume occupied by one mole of the two interacting entities, oxygen and porphyrin.

The volume v of the sphere of action or the volume occupied by one molecule is equal to:

$$v = 1000 \, V / N \qquad (9.6)$$

where N is the Avogadro Number (6.022×10^{23}) and 1000 is the number of cm^3 in 1 liter.

V can be determined from the asymptote drawn at low oxygen concentration (Fig. 9.20, slope 1). In fact, the asymptote supposes that exponential factor is absent and thus the Stern-Volmer plot is linear and can be described with the classical Stern-Volmer Equation (Eq. 4.6).

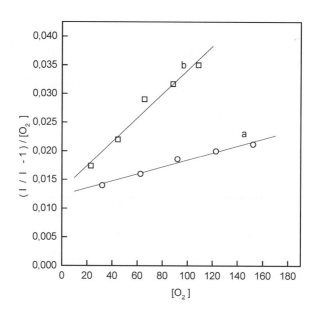

Figure 9.19.Analysis of the fluorescence intensity quenching by oxygen of myoglobin porphyrin with Eq. 4.29. Experiments were performed at 15 (a) and 35°C (b). Only dynamic quenching constants were obtained, equal to 13.2 and 12.3 M^{-1} at 35 et 15°C, respectively.

Table 9.7 shows the values of V calculated at different oxygen concentrations. One can notice that the values are very close one to the others. The mean value of V obtained from the six measurements is 1.7245 M^{-1}.

Eq. 9.6 yields a value of v equal to 286×10^{-23} cm^3 or 2860 $Å^3$. The radius r of this volume is equal to 8.8 Å. This value is in the same range of the sum of the radius of oxygen (2Å) and of that of porphyrin (6Å).

Figures 9.19 and 9.20 along with the corresponding calculations are the proof that oxygen does not bind to Mb^{desFe} since iron is absent. Only dynamic quenching occurs accompanied with an accumulation of oxygen molecules in the heme pocket.

Sphere of action was also observed for the interaction between oxygen and perylene in dodecane (Lakowicz and Weber. 1973) To the difference of Mb^{desFe}, both the fluorophore (the perylene) and the quencher (oxygen) diffuses in the medium. Porphyrin is located within the heme-pocket of apomyoglobin and does not diffuse independently from the protein. However, since quenching of porphyrin fluorescence is the result of dynamic collision with oxygen, the sphere of action calculated characterizes oxygen-porphyrin interaction.

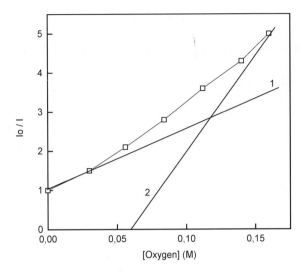

Figure 9.20. Determination of the sphere of action of fluorescence quenching with oxygen of PPIX in myoglobin. Slope 1 yields a value for r = 8.8 Å. (R of porphyrin + R of oxygen = 6 + 2 = 8 Å.) Slope 2 yields a value for r resulting from oxygen diffusion in the whole protein. At O_2 = 0.056 M, r = 18.2 Å. If myoglobin is considered as a sphere, its radius is equal to 19.4 Å.

Table 9.7. Volume of one mole of porphyrin-oxygen collision at different oxygen concentrations and at 20°C.

$[O_2]$ (M)	V (M^{-1})
0.03	1.13
0.056	1.785
0.084	1.835
0.112	1.93
0.14	1.783
0.16	1.882

Is it possible to measure the sphere of action of the myoglobin-oxygen interaction? This can be done if we plot the asymptote at high oxygen concentration (Fig. 9.20). The volume of the sphere of action can be calculated at the intercept of the asymptote at the X axis for $I_0 / I = 1$. The graph shows that the intercept occurs at $[O_2]$ equal to 0.058 M. The value of the radius of the sphere of action is calculated to be equal to 18.2 Å. This value is within the same range of that calculated theoretically for the sum of myoglobin and oxygen radius (19.4 Å + 2 Å = 21.4 Å).

Table 9.8 shows the values of V and of the radius of the sphere of action calculated at different oxygen concentration. We can notice that the values differ from one concentration to another. At the highest oxygen concentrations, the values are smaller. This clearly indicates that upon increasing oxygen concentration, fluorescence lifetime decreases and the studied protein area is more and more small. This result is in complete opposition to that obtained when the asymptote was drawn at low oxygen concentration. In this case, at all oxygen concentrations, one can determine very close volume of sphere of action.

Table 9.8. Different parameters of the sphere of action calculated from the asymptote plotted at high oxygen concentrations.

$[O_2]$ (M)	V (M^{-1})	v ($Å^3$)	r (Å)
0.056	15.125	25110	18.16
0.084	4	6642	11.66
0.112	1.325	2200	8
0.14	0.168	279	4

7. Effect of sucrose on the bimolecular diffusion constant

Since fluorescence quenching of porphyrin with oxygen is a dynamic process, we calculated the bimolecular diffusion constant at two oxygen concentrations. At low oxygen concentration, fluorescence lifetime is equal to 17.8 ns. Thus, we are observing the fluorescence intensity quenching of the porphyrin with oxygen molecules located by and far from the heme pocket, i.e., we are studying the diffusion of oxygen molecules in the protein matrix. Therefore, the value of k_q will depend on the dynamics of the studied area.

Upon increasing oxygen concentration, fluorescence lifetime decreases and we are going to monitor fluorescence quenching by oxygen molecules that are closer to the porphyrin. Thus, the studied domain is much more restricted than in the case of longer fluorescence lifetimes. At very high oxygen concentrations, fluorescence lifetime is short and the monitored domain will be that of the heme pocket.

We have simplified the analysis to two domains, the whole protein (τ_o = 17.8 ns) and a more restricted area, the heme pocket and some of the protein matrix surrounding it (τ_o = 10 ns at 68 mM of oxygen at 20°C and 8.3 ns at 53 mM of oxygen at 35°C) (Only, plots at 20°C are shown, figure 9.20). From the slopes at low oxygen concentrations, we found values of k_q equal to 8.4 x 10^8 M^{-1} s^{-1} and 13 x 10^8 M^{-1} s^{-1} at 20 and 35°C, respectively. The slopes at high quencher concentrations give values of k_q equal to 3.65 x 10^9 M^{-1} s^{-1} and 6 x 10^9 M^{-1} s^{-1} at 20 and 35°C, respectively. The kinetic constant measured at low oxygen concentration is an indication of oxygen migration through the protein matrix far from the heme region, only. In fact, diffusion of oxygen within the heme pocket (Figure 9.17) and in water is temperature independent.

The results obtained from figure 9.20 indicate that for both slopes (at low and high oxygen concentrations), the values of k_q at 20°C are 36 to 40% lower than those measured at 35°C. Thus, the kinetic constants measured at high oxygen concentrations (low fluorescence lifetimes) reflect also the diffusion of oxygen through the protein matrix, within an area smaller than that observed at low oxygen concentrations (higher lifetimes).

Studies performed in presence of 40% sucrose (Jameson et al. 1984) showed that at low oxygen concentration (τ_o = 18.45 ns), k_q is equal to 0.23 x 10^9 M^{-1} s^{-1} while at higher oxygen concentration (τ_o is taken as 15.375 ns calculated at 116.7 mM of oxygen), k_q is equal to 1.12 x 10^9 M^{-1} s^{-1}. We can notice that at low oxygen concentration k_q value is 68% lower than that observed in the same, conditions in the absence of sucrose. At high oxygen concentrations, k_q decrease is only 25% compared to the value measured at the same conditions of oxygen concentrations in absence of sucrose.

Sucrose increases the viscosity of the medium and thus induces a decrease in the fluctuations of the protein matrix, i.e., a decrease in the mobility of oxygen molecules through the protein matrix. In this case, we should expect a value for k_q in the presence of sucrose lower than that calculated in its absence. We can notice that this decrease is not identical when analyses were made at low and at high oxygen concentrations. k_q decrease is much more important at low than at high oxygen concentrations. Protein fluctuations seem to be much more sensitive to modifications in viscosity at low than at high oxygen concentration, i.e., we are not observing the same dynamic domain, otherwise decrease in k_q value should be identical. At low oxygen concentrations (high fluorescence lifetime), we are studying oxygen diffusion mainly in the protein matrix. At high oxygen concentrations (low fluorescence lifetime) we are observing oxygen diffusion in a domain of restricted area limited to the heme-pocket and the protein matrix surrounding it. At very high oxygen concentrations, i.e., at very low fluorescence lifetime (case not reached in experiments with Mb^{desFe}), the only domain that can be observed is the heme-pocket. In this case, oxygen diffusion becomes temperature – independent. This was observed for $Mb^{Fe \rightarrow Zn}$ (Fig. 9.18).

Fluorescence quenching with oxygen of $Mb^{Fe \rightarrow Zn}$ was also studied by separating the different rate constants, the oxygen entry rate (k^+), the exit rate (k^-) and migration rate (χ). The three constants measured at different temperatures are displayed in Table 9.9.

Table 9.9. Oxygen entry rate (k^+), exit rate (k^-) and migration rate (χ) determined on $Mb^{Fe \rightarrow Zn}$.

t (°C)	k^+ ($10^9 M^{-1} s^{-1}$)	k^- ($10^9 s^{-1}$)	χ ($10^9 s^{-1}$)
5	0.20 ± 0.03	0.08 ± 0.01	2.6 ± 0.2
10	0.26 ± 0.03	0.09 ± 0.02	2.6 ± 0.2
15	0.31 ± 0.05	0.10 ± 0.02	2.6 ± 0.2
20	0.37 ± 0.05	0.11 ± 0.02	2.6 ± 0.2
25	0.44 ± 0.05	0.12 ± 0.02	2.6 ± 0.2
30	0.55 ± 0.05	0.14 ± 0.02	2.6 ± 0.2

Source: Carrero, J., Jameson, D. M. and Gratton. E. 1995. Biophysical Chemistry. 54, 143 – 154.

We can notice that only migration constant does not vary with the temperature. Since fluorescence lifetime is short and thus the studied area concerns the heme pocket and its close surrounding protein matrix, the migration constant describes oxygen diffusion mainly within this pocket. The absence of any significant motions will yield a migration constant that is temperature-independent. This is not the case for the entry and exit constants, since we notice that they are temperature-dependent and thus they vary with the protein fluctuations. Since oxygen diffusion in proteins is anisotropic and does not take one precise path, entry and exit rate constants should be considered as mean values of the different paths that oxygen molecules take to enter and to exit from the protein. Also, since the studied area is limited in space as the result of the short fluorescence lifetime, the exit and entry rate constants concern the protein matrix in proximity of the heme pocket.

In conclusion, the internal mobility of myoglobin and / or hemoglobin and its subunits appears to govern the ligand diffusion in the interior of the heme proteins. Fluorophores located in the same area such as the heme pocket and having fluorescence lifetimes different one from each other can help to investigate different areas of hemoproteins. This method is efficient and would allow drawing "a map" describing the dynamics of the different areas of a protein.

Chapter 10

FLUORESCENCE FINGERPRINTS OF ANIMAL AND VEGETAL SPECIES AND / OR VARIETIES

1. Fluorescence fingerprints of Eisenia fetida and Eisenia andrei

1a. Introduction

Performing an ecological and environmental diagnosis of an ecosystem supposes that one applies consistent methodologies on population sensitive to modifications that occur within the environment. Among these methodologies, one can mention the diagnosis at the cellular and / or individual level and the analysis of the functioning of the ecosystem. All the approaches and tests used rest on the choice of the right species. Earthworms are the most important biotic components in the soil, and thus they are commonly used in studies of toxicity.

Eisenia andrei (Bouché 1972) and *Eisenia fetida* (Savigny 1826) are recommended by the OECD for tests of acute and sub-acute toxicity of soil. These worms are easily cultured in the laboratory (Tomlin and Miller, 1989) and an extensive database on effects of all classes of chemical on these two species is also available (Edwards and Bohlen, 1996). These worms feed themselves at the surface of the soils and thus reflect the impact of the recent contributions of soil pollutants. The two species display similar body length and segment number as well as showing resemblance in the shape of the clitellum and tubercula pubertatis. They can be distinguished in the adult form due to their gross morphology mainly their pigmentation (Fig. 10.1). The taxonomic status of *Eisenia andrei* and *Eisenia fetida* was confirmed by Jaenike (1982) employing electrophoretic techniques. Until recently they have usually been considered as subspecies according to their different body pigmentation. André in 1963 described *Eisenia fetida* form *typica*, with a characteristic stripped pattern and *Eisenia fetida* form *unicolor* with a uniform reddish color. Bouché in 1972 considered that the term *unicolor* had a low systematic value, and thus designated these forms *Eisenia fetida fetida* and *Eisenia fetida andrei*. André was the first to demonstrate the specific status of these two forms by recording signs of reproductive isolation between them but he did not give them the status of separate species. This status was given later after biochemical characterizations of these species (Jaenike, 1982 ; Roch et al. 1980 ; Oien and Stenersen, 1984 ; Engelstad and Stenersen, 1991). Although the two species have been distinguished at the molecular level, they are still used indifferently, according to the norm ISO /TC 190 / SC4 in ecotoxicological studies.

NMR studies performed on tissue extracts on the coelomic fluid of *Eisenia andrei* and *Eisenia fetida* allowed to distinguish the two species one from each other (Bundy et al. 2002). We have seen in all over this book that fluorescence spectroscopy can detect differences that occur on the structural and dynamical levels purified macromolecules. The following data will show that it is possible to use fluorescence spectroscopy to detect differences in the cells at the structural and the metabolical levels.

Figure 10.1. Photos of adults *Eisenia andrei* (a) and *Eisenia fetida* (b). The two worms can be distinguished due to their different pigmentations. Sources of Figures 10.1 to 10.4 : Albani, J. R., Demuynck, S., Grumiaux, F. and Leprêtre, A. 2003. Photochemistry and Photobiology. 78, 599-602. Authorization of reprint accorded by the American Society for Photochemistry.

Fluorescence studies are performed on the coelomic liquid of *Eisenia andrei* and *Eisenia fetida*. The results described here show that the coelomic fluid of *Eisenia andrei* displays a characteristic fluorescence absent in the coelomic fluid of *Eisenia fetida*. Also, the data indicate that the two species do not metabolize the same types of molecules and thus can be differentiated quickly and easily at the molecular / metabolic level. The experiments, we are going to describe, allow easy and unambiguous identifications of both species.

1b. Results

Figure 10.2 displays the fluorescence emission spectra of coelomic fluids of *Eisenia andrei* and *Eisenia fetida* (spectra a and b, respectively). We notice that the fluorescence occurs only from *Eisenia andrei*. The fluorescence observed from *Eisenia fetida* is almost non-existent and in the most of the studied earthworms, it was identical to the diffusion of the buffer. Thus, *Eisenia andrei* and *Eisenia fetida* are physiologically and metabolically different one from each other since the coelomic fluid of *Eisenia andrei* contains a fluorescence substance that is absent within the coelomic fluid of *Eisenia fetida*.

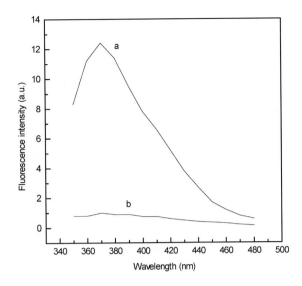

Figure 10.2. Fluorescence emission spectra of coelomic fluid of *Esenia fetida* (b) and *Esenia andrei* (a) obtained at λ_{ex} = 320 nm. λ_{em} = 370 nm. Temperature = 20°C and pH 8.6.

The fluorescence excitation spectrum of a fluorophore characterizes the electron distribution of the molecule in the ground state. Thus, it helps to identify the structure and / or the nature of the emitting molecule. The fluorescence excitation spectrum recorded on the coelomic fluid of *Eisenia andrei* shows that the fluorophore absorbs at a peak equal to 314 nm at pH 8 (Fig. 10.3). Fluorescence properties of the emitting fluorophore are characteristic of the 4-methylumbelliferyl β-D-glucoronide (MUGlcU) (Molecular Probes, 1992-1994). The absence of fluorescence in *Eisenia fetida* indicates that the metabolism of this substrate is inhibited by one or many enzymes present only in the coelomic fluid of *Eisenia fetida.*

Excitation at 280 nm of the coelomic fluid yields a fluorescence emission at 333 nm characteristic of dissolved protein (data not shown). Addition of 2-p-toluidinyl-naphthalene-6-sulfonate (TNS) induces a decrease in the fluorescence emission of the proteins and an increase of TNS fluorescence indicating an interaction between the extrinsic fluorophore and the proteins.

In presence of coelomic fluid, TNS shows a significant fluorescence. However, the position of the peak varies with the origin of the coelomic fluid. In fact, the emission peak of TNS is equal to 397 nm and 410 nm for *Eisenia andrei* and *Eisenia fetida,* respectively (Fig. 10.4). This difference is the result of the contribution of the emission of the coelomic fluid to the recorded emission spectrum. In fact, since there is no fluorescence occurring from the coelomic fluid of *Eisenia fetida*, the observed emission in presence of TNS is from the extrinsic fluorophore only. Meanwhile, the spectrum recorded for *Eisenia andrei* is the result of the contribution of both TNS and the natural fluorophore present in the coelomic fluid. Experiments with TNS confirm the fact that coelomic fluids of *Eisenia andrei* and *Eisenia fetida* differ, i.e., the metabolism of the two earthworms is not the same.

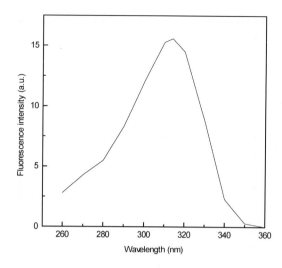

Figure 10.3. Fluorescence excitation spectrum of coelomic fluid of *Eisenia andrei*. λ_{em} = 380 nm. The peak of the spectrum is located at 314 nm. Temperature = 20°C and pH = 8.

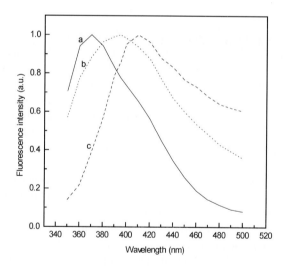

Figure 10.4. Normalized fluorescence emission spectra of coelomic fluid of *Eisenia andrei* λ_{max} = 370 nm) (a), of TNS bound to *Esenia andrei* λ_{max} = 397 nm (b) and of TNS bound to *Esenia fetida* λ_{max} = 410 nm (c). λ_{ex} = 320 nm. Temperature = 20°C and pH = 8.

In conclusion, fluorescence data shown indicates clearly that *Eisenia andrei* and *Eisenia fetida* display different molecular composition in their coelomic fluid and thus have different metabolisms. This work reveals for the first time the possibility of using fluorescence spectroscopy as a tool to perform taxonomic studies.

2. Structural characterization of varieties of crops among a species and of genetically modified organisms: a fluorescence study

2a. Pioneering work of Zandomeneghi

In general, fluorescence spectroscopy is applied to solution rather to solid. One main reason is that in solution it is easier to work on homogeneous medim. Wrighton et al (1974) and then Ramasamy et al (1986) were among the firsts to study fluorescence of powders and solid. In 1999, Zandomeneghi published a paper where emission spectra of cereal powders have been described. The author shows that there are at least three different types of fluorophores that cause the emission and reabsorption phenomena observed when the aromatic residues of the embodied proteins are excited with near-UV light (275 nm). Energy transfer occurs from the aromatic amino acids to other fluorophores located in the flour and which emit with a band centered at 430 nm. The maximum of the fluorescence excitation spectrum of these fluorophores was found at 335 nm and is found practically in all of the cereal flours investigated. Also, the authors found that the 430 nm emission is partly absorbed by other chromophores and emitted at longer wavelengths. The authors attributed fluorescence emissions in the visible to compounds such as 4-aminobenzoic acid, pyridoxine, tocopherols, etc...

Reflectance spectra show four absorption bands centered at about 280, 330, 400 and 450 nm. The two most intense bands, i.e., those at 280 and 330 nm, and that at 450 nm are revealed by the fluorescence excitation spectra and thus characterizes emitting fluorophores present naturally in the flours.

In another paper, Zandomeneghi et al (2003) put into evidence that riboflavin is one the fluorophores emitting in the visible. In fact, excitation at 430 nm induces a fluorescence emission spectrum with a maximum located at 520 nm accompanied with another peak around 475 nm. The emission peak at 520 nm characterizes that of riboflavin while that of 475 nm is the result of an emission from a fluorophore other than riboflavin. In fact, increasing the excitation wavelength to 470 or 480 nm leads to the disappearance of the peak at 475 nm (Fig. 10.5).

Then, the authors recorded the fluorescence emission of Idra flour samples enriched with riboflavin (1.47, 2.94, 6.25, 12.48 and 25 µg/g) (Fig. 10.6) then plotted the integral of the fluorescence spectra from 520 to 775 nm against the riboflavin concentration added to the flour. A quadratic function was obtained allowing the determination of the concentration of riboflavin in native Idra flour. This riboflavin concentration was found excitation wavelength dependent. For example, scattered light is very important from 480 to 500 nm excitation wavelengths. The most suitable excitation wavelength to determine correctly the riboflavin concentration without any interference was found equal to 470 nm (Fig. 10.7).

The works initiated by Zandomeneghi showed that fluorescence differs from a species to another. For example, fluorescence intensity in the U.V. of corn is not similar to that recorded for rice. Also, another interesting finding is the fact that because of the different natural emitting species, variation of the excitation wavelength leads to different emission spectra. These two fluorescence features characterized by the works of Zandomeneghi and his collaborators allowed us to initiate a new type of work which is the characterization of different varieties within the same species and the differentiation of a transgenic crop from a non transgenic one.

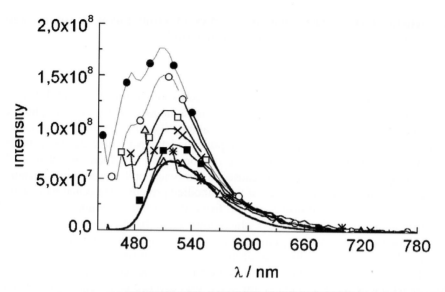

Figure 10.5. Fluorescence spectra of Hydra wheat flour on 430 (·), 440 (○), 450 (□), 460 (×), 470 (■), 480 (△), 500 (*) nm excitation; and spectrum of RBF in EtOH solution on 450 nm excitation (-). Source: Zandomeneghi M., Carbonaro, L., Calucci, L., Pinzino, C., Galleschi, L. and Ghiringhelli, S. 2003, Journal of Agricultural and Food Chemistry 51, 2888-2895. Authorization of reprint accorded by the American Chemical Society.

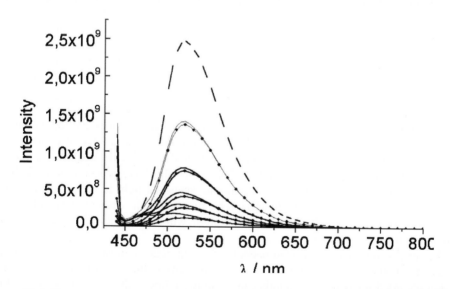

Figure 10.6. Fluorescence spectra on 430 nm excitation of Idra wheat flour added with variable amounts of RBF; from top to bottom: dashed line, 25 μg/g; solid lines, 12.48, 6.25, 2.94, 1.47, and 0 μg/g. Spectra calculated as a fraction of the most enriched sample (-·- lines) are reported too. Source: Zandomeneghi M., Carbonaro, L., Calucci, L., Pinzino, C., Galleschi, L. and Ghiringhelli, S. 2003, Journal of Agricultural and Food Chemistry 51, 2888-2895. Authorization of reprint accorded by the American Chemical Society.

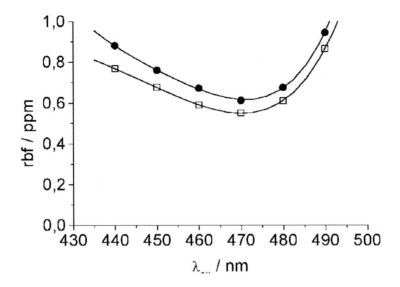

Figure 10.7. Plots of the contents of RBF (*µ*g/g) of native Idra flour against the excitation wavelength, as calculated on the basis of the data and the correspondent quadratic fits reported in figure 10.6 from fluorescence spectra (•); (□): from fluorescence spectra calculated as a fraction of the most enriched sample. Source: Zandomeneghi M., Carbonaro, L., Calucci, L., Pinzino, C., Galleschi, L. and Ghiringhelli, S. 2003, Journal of Agricultural and Food Chemistry 51, 2888-2895. Authorization of reprint accorded by the American Chemical Society.

2b. Structural characterization of crops

Emission of intrinsic fluorophores of crops depends on the structure of their microenvironments (molecular and / or cellular levels). We are going to describe here the emission of different varieties of corns (standard known as Safran, the amylose extender mutant and sweet) and of different varieties of white rice (Surinam, Thai, Sherbati and Basmati). Our results indicate that within each species, each variety has a characteristic fluorescence. Also we compared the emission of non genetically modified and genetically modified soya. Genetically modified organisms display fluorescence emission spectra different from those of the same non-GMO variety.

Figure 10.8 displays the emission spectra of standard (75% amylopectine and 25% amylose) (spectrum a) and simple mutant AE (25% amylopectine and 75% amylose) (spectrum b) corns (λ_{ex} = 320 nm). The emission peaks are located at 420 and 535 for the standard and 425 and 530 nm for the AE mutant. In this corn, the AE gene (Amylose Extender) becomes dominant, increasing the proportion of amylose.

The sweet corn « Sweet Sunglow Hybrid » (U.S.A.) displays a fluorescence emission spectrum with two peaks located at 438 and 529 nm (data not shown). All tested corns present two emission peaks.

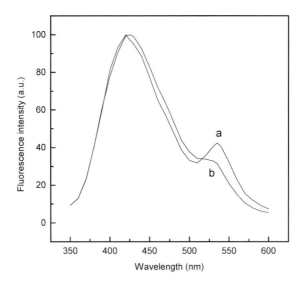

Figure 10.8. Normalized fluorescence emission spectra of standard corn (a) and of the mutant AE corn (b). $\lambda_{ex} = 320$ nm. $\lambda_{em} = 420$ and 535 nm for the standard corn and 425 and 530 nm for the mutant.

In order to know whether the two peaks originate from one group of fluorophore or two different groups, we recorded the fluorescence excitation spectra at two emission wavelengths, 540 and 435 nm.

The fluorescence excitation spectrum characterizes the state of the molecule at the fundamental state. Figure 10.9 shows the fluorescence excitation spectra of the standard (spectrum 3), the AE (spectrum 2) and a hybrid sweet (1) corn, obtained with emission wavelength of 435 nm (spectra a) and of 540 nm (spectra b).

The following conclusions can be drawn from figure 10.9:

- For each emission wavelength (435 or 540 nm), we have a specific excitation spectrum. In other terms, we have two fluorophores or two groups of fluorophores that emit differently. Excitation spectra of region (a) are characteristics of fluorophores that are the origins of the emission spectra with a peak located around 420 nm, while spectra of region (b) characterize fluorophores that are emitting with a maximum around 530 nm.

- All tested corns possess the two excitation spectra observed in figure 10.9. The positions of the maxima of these spectra are identical or close to those observed in figure 10.9 (data not shown). Although the genetic mutation affected the composition in amylose and in amylopectin and thus the structure of the starch, the global shapes of the excitation spectra are not affected, i.e., the structure of the fluorophore(s) in corn are not greatly modified. Therefore, all change in fluorescence observed between the different corns is the consequence of the structural modification within the corn owed to the genetic mutation.

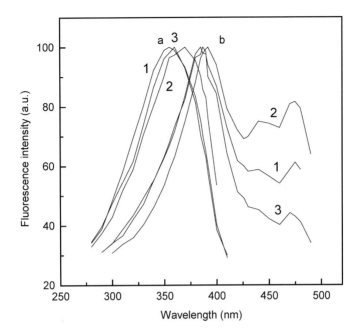

Figure 10.9. Fluorescence excitation spectra of the standard (spectrum 3), the AE (spectrum 2) and a hybrid sweet (spectrum 1) corns, obtained with emission wavelengths of 435 nm (spectra a) and of 540 nm (spectra b). λ_{max} = 360 and 387 nm for the standard, 355 and 385 nm for the AE and 370 and 392 nm for the hybrid sweet corns.

Figure 10.10 shows the variation of the position of the emission maximum with the excitation wavelength of the following corns: the Standard (a), the mutant AE (b), the hybrid American E.H.seeds (c), the French hybrid corn (Epi d'or) (d), a sweet corn cultivated in Seattle (e), and of a French non hybrid sweet corn (Epi d'or) (f). We notice that the standard corn emits at 412-414 nm at λ_{ex} 280-290 nm. The mutant AE corn displays an emission maximum located at 425 nm that remains stable at different excitation wavelengths and all sweet corns possess a fluorescence peak around 430-435 nm.

Figure 10.11 displays the fluorescence emission spectra of the standard corn when we received it in the absence of humidity (spectrum a) and after three months kept in the laboratory at normal conditions of temperature (20°C) and humidity (not measured) (spectrum b). One can clearly observe that the intensities of the two peaks is modified with time. A fresh corn not affected by the humidity of the surrounding medium yields a peak in the 420-430 nm wavelength range of higher intensity compared to the peak observed in the 530-535 nm wavelength range. When the internal humidity of the corn increases, we observe a drastic modification in the intensities of the peaks. We believe that intensity variation or / and peak ratios are proportional to the humidity percentage. The spectra in figure 10.11 indicate clearly that it is possible to find out by a simple fluorescence measurement whether a corn is of a good quality or not.

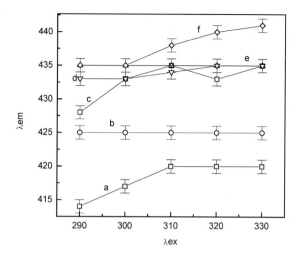

Figure 10.10. Variation of the position of the emission maximum with the excitation wavelength. of the following corns : the Standard (a), the mutant AE (b), the hybrid American E.H.seeds (c), the French hybrid corn (épi d'or) (d), a sweet corn cultivated in Seattle (e), and of a French non hybrid sweet corn (épi d'or) (f). The standard corn starts at 412-414 nm at λ_{ex} 280-290 nm. The mutant corn displays an emission maximum located at 425 nm that remains stable at different excitation wavelengths and all sweet corn possess a fluorescence peak around 430-435 nm.

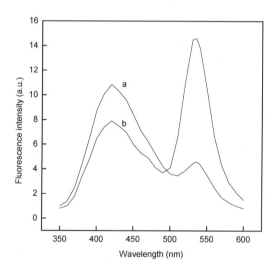

Figure 10.11. Fluorescence emission spectra of standard corn in the absence of external humidity (spectrum a) and after three months kept in the laboratory at normal conditions of temperature (20°C) and humidity (spectrum b). λ_{ex} = 320 nm.

Figure 10.12 shows fluorescence emission spectra of a Surinam variety rice as a function of wavelength. Although, the shapes of the spectra are identical for all varieties of rice, the positions of the emission spectra differ from a variety to another.

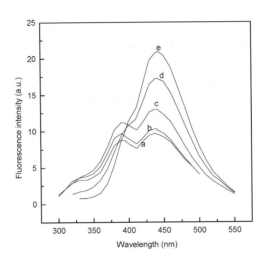

Figure 10.12. Fluorescence emission spectra of Surinam white rice recorded at $\lambda_{ex} = 280$ to 320 nm, spectra (a) to (e), respectively. The emission peaks are located at 390 nm and 438 nm at $\lambda_{ex} = 280$ to 300 nm.

Figure 10.13 displays the variation of the emission maximum as a function of the excitation wavelength for a Thaï variety rice (a) and a basmati rice (b).

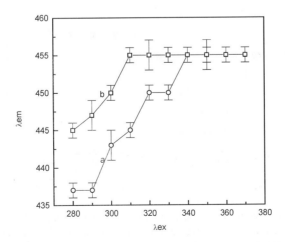

Figure 10.13. variation of the emission maximum as a function of the excitation wavelength for a Thaï variety rice (a) and a basmati rice (b).

Figures 10.14 and 10.15 displays the variation of the emission wavelength of different indian Basmati rice in presence of two Basmati rice bought from the French market.

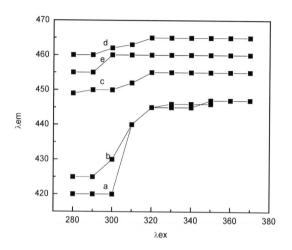

Figure 10.14. Variation of the emission maximum of different indian Basmati rice. (a) : Pusa. (b): Dehradum. (c): Taraori.   (d) : Basmati 370.   (e) : Basmati (I). The last Basmati is a mixture of pure Basmatis or a mixture of a Basmati with a non Basmati rice.

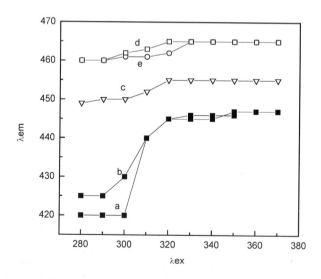

Figure 10.15. Variation of the emission maximum of different indian Basmati rice. (a) : Pusa. (b): Dehradum. (c): Taraori.   (d) : Basmati 370. (e) : Basmati (II). The last Basmati seems to correspond to a pure Basmati variety, Basmati 370.

Figure 10.16 displays the fluorescence emission spectra of a non-transgenic (a) and a transgenic (b) soya. We notice that there is a shift in the position of the emission peaks mainly of that located at 513 nm for the non-GMO soya. The position of the emission peak shifts to 520 nm in the GMO-soya. The fluorescence excitation spectra of both soya are identical, they have the same shape and the same maxima, located at 398 and 438 nm (data not shown).

Therefore, addition of a gene to soya does not affect the structure of the fluorophores present in the soya.

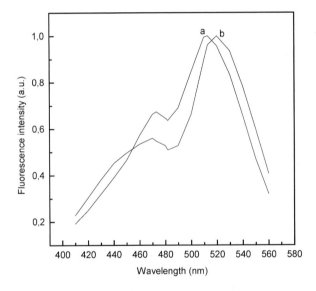

Figure 10.16. Normalized fluorescence emission spectra of non-GMO (a) and GMO soya (b). The positions of the peaks are 513 and 473 nm (non transgenic soya) and 520 and 470 nm (transgenic soya). λ_{ex} = 380 nm.

One should indicate here that we found for a non-transgenic soya cultivated in the north of France a fluorescence emission spectrum identical to that we found for the transgenic soya. Therefore, one should be careful in interpreting the results obtained with soya. Varying the culture conditions (type of soil for example) can induce different fluorescence spectra.

The method we are describing here is based on the intrinsic fluorophore properties present in crops. As the fluorescence excitation spectra indicate, these fluorophores preserve the same structure following a genetic mutation. Therefore, all changes in fluorescence parameters are the result of the structural modification that takes place within the crop. These structural modifications are obvious in corn and concern mainly the starch. Concerning genetically modified organisms, when a gene is added to a host, a structural modification can occur within the host. This structural modification may concern some local parts of the cells of the host but also a local or global cells reorganization of the host. As it is already known, the physiological properties of a cell and of a macromolecule are directly related to their structure. Therefore, inducing a genetic mutation or addition of a new gene within a host can confer to the host new physiological properties accompanied with a local modified structure.

Figure 10.17 displays the fluorescence emission spectra of quinoa in grains (spectrum a) and grounded (spectrum b). One can notice that the peak at 513 nm is not very obvious in the spectrum of the grains. Quenching of the fluorescence emission at 513 nm is important in the grain. When the grain in grounded, quenching efficiency observed in the grain is weakening inducing a significant emission of the peak at 512 nm.

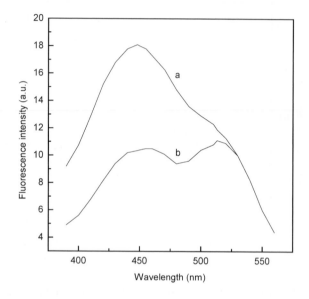

Figure 10.17. Fluorescence emission spectra of quinoa (biological culture) in grains (a) and grounded (b). λ_{ex} = 360 nm and λ_{em} = 448 nm (grains) and 455 and 513 nm (grounded).

Our results show also that the profile of the fluorescence spectrum of each of the four different species studied in this work is different from the other three. However, within each species, the profiles of the fluorescence emission spectra are identical. Therefore, one application of our method is the possibility to get a fluorescence fingerprint of a precise species and its varieties and therefore to follow the trace of this species with time.

In conclusion, the method we described here shows that biological species and varieties within each species can each have its own fluorescence fingerprint. Our method can be complementary to the polymerase chain reaction analysis and is appropriate for large-scale laboratory and industry analyses.

References

Ackroyd, R., Kelty, C., Brown, N. and Reed, M, 2001, The history of photodetection and photodynamic therapy. Photochemistry and Photobiology 74, 656-669.

Alcala, J. R., Gratton, E. and Jameson, D. M, 1985, A multifrequency phase fluorometer using the harmonic content of a mode-locked laser. Analytical Instrumentation. 14, 225-240.

Albani, J, 1985, Fluorescence studies on the interaction between two cytochromes extracted from the yeast, Hansenula anomala. Archives of Biochemistry and Biophysics 243, 292-297.

Albani, J., Alpert, B., Krajcarski, D. and Szabo, A. G, 1985, A fluorescence decay time study of tryptophan in isolated hemoglobin subunits. FEBS Letters. 182, 302-304.

Albani, J. and Alpert, B, 1986, Porphyrin motions in the myoglobin heme-pocket. Chemical Physical Letters. 131, 147-152.

Albani, J. and Alpert, B, 1987, Fluctuation domains in myoglobin. Fluorescence quenching studies. European Journal of Biochemistry 162, 175-178.

Albani, J, 1992, Motions studies of the human α_1-acid glycoprotein (orosomucoid) followed by red-edge excitation spectra and polarization of 2-p-toluidinylnaphthalene-6-sulfonate (TNS) and of tryptophan residues. Biophysical Chemistry. 44, 129 - 137.

Albani, J, 1993, A study of the interaction between two proteins, one containing a flavin mononucleotide. Journal of Biochemical and Biophysical Methods. 26, 105-112.

Albani, J. R. 1994 Effect of sialic acids and β-drug adrenergic blocker, propranolol, on dynamics of human alpha 1 acid glycoprotein. A fluorescence study. J. Biochem. 116, 625-630.

Albani, J., Vos, R., Willaert, K. and Engelborghs, Y, 1995, Interaction between human α_1-acid glycoprotein (orosomucoid) and 2-p-toluidinylnaphthalene-6-sulfonate. Photochemistry and Photobiology 62, 30-34.

Albani, J. R, 1996, Dynamics of Lens culinaris agglutinin studied by red-edge excitation spectra and anisotropy measurements of 2-p-toluidinylnaphthalene-6-sulfonate (TNS) and of tryptophan residues. Journal of fluorescence. 6, 199-208.

Albani, J. R. 1996, Motions of tryptophan residues in asialylated human α_1-acid glycoprotein. Biochimica Biophysica Acta. 1291, 215-220.

Albani, J. R, 1997, Binding effect of progesterone on the dynamics of α_1- acid glycoprotein. Biochimica Biophysica Acta 1336, 349-359.

Albani, J. R., 1997, Interaction between cytochrome b_2 core and flavodehydrogenase from the yeast Hansenula anomala. Photochem . Photobiol. 66, 72 – 75.

Albani, J. R. , Debray, H., Vincent, M. and Gallay, J. 1997, Role of the carbohydrate moiety and of the alpha L-fucose in the stabilization and the dynamics of the Lens culinaris agglutinin - glycoprotein complex. A fluorescence study. Journal of Fluorescence. 7, 293-298.

Albani, J. R, 1998, Lens culinaris agglutinin – lactotransferrin and serotransferrin complexes, followed by fluorescence intensity quenching of fluorescein (FITC) with iodide and temperature. Biochimica Biophysica Acta 1425, 405-410.

Albani, J. R. 1998, Dynamics of Trp residues in crystals of human α_1- acid glycoprotein (orosomucoid) followed by red-edge excitation spectra. Journal of Fluorescence, 8, 213-224.

Albani, J. R, 1998, Corelation between dynamics, structure and spectral properties of human α_1-acid glycoprotein (orosomucoid): a fluorescence approach. Spectrochimica Acta Part A 54, 175-183.

Albani, J. R. and Plancke, Y. D, 1998 and1999, Interaction between calcofluor white and carbohydrates of α_1- acid glycoprotein. Carbohydrate Research 314, 169-175 and 318, 194-200

Albani, J. R, 1999, New insights in the conformation of α_1-acid glycoprotein (orosomucoid). Quenching resolved emission anisotropy studies. Spectrochimica Acta, Part A. 55, 2353-2360.

Albani, J. R., Sillen, A., Coddeville, B., Plancke, Y.D., Engelborghs, Y, 1999, Dynamics of carbohydrate residues of α_1- acid glycoprotein (orosomucoid) followed by red-edge excitation spectra and emission anisotropy studies of Calcofluor White. Carbohydrate Research 322, 87-94.

Albani, J. R., Sillen, A., Plancke, Y.D., Coddeville, B., Engelborghs, Y, 2000, Interaction between carbohydrate residues of α_1- acid glycoprotein (orosomucoid) and saturating concentrations of Calcofluor White. A fluorescence study. Carbohydrate Research 327, 333-340.

Albani. J. R, 2001, Effect of binding of calcofluor white on the carbohydrate residues of α_1-acid glycoprotein (orosomucoid) on the structure and dynamics of the protein moiety. A fluorescence study. Carbohydrate Research. 334, 141-151.

Albani, J. R., *Absorption et Fluorescence : Principes et Applications.* Editions Lavoisier – Tec & Doc. Paris, 2001.

Albani, J. R, 2002, Interaction between carbohydrate residues of α_1- acid glycoprotein (orosomucoid) and progesterone. A fluorescence study. Carbohydrate Research 337, 1405-1410.

Albani, J. R, 2003, Relation between the secondary structure of carbohydrate residues of α_1-acid glycoprotein (orosomucoid) and the fluorescence of the protein. Carbohydrate Research 338, 1097-1101.

Albani, J. R. 2003, Förster energy-transfer studies between Trp residues of α_1- acid glycoprotein (orosomucoid) and the glycosylation site of the protein. Carbohydrate Research 338, 2233-2236..

Albani, J. R., Demuynck, S., Grumiaux, F. and Leprêtre, A, 2003, Fluorescence fingerprints of *Eisenia fetida* and *Eisenia andrei.* Photochemistry and Photobiology. 78, 599-602..

Albani, J. R, 2004, Tertiary structure of human α_1-acid glycoprotein (orosomucoid). Straightforward fluorescence experiments revealing the presence of a binding pocket. Carbohysrate Research. 339, 607-612.

Albani, J. R., 2004, Effect of the secondary structure of carbohydrate residues of α_1-acid glycoprotein (orosomucoid) on the local dynamics of Trp residues. Chemistry & Biodiversity. 1, 152-160.

Alpert, B. and Lopez-Delgado, R, 1976, Fluorescence lifetimes of haem proteins excited Into the tryptophan absorption band with synchrotron radiation. Nature, 263, 445 - 446.

Åman, P., McNeil, M., Franzén, L-E., Darvill, A. G. and Albersheim, P, 1981, Structural elucidation, using h.p.l.c.-m.s. and g.l.c.-m.s., of the acidic polysaccharide secreted by *rhizobium meliloti* strain 1021[*1, *2]. Carbohydrate Research 95, 263-282.

Amisha Kamal, J. K. and Behere, D. V. Steady-state and picosecond time-resolved fluorescence studies on native and apo seed coat soybean peroxidase. Biochemical Biophysical Research Communication 289, 427-433.

Andersen, K.V. and Poulsen, F. M, 1993, The three-dimensional structure of acyl-coenzyme A binding protein from bovine liver: structural refinement using heteronuclear multidimensional NMR spectroscopy. Journal of Biomolecular NMR 3, 271-284.

André, F, 1963, Contribution à l'analyse expérimentale de la reproduction des lombriciens. Bulletin Biologique de la France et de la Belgique. 97, 3-101.

Athes, V., Lange, R. and Combes, D., 1998, Influence of polyols on the structural properties of Kluyveromyces lactis beta-galactosidase under high hydrostatic pressure. European Journal of Biochemistry 255, 206-212.

Badea, M. G. and Brand, L. 1979, Time-resolved fluorescence measurements. Methods in Enzymology. 61, 378-425.

Baker, E. N. and Lindley, P. F. 1992, New perspectives on the structure and function of transferrins. Journal of Inorganic Biochemistry 47, 147-160.

Balabanli, B., Kamisaki, Y., Martin, E. and Murad, F. 1999, Requirements for heme and thiols for the nonenzymatic modification of nitrotyrosine. Proceedings of the Natural Academy of Sciences. U.S.A. 96, 13136 - 13141.

Barbato, G., Ikura, M., Kay, L. E., Pastor, R. W. and Bax, A. 1992, Backbone dynamics of calmodulin studied by ^{15}N relaxation using inverse detected two-dimensional NMR spectroscopy: the central helix is flexible. Biochemistry, 31, 5269-5278.

Barik, A., Priyadarsini, K. I. and Mohan, H. 2003, Photophysical Studies on Binding of Curcumin to Bovine Serum Albumin. Photochemistry and Photobiology. 77, 597–603.

Barrick, D., and Baldwin, R. L. 1993, Three-state analysis of sperm whale apomyoglobin unfolding. Biochemistry 32, 3790-3796

Beckingham, J. A., Bottomley, S. P., Hinton, R., Sutton, B. J. and Gore, M. G, 1999, Interactions between a single immunoglobulin-binding domain of protein L from *Peptostreptococcus magnus* and a human k light chain. Biochemical Journal 340, 193-199.

Beratan, D. N., Betts, J. N. and Onuchic, J. N, 1991, Protein electron transfer rates set by the bridging secondary and tertiary structure. Science. 252, 1285-1288.

Bilsel, O., Yang,, L., Zitzewitz, J. A., Beechem, J. M. and Matthews, C. R. 1999, Time-Resolved Fluorescence Anisotropy Study of the Refolding Reaction of the α-Subunit of Tryptophan Synthase Reveals Nonmonotonic Behavior of the Rotational Correlation Time. Biochemistry 38, 4177 – 4187.

Bismuto, E., Irace, G., D'Auria S., Rossi, M. and Nucci, R., 1997, Multitryptophan-fluorescence-emission decay of beta-glycosidase from the extremely thermophilic archaeon Sulfolobus solfataricus. European Journal of Biochemistry 244, 53-58.

Blat, Y. and Eisenbach, M. 1996, Oligomerization of the phosphatase CheZ upon interaction with the phosphorylated form of CheY. The signal protein of bacterial chemotaxis. J. Biol. Chem. 271, 1226-1231.

Bouché, M. B, 1972, Lombriciens de France, écologie et systématique. Annales Zoologie et Ecologie Animale. (Numéro spécial.) 72, 1-671.

Bourne, Y., Van Tilbeurgh, H. and Cambillau, C., 1993, Protein–carbohydrate interactions. Current Opinion of Structural Biology 3, 681-686.

Brinkman-van der Linden EC, van Ommen EC, van Dijk W, 1996, Glycosylation of alpha 1-acid glycoprotein in septic shock: changes in degree of branching and in expression of sialyl Lewis(x) groups. Glycoconjugate Journal 13, 27 – 31.

Britz-McKibbin, P., Markuszewski, M. J., Iyanagi, T., Matsuda, K., Nishioka, T. and Terabe, S. 2003, Picomolar analysis of flavins in biological samples by dynamic pH junction-sweeping capillary electrophoresis with laser-induced fluorescence detection. Analytical Biochemistry 313, 89-96

Bundy, J.G., Spurgeon, D. J., Svendsen, C., Hankard, P. K., Osborn, D., Lindon, J. C. and Nicholson, J. K, 2002, Earthworm species of the genus *Eisenia* can be phenotypically differentiated by metabolic profiling. *FEBS letters*. 521, 115-120.

Burstein, E.A, Vedenkina, N.S, and Ivkova. M. N, 1973, Fluorescence and the location of tryptophan residues in protein molecules. Photochemistry and Photobiology 18, 263–279.

Burstein, E.A. and Emelyanenko, V. L, 1996, Log-normal description of fluorescence spectra of organic fluorophores. Photochemistry and Photobiology 64, 316-320.

Canguly, M. and Westphal, U. 1968, Steroid.protein interactions. XVII. Influence of solvent environment on interaction between human α_1–acid glycoprotein and progesterone. Journal of Biological Chemistry 243, 6130-6139.

Cantor, C. R. and Schimmel, P. R., Biophysical Chemistry, 1980, Editions W. H. Freeman and Company, New York.

Capeillere-Blandin, C., 1982, Transient kinetics of the one-electron transfer reaction between reduced flavocytochrome b_2 and oxidized cytochrome c. Evidence for the existence of a protein complex in the reaction. European Journal of Biochemistry 128, 533-542.

Capeillère-Blandin, C. and Albani, J. 1987, Cytochrome b_2, an electron carrier between fiavocytochrome b, and cytochrome c: Rapid kinetic characterisation of the electron transfer parameters with ionic strength dependence. Biochemical Journal 245, 159-165.

Carpita, N. C. and Gibeaut, D. M. 1993, Structural models of primary cell walls in flowering plants: consistency of molecular structure with the physical properties of the walls during growth. Plant Journal 3, 1–30.

Carrero, J., Jameson, D. M. and Gratton, G, 1995, Oxygen distribution and diffusion into myoglobin revealed by quenching of zincprotoporphyrin IX fluorescence. Biophysical Chemistry 54, 143 – 154.

Carter, C. W. Jr, Kraut, J., Freer, S. T., Nguyen, H.-X., Alden, R. A. and Bartsch, R. G, 1974, Two-Angstrom crystal structure of oxidized Chromatium high potential iron protein. Journal of Biological Chemistry 249, 4212-4225.

Casimiro, D. R., Wong, L. L., Colon, J. L., Zewert, T. E., Richards, J. H., Chang, I-J., Winkler, J. R. and Gray, H. B, 1993, Electron transfer in ruthenium / zinc porphyrin derivatives of recombinant human myoglobins. Analysis of tunneling pathways in myoglobin and cytochrome c. Journal of American Chemical Society 115, 1485-1489.

Chang, Y-C. and Ludescher, R. D, 1994, Local conformation of rabbit skeletal myosin rod filaments probed by intrinsic tryptophan fluorescence. Biochemistry 33, 2313 – 2321.

Cheng, Y and Prusoff, W. H, 1973, Relationship between the inhibition constant (K_i) and the concentration of inhibitor which causes 50 per cent inhibition (I50) of an enzymatic reaction. Biochemical Pharmacology 22, 3099-3108.

Chiu, K. M., Mortensen, R. F., Osmand, A. P and Gewurz, H, 1977, Interactions of alpha1-acid glycoprotein with the immune system. I. Purification and effects upon lymphocyte responsiveness. Immunology 32, 997-1005.

Christie, J. M., Salomon, M., Nozue, K., Wada, M. and Briggs, W. R. LOV (light, oxygen, or voltage) domains of the blue-light photoreceptor phototropin (nph1): binding sites for the chromophore flavin mononucleotide. Proc. Natl. Acad. Scien. U.S.A. 1999, 96, 8779-8783.

Cinelli, R. A., G., Ferrari, A., Pellegrini, V., Tyagi, M., Giacca, M. and Beltram F. 2004, The enhanced green fluorescent protein as a tool for the analysis of protein dynamics and localization: Local fluorescence study at the single-molecule level. Photochemistry and Photobiology 71, 771–776.

Clayton, A. H. A. and Sawyer, W. H., 1999, Tryptophan rotamer distributions in amphipathic peptides at a lipid surface. Biophysical Journal, 76, 3235-3242.

Collins, J. H., Cox, J. A. and Theibert, J. L, 1988, Amino acid sequence of a sarcoplasmic calcium-binding protein from the sandworm Nereis diversicolor. Journal of Biological Chemistry 263, 15378-15385.

Daff, S., Sharp, R. E., Short, D. M., Bell, C., White, P., Manson, F. D., Reid, G. A. and Chapman, S. K, 1996. Interaction of cytochrome c with flavocytochrome b_2. Biochemistry 35, 6351-6357.

Dahms, T. E. S., Willis, K. J. And Szabo, A. G., 1995, Conformational heterogeneity of tryptophan in a protein crystal. Journal of The American Chemical Society. 117, 2321-2326.

Das, T. K. and Mazumdar, S. 1995, pH-induced conformational perturbation in horseradish peroxidase. Picosecond tryptophan fluorescence studies on native and cyanide-modified enzymes. Eur. J. Biochem. 227, 823-828

Das, T. K. and Mazumdar, S. 1995. Conformational substates of apoprotein of Horsedadish peroxidase in aqueous solution – a fluorescence dynamics study. J. Phys. Chem. 99, 13283-13290.

Davis, T. N., Urdea, M. S., Masiarz, F.R. and Thorner, J. (1986) Isolation of the yeast calmodulin gene: calmodulin is an essential protein. Cell 47, 423-431.

De, S. and Girigoswami, A. 2004, Fluorescence resonance energy transfer-a spectroscopic probe for organized surfactant media. Journal of Colloidal Interface Science 271, 485-495.

De Beuckeleer, K., Volckaert, G. and Engelborghs, Y, 1999, Time resolved fluorescence and phosphorescence properties of the individual tryptophan residues of barnase : Evidence for protein-protein interactions. Proteins : Structure, Function and Genetics. 36, 42-53.

De Ceukeleire, M., Albani, J. R, 2002, Interaction between carbohydrate residues of α_1- acid glycoprotein (orosomucoid) and progesterone. A fluorescence study. Carbohydrate Research 337, 1405-1410.

Demchenko, A. P, 1985, Fluorescence molecular relaxation studies of protein dynamics. FEBS Letters. 182, 99-102.

Demchenko, A.P, 2002, The red-edge effects: 30 years of exploration. Luminescence, 17, 19-42.

Dente L, Pizza MG, Metspalu A, Cortese R, 1987, Structure and expression of the genes coding for human alpha 1-acid glycoprotein. EMBO J. 6, 2289-2296.

Devaraj, S., Xu, D. Y. and Jialal, I, 2003, C-reactive protein increases plasminogen activator inhibitor-1 expression and activity in human aortic endothelial cells: implications for the metabolic syndrome and atherothrombosis. Circulation. 107, 398-404.

Dickerson, R. E. and Geis, I, 1983, Hemoglobin : structure, function, evolution and pathology. The Benjamin / Cummings Publishing Company, Inc.

Di Venere, A., Mei, G., Gilardi, G., Rosato, N., De Matteis, F., McKay, R., Gratton, E. and Finazzi Agro, A. 1998. Resolution of the heterogeneous fluorescence in multi-tryptophan proteins : ascorbate oxidase. European Journal of Biochemistry 257, 337-343.

Donovan, J. W. 1969, Changes in Ultraviolet Absorption Produced by Alteration of Protein Conformation. Journal of Biological Chemistry 244, 1961-1967.

Drew, J., Thistletwhaite, P. and Woolfe, G, 1983, Excited-state relaxation in 1-amino-8-naphtthalene-sulfonate. Chemical Physical Letters 96, 296- 301.

Du, H., Fuh, R. A., Li, J., Corkan, A. and Lindsey, J. S. 1998. PhotochemCAD: A computer-aided design and research tool in photochemistry. Photochemistry and Photobiology 68, 141-142.

Easter J.H, DeToma R.P and Brand L, 1976, Nanosecond time-resolved emission spectroscopy of a fluorescence probe adsorbed to L-alpha-egg lecithin vesicles. Biophysical Journal 16, 571-583.

Ebu Isaac, V., Patel, L., Curran, T., Abate-Shen, C, 1995, Use of fluorescence energy transfer to estimate intramolecular distances in the Msx-1 homeodomain. Biochemistry. 34, 15276-15281.

Edwards, C. A. and Bohlen, P. J, 1996, *Biology and ecology of earthworms*, third ed. Chapman and Hall, London.

Eftink M, 1983, Quenching-resolved emission anisotropy studies with single and multitryptophan-containing proteins. Biophysical Journal 43, 323-334.

Eftink, M. R. and Ghiron, C. A, 1976, Exposure of tryptophanyl residues in proteins. Quantitative determination by fluorescence quenching studies. Biochemistry 15, 672 – 680

Engelstad F. and Stenersen, J, 1991, Acetylesterase pattern in the earthworm genus *Eisenia* (Oligochaeta, Lumbricidae) : implications for laboratory use and taxonomic status. Soil Biology and Biochemistry. 23, 243-247.

Falk, J. E. 1975. in « Porphyrins and metalloporphyrins". Edited by Kevin M. Smith. Elsevier Scientific Publishing Company.

Feitelson, J. and Spiro, T. G. 1986, Bonding in zinc proto- and mesoporphyrin substituted myoglobin and model compounds studied by resonance raman spectroscopy. Inorganic Chemistry 25, 861-865.

Fisher, C. A., Narayanaswami, V. and Ryan, R. O, 2000, Journal of Biological Chemistry 275, 33601 - 33606.

Flower, D. R, 1995, Multiple molecular recognition properties of the lipocalin protein family. Journal of Molecular Recognition 8, 185- 195.

Flower, D. R, 1996, The lipocalin protein family: structure and function. Biochemical Journal 318, 1-14.

Foriers, A., Wunilmart, C., Sharon, C. and Strosberg, A.D, 1977, Extensive sequence homologies among lectins from leguminous plants. Biochemical Biophysical Research Communication 75, 980-986.

Forster, T. (1948) Intermolecular energy migration and fluorescence. Annual of Physics (Leipzig) 2, 55-75.

Frank, I. M. and Vavilov, S. I, 1931. Über die Wirkungssphäre der Auslöschungsvargänge in den flureszierenden flüssigkeiten. Z. Phys. 69, 100 – 110.

Franzen, L. E., Svensson, S. and Larm, O, 1980, Structural studies on the carbohydrate portion of human antithrombin III. Journal of Biological Chemistry. 255, 5090-5093.

Frauenfelder, H., Petsko, C.A., and Tsernoglou, D, 1979, Temperature-dependent X-ray diffraction of a probe of protein structural dynamics. Nature 280, 558-563.

Froschle, M., Ulmer, W. and Jany, K. D. 1984, Tyrosine modification of glucose dehydrogenase from bacillus megaterium . Effect of tetranitromethane on the enzyme in the tetrameric and monomeric state. European Journal of Biochemistry 142, 533-540 .

Fujimoto, K., Shimizu, H. and Inouye, M. 2004, Unambiguous Detection of Target DNAs by Excimer-Monomer Switching Molecular Beacons. Journal of Organic Chemistry, 69, 3271 –3275.

Fussle, R., Bhakdi, S., Sziegoleit, A., Tranum-Jensen, J., Kranz, T. and Wellensiek, H. J. 1981, On the mechanism of membrane damage by Staphylococcus aureus α-toxin. Journal of Cell Biology 91, 83-94.

Gajhede, M., Schuller, D. J., Henriksen, A., Smith, A. T. and Poulos, T. L. Crystal structure of horseradish peroxidase C at 2.15 Å resolution. 1997. Nature, Structural Biology. 4, 1032-1038.

Gambacorti-Passerini C, Barni R, le Coutre P, Zucchetti M, Cabrita G, Cleris L, Rossi F, Gianazza E, Brueggen J, Cozens R, Pioltelli P, Pogliani E, Corneo, G, Formelli F, D'Incalci M. 2000, Role of alpha1 acid glycoprotein in the in vivo resistance of human BCR-ABL(+) leukemic cells to the abl inhibitor STI571. Journal of the National Cancer Institute. 92, 1641-1650.

Gervais. M., Groudinsky, O., Risler, Y. and Labeyrie, F., 1977, Dissection of flavocytochrome b₂ – a bifunctional enzyme – into a cytochrome   core and a flavoprotein molecule. Biochemical Biophysical Research Communication 77, 1543-1551.

Gervais, M., Labeyrie, F., Risler, Y. and Vergnes, O, 1980, A flavin-mononucleotide-binding site in *Hansenula anomala* nicked flavocytochrome b₂, requiring the association of two domains. European Journal of Biochemistry 111, 17-31.

Gibney, B. R., and Dutton, P. L, 1999, Histidine Placement in De Novo Designed Heme Proteins. Protein Science 8, 1888-1898.

Gill, S. C. and von Hippel, P. H. 1989, Calculation of protein extinction coefficients from amino acid sequence data. Analytical Biochemistry, 182, 319-326. Erratum in: Anal. Biochem. 1990, 189, 283.

Goto, Y., and Fink, A. L. 1990, Phase diagram for acidic conformational states of apomyoglobin. Journal of Molecular Biology 214, 803-805

Gratton, E., US. Patent 4,840,485. Frequency-domain Cross-Correlation Fluorometry with Phase- Locked Loop Frequency Synthesizers. June 20, 1989.

Gray, G. S., and Kehoe, M. 1984, Primary sequence of the alpha-toxin gene from Staphylococcus aureus wood. Infection and Immunity 46, 615-618.

Grossniklaus H.E., Waring, G.O., 4th, Akor, C., Castellano-Sanchez, A. A. and Bennett, K, 2003, Evaluation of hematoxylin and eosin and special stains for the detection of acanthamoeba keratitis in penetrating keratoplasties. American. Journal of Ophthalmology 136, 520-526.

Gryczynski, Z., Lubkowski, J. and Bucci, E. 1995, Heme-Protein Interactions in Horse Heart Myoglobin at Neutral pH and Exposed to Acid Investigated by Time-resolved Fluorescence in the Pico- to Nanosecond Time Range. Journal of Biological Chemistry 270, 19232-19237.

Gryczynski, Z and Bucci, E. 1998. Time resolved emissions in the picosecond range of single tryptophan recombinant myoglobins reveal the presence of long range heme protein interactions. Biophysical Chemistry 74, 187-196

Hagihara, Y., Aimoto, S., Fink, A. L. and Goto, Y. 1993, Guanidine hydrochloride-induced folding of proteins. Journal of Molecular Biology 231, 180-184

Harada, S., Kitadokoro, K., Kinoshita, T., Kai, Y. and Kasai, N. 1991, Crystallization and main-chain structure of neutral protease from Streptomyces caespitosus. Journal of Biochemistry 110, 46-49.

Hatton, M. W. C., Marz, L., Berry, L. R., Debanne, M. T. and Regoeczi, E., 1979, Bi- and tri-antennary human transferrin glycopeptides and their affinities for the hepatic lectin specific for asialo-glycoproteins. Biochemical Journal 181, 633 – 638.

Heim, R., Cubitt, A. B., Tsien, Y. 1995, Improved green fluorescence. Nature. 373, 663-664.

Heim, R, 1999, Green fluorescent protein forms for energy transfer. Methods in Enzymology, 302, 408-423.

Heyduk, T, and Heyduk, E, 2002, Molecular beacons for detecting DNA binding proteins. Nature Biotechnology, 20, 171–176.

Hirsch, R. E., Zukin, S. R. and Nagel, R. L, 1980, Intrinsic fluorescence emission of intact oxy hemoglobins. Biochemical Biophysical Research Communications. 93, 432-439.

Hirsch, R. E. and Nagel, R. L, 1981, Conformational studies of hemoglobins using intrinsic fluorescence measurements. Journal of Biological Chemistry 256, 1080-1083.

Hirshfield, K. M., Toptygin, D., Grandhige, G., Kim, H., Packard, B. Z. and Brand, L, 1996, Steady-state and time-resolved fluorescence measurements for studying molecular interactions: interaction of a calcium-binding probe with proteins. Biophysical Chemistry 62, 25-38

Hof, M., Vajda, S., Fidler, V, and Karpenko, V, 1996, Picosecond tryptophan fluorescence of human blood serum orosomucoid. Collect. Czech. Chem. Commun. 61, 808-818.

Hutterer, R., Schneider, F. W., Sprinz, H. and Hof, M. 1996, Binding and relaxation behaviour of prodan and patman in phospholipid vesicles: a fluorescence and ^{1}H NMR study. Biophysical Chemistry 61, 151-160

Hyde, C. C., Ahmed, S. A., Padlan, E. A., Miles, E. W. and Davies, D. R. 1988, Three-dimensional structure of the tryptophan synthase alpha 2 beta 2 multienzyme complex from Salmonella typhimurium. Journal of Biological Chemistry 263, 17857-17871.

Ichikawa, T. and Terada, H, 1977,. Second derivative spectrophotometry as an effective tool for examining phenylalanine residues in proteins. Biochimica Biophysica Acta 494, 267 – 270.

Ichikawa, T. and Terada, H, 1979, Estimation of state and amount of phenylalanine residues in proteins by second derivative spectrophotometry. Biochimica Biophysica Acta. 580, 120-128.

Itoh, M., Mizukoshi, H., Fuke, K., Matsukawa, S., Mawatari, K., Yoneyama, Y. Sumitani, M. and Yoshihara, Y, 1981, Tryptophan fluorescence of human hemoglobin. I. Significant change of fluorescence intensity and lifetimes in the T-R transition. Biochemical Biophysical Research Communications 100, 1259-1265.

Jablonski, A., 1935, Über den Mechanismus des Photolumineszenz von Farbstoffphosphoren. Z. Phys. 94, 38-64.

Jaenike, J. 1982, "*Eisenia foetida*" is two biological species. Megadrilogica 4, 6-7.

James, D. R., Siemiarczuk, A. and Ware, W. R., 1992, Stroboscopic optical boxcar technique for the determination of fluorescence lifetimes. Review of Scientific Instrument. 63, 1710-1716.

Jameson, D. M., Gratton, E., G., Weber, G. and Alpert, B, 1984, Oxygen distribution and migration within Hb^{desFe} and Hb^{desFe}. Multifrequency phase and modulation fluorometry study. Biophysical Journal 45, 795 – 803.

Jorgensen, H. G., Elliott, M. A., Allan, E. K., Carr, C. E., Holyoake, T. L., Smith, K. D, 2002, α_1-acid glycoprotein expressed in the plasma of chronic myeloid leukemia patients does not mediate significant in vitro resistance to STI571. Blood. 99, 713-715.

Jullien, M., Garel, J-R., Merola, F. and Brochon J-C, 1986, Quenching by acrylamide and temperature of a fluorescent probe attached to the active site of ribonuclease. European Biophysical Journal 13, 131-137

Kabsch, W., H. G. Mannherrz, D. Shuck, E. F. Pai, and K. C. Holmes. 1990. Atomic structure of actin: DNase I complex. Nature 347, 37-44

Karplus, M. and McCammon, J. A. 1983, Dynamics of proteins: Elements and function. Annual Review of Biochemistry 53, 263-300

Kauzmann, W., 1957, Quantum Chemistry, Academic Press.

Kehoe, M., Duncan, I., Foster, T., Fairweather, N. and Dougan, G, 1983, Cloning, expression, and mapping of the Staphylococcus aureus alpha-hemolysin determinant in *Escherichia coli* K-12. Infection and Immunity 4, 1105-1111

Khan, K.K., Mazumdar, S., Modi, S., Sutcliffe, M., Roberts, G. C. K. and Mitra, S. 1997, Steady-state and picosecond-time-resolved fluorescence studies on the recombinant heme domain of *Bacillus megaterium* cytochrome P-450. European Journal of Biochemistry 244, 361-370.

Khan, M. J., Joginadha Swamy, M., Krishna Sastry, M. V., Umadevi Sajjan, S., Patanjali, S. R., Rao, P., Swarnalatha, G. V., Banerjee, P. and Surolia, A. 1988, Saccharide binding to three Gal/GalNac specific lectins : Fluorescence, spectroscopic and stopped-flow kinetic studies. Glycoconjugate Journal 5, 75-84.

Kirby, E. P. and Steiner, R. F. 1970, The tryptophan microenvironments in apomyoglobin. Journal of Biological Chemistry 245, 6300-6630

Kirley, T. L., Spargue, E. D. and Halsall. H. B. 1982, The binding of spin-labeled propranolol and spin labeled progesterone by orosomucoid. Biophysical Chemistry 15, 209-216.

Kiss, R. S., Kay, C. M. and Ryan, R. O. 1999, amphipathic α - helix bundle organization of lipid – free chicken apolipoprotein A-I. Biochemistry 38, 4327-4334.

Kita, A., Kasai, N., Kasai, S., Nakaya, T. and Miki, K. 1991, Crystallization and preliminary X-ray diffraction studies of a flavoprotein, FP390, from a luminescent bacterium, Photobacterium phosphoreum. Journal of Biochemistry 110, 748-750.

Kopecky, V. Jr., Ettrich, R., Hofbauerova, K. and Baumruk. V, 2003, Structure of human α_1-acid glycoprotein and its high-affinity binding site. Biochemical Biophysical Research Communications 300, 41-46.

Kornfeld, K., Reitman, M-L. and Kornfeld, R, 1981, The carbohydrate-binding specificity of pea and lentil lectins. Fucose is an important determinant. Journal of Biological Chemistry 256, 6633-6640.

Kraulis, P. J. 1991, MOLSCRIPT: a program to produce both detailed and schematic plots of protein structures. Journal of Applied Crystallography 24, 946-950.

Krishna, M. M. and Periasamy, N, 1997, Spectrally constrained global analysis of fluorescence decays in biomembrane systems. Analytical Biochemistry 253, 1-7.

Kubota, Y., Matado, Y., Shi-Genuare, Y. and Fujisaki, Y, 1979, Fluorescence quenching of 10-methylacridinium chloride by nucleotides. Photochemistry and Photobiology 29, 1099-1106.

Kuipers, O. P., Vincent, M., Brochon, J-C., Verheij, H. M., De Haas, G. H. and Gallay, J. 1991, Insight into the conformational dynamics of specific regions of porcine pancreatic phospholipase A2 from a time-resolved fluorescence study of a genetically inserted single tryptophan residue. Biochemistry 30, 8771 – 8785.

Kungl, A. J., Visser, N. V., van Hoek, A., Visser, A. J., Billich, A., Schilk, A., Gstach, H. and Auer, M, 1998, Time-resolved fluorescence anisotropy of HIV-1 protease inhibitor complexes correlates with inhibitory activity. Biochemistry 37, 2778-2786.

Kunz, L. and MacRobert, A. J, 2002, Intracellular photobleaching of 5,10,15,20-tetrakis(m-hydroxyphenyl) chlorin (Foscan) exhibits a complex dependence on oxygen level and fluence rate. Photochemistry and Photobiology 75, 28-35.

Kuppens S, Hellings M, Jordens J, Verheyden S, Engelborghs Y, 2003, Conformational states of the switch I region of Ha-ras-p21 in hinge residue mutants studied by fluorescence lifetime and fluorescence anisotropy measurements. Protein Science 12, 930-938.

Kute T and Westphal U, 1976, Steroid-protein interactions. XXXIV. Chemical modification of alpha1-acid glycoprotein for characterization of the progesterone binding site. Biochimica Biophysica Acta 420, 195-213

Labeyrie, F. and Baudras, A, 1972, Differences in quaternary structure and constitutive chains between two homologous forms of cytochrome b_2 (L-lactate: cytochrome c oxidoreductase). European Journal of Biochemistry 25, 33-40.

Ladokhin, A. S. and White, S. H, 2001, Alphas and Taus of Tryptophan Fluorescence in Membranes. Biophysical Journal. 181, 1825-1827.

Lahiri, B., Bagdasarian, A., Mitchell, B., Talamo, R. C., Colman, R. W. and Rosenberg, R. D, 1976, Antithrombin-Heparin Cofactor: An inhibitor of plasma kallikrein. Archives of Biochemistry and Biophysics 175, 737- 747.

Lakowicz, J. R. and Weber, G, 1973, Quenching of fluorescence by oxygen. A probe for structural fluctuation in macromolecules. Biochemistry. 12, 4161- 4170.

Lakowicz and Maliwal, 1983, Oxygen quenching and fluorescence depolarization of tyrosine residues in proteins. Journal of Biological Chemistry 258, 4794-4801.

Lakowicz, J. R., Maliwal, B., Cherek, H. and Balter, A. 1983, Rotational freedom of tryptophan residues in proteins and peptides. Biochemistry. 22, 1741 – 1752.

Lakowicz, J. R., Gratton, E., Cherek, H., Maliwal, B. P. and Laczko, G, 1984, Determination of time-resolved fluorescence emission spectra and anisotropies at a fluorophore-protein complex using frequency-domain phase-modulation fluorometry. Journal of Biological Chemistry 259, 10967-10972.

Lakowicz, J. R., Laczko, G. and Gryczinski, I, 1985, 2-GHz frequency-domain fluorometer. Review of Scientific Instrumunents 57, 2499-2506.

Lakowicz, J. R., 1983, 1999, Principles of Fluorescence Spectroscopy, Kluwer Academic/Plenum Publishers, New York.

Lawson D. M, Derewenda, U., Serre, L., Ferri, S., Szittner, R., Wei, Y., Meighen, E. A. and Derewenda, Z. S. 1994, Structure of a myristoyl-ACP-specific thioesterase from Vibrio harveyi. Biochemistry 33, 9382-9388.

Lee, J-S., Liao, J-H., Wu, S-H., Chiou, S. H. 1997, α–Crystallin acting as a molecular chaperonin against photodamage by UV irradiation. Journal of Protein Chemistry 16, 283-289.

Lee S, Y, 2000, NMR Studies of calmodulin from *S. cerevisiae*. Ph. D. thesis. University of Washington.

Lee, S. Y. and Klevit, R. E., 2000, The whole is not the simple sum of its parts in calmodulin from *S. cerevisiae*. Biochemistry 39, 4225-4230.

Leenders. R., Van Hoek, A., Van Iersel. M., Veeger, C. and Visser, A. J. W. G, 1993, Flavin dynamics in oxidized *Clostridium beijerinckii* flavodoxin as assessed by time-resolved polarized fluorescence. European Journal of Biochemistry 218, 977-984.

Lehrer S.S, 1971, Solute perturbation of protein fluorescence. The quenching of the tryptophyl fluorescence of model compounds and of lysozyme by iodide ion. Biochemistry 10, 3254-3263.

Le Tilly, V. and Royer, C, 1993, Fluorescence anisotropy assays implicate protein-protein interactions in regulating trp repressor DNA binding. Biochemistry 32, 7753-7758.

Li, J., Szittner, R. and Meighen, E. A. 1998, Tryptophan Fluorescence of the lux-Specific *Vibrio harveyi* Acyl-ACP Thioesterase and Its Tryptophan Mutants: Structural Properties and Ligand-Induced Conformational Change. Biochemistry 37, 16130 – 16138.

Liang, Z-X., Nocek, J. M., Kurnikov, I. V., Beratan, D. N. and Hoffman, B. M, 2000, Electrostatic control of electron transfer between myoglobin and cytochrome b_5 : effect of methylating the heme propionates of zn-myoglobin. Journal of The American Chemical Society 122, 3552-3553.

Liang, Z-X., Nocek, J. M., Huang, K., Kurnikov, I. V., Beratan, D. N. and Hoffman, B. M, 2002, Dynamic docking and electron transfer between Zn-myoglobin and cytochrome b_5. Journal of The American Chemical Society 124, 6849-6859.

Libich, D.S, Hill, C.M, Bates, I.R, Hallett, F.R, Armstrong, S, Siemiarczuk, A and Harauz, G, 2003, Interaction of the 18.5-kD isoform of myelin basic protein with Ca2+ -calmodulin: effects of deimination assessed by intrinsic Trp fluorescence spectroscopy, dynamic light scattering, and circular dichroism. Protein Science 12, 1507-1521.

Liener, I. E., 1976, Phytohemagglutinins (Phytolectins) Annual Review of Plant Physiology 27, 291-319.

Lin, J. K. and S. Y. Lin-Shiau, 2001, Mechanisms of cancer chemoprevention by curcumin. Proc. Natl. Sci. Commun., BOC(B). 25, 59–66.

Livesey, A. K. and Brochon, J-C, 1987, Analyzing the distribution of decay constants in pulse-fluorimetry using the maximum entropy method. Biophysical Journal 52, 693-706.

Loebel, D., Scaloni, A., Paolini, S., Fini, C., Ferrara, L., Breer, H. & Pelosi, P, 2000, Cloning, post-translational modifications, heterologous expression and ligand-binding of boar salivary lipocalin. Biochemical Journal. 350 Part 2, 369-379.

LopezÝArbeloa, F., RuizÝOjeda, P. and LopezÝArbeloa, I. 1989. Fluorescence self-quenching of the molecular forms of rhodamine B in aqueous and ethanolic solutions. Journal of Luminescence 44, 105-112.

Loris, R., Steyaert, J., Maes, D., Lisgarten, J., Pickersgill, R. and Wyns, L. 1993, Crystal structure determination and refinement at 2.3-Å resolution of the lentil lectin. Biochemistry 32, 8772-8781.

Loskutoff, D. J. and Schleef, R. R, 1988, Plasminogen activators and their inhibitors. Methods in Enzymology 163, 293–302

Maeda H and Ishida N, 1967, Specificity of binding of hexopyranosyl polysaccharides with fluorescent brightener. Journal of Biochemistry 62, 276-278.

Margalit, R., Shaklai, N. and Cohen, S, 1983, Fluorometric studies on the dimerization equilibrium of protoporphyrin IX and its haemato derivative. Biochemical Journal 209, 547-552.

MacPhee, C. A., Howlett, G. J., Sawyer, W. H. and Clayton, A. H. A. 1999, Helix-Helix Association of a Lipid-Bound Amphipathic α-Helix Derived from Apolipoprotein C-II. Biochemistry 38, 10878-10884.

Mackiewicz A and Mackiewicz K, 1995, Glycoforms of serum α_1-acid glycoprotein as markers of inflammation and cancer. Glycoconjugate Journal 12, 241-247.

Majoul, I., Straub, M., Duden, R., Hell, S. W., Soling, H. D, 2002, Fluorescence resonance energy transfer analysis of protein-protein interactions in single living cells by multifocal multiphoton microscopy. Journal of Biotechnology 82, 267-277

Malicka, J., Gryczynski, I., Fang, J., Kusba, J. and Lakowicz, J. R. 2003, Increased resonance energy transfer between fluorophores bound to DNA in proximity to metallic silver particles. Analitical Biochemistry 315, 160-169.

Maliwal, B.P., Kusba, J. and Lakowicz, J. R, 1995, Fluorescence energy transfer in one dimension: Frequency-domain fluorescence study of DNA-fluorophore complex. Biopolymers, 35, 245-255.

Magnusson, S. Primary structure of antithrombin III (Heparin Cofactor). Partial homology between α_1 antitrypsin and antithrombin-III , in Collen DW, Wiman B, Verstraete M (eds): The Physiological Inhibitions of Blood Coagulation and Fibrinolysis. Amsterdam, The Netherlands, Elsevier/North-Holland Biomedical Press , 1979 , p 43.

Marcus, R. A., and Sutin, N, 1985, Electron Transfers in Chemistry and Biology. Biochimica Biophysica Acta. 811, 265-322.

Matthews B. W. 1968, Solvent content of protein crystals. Journal of Molecular Biology 33, 491-497.

Matthews, C. R., and Crisanti, M. M, 1981, Urea-induced unfolding of the α-subunit of tryptophan synthase: evidence for a multistate process. Biochemistry 20, 784-792.

Matulis, D and Lovrien, R. 1998, 1-Anilino-8-naphthalene sulfonate anion-protein binding depends primarily on ion pair formation. Biophysical Journal 74, 422-429.

Matulis, D., Baurmann, C. G., Bloomfield, V. A. and Lovrien, R. E. 1999, 1-anilino-8-naphthalene sulfonate as a protein conformational tightening agent. Biopolymers, 49, 451-458.

Marz, L., Hatton, M. W., Berry, L. R. and Regoeczi, E, 1982, The structural heterogeneity of the carbohydrate moiety of desialylated human transferrin. Canadian Journal of Biochemistry 60, 624-630.

McClure, W.O. and Edelman, G.M, 1966, Fluorescent probes for conformational states of proteins. 1. Mechanism of fluorescence of 2-p-toluidinylnaphthalene.6-sulfonate. A hydrophobie probe. Biochemistry 5, 1908-1918.

McPherson, A., Mickelson, K. E. and Westphal, U. 1980. Crystallization of corticosteroid binding globulin (CBG) and alpha 1-acid glycoprotein (AAG). Journal of Steroid Biochemistry 13, 991-992.

McPherson, A., Friedman, M. L. and Halsall, B. 1984, Crystallization of α_1-acid lipoprotein. Biochemical Biophysical Research Communications 124, 619-624.

Meagher, J. L., Beechem, J. M., Olson, S. T. and and Gettins, P. G. W. 1998, Deconvolution of the fluorescence emission spectrum of human antithrombin and identification of the tryptophan residues that are responsive to heparin binding. Journal of Biological Chemistry 273, 23283-23289.

Megan, L. J., Kurnikov, I. V. and. Beratan, D. N, 2002, The nature of tunneling pathway and average packing density models for protein-mediated electron transfer. Journal of Physical Chemistry A. 106, 2002-2006.

Mehta, K., P. Pantazis, T. McQueen and B. Agarwal, 1997, Antiproliferative effect of curcumin against human breast tumor cell lines. Anticancer Drugs. 8, 471–480.

Mei, G., Di Venere, A., Campeggi, F.M., Gilardi, G., Rosato, N., De Matteis, F. and Finazzi-Agrò, A. 1999, The effect of pressure and guanidine hydrochloride on azurins mutated in the hydrophobic core. European Journal of Biochemistry 265, 619-626.

Menache, D. 1991, Antithrombin III: Introduction. Semin Hematol 28:1-2.

Menache, D., Grossman, B. and Jackson, C, 1992, Antithrombin III: Physiology, deficiency, and replacement therapy. Transfusion 32, 580-588.

Menestrina, G. 1986, Ionic channel formed by Staphylococcus aureus α-toxin: voltage-dependent inhibition by divalent and trivalent cations. Journal of Membrane Biology 90, 177-190.

Messerschmidt, A., Rossi, A., Landenstein, R., Huber, R., Bolognesi, M., Gatti, G., Marchesini, A., Petruzzelli, R. and Finazzi Agro, A. 1989. X-ray crystal structure of the blue oxidase ascorbate oxidase from zucchini. Journal of Molecular Biology 206, 513 - 529.

Messerschmidt, A., Landenstein, R., Huber, R., Bolognesi, M., Avigliano, L., Petruzzelli, R., Rossi, A. and Finazzi Agro, A. 1992. Refined crystal structure of ascorbate oxidase at 1.9 Å resolution. Journal of Molecular Biology 224, 179-205.

Miki, K., Ezoe, T., Masui, A., Yoshisaka, T., Mimuro, M., Fujiwara-Arasaki, T. and Kasai, N, 1990, Crystallization and preliminary X-ray diffraction studies of C-phycocyanin from a red alga, Porphyra tenera. Journal of Biochemistry 108, 646-649.

Molecular Probes. *Handbook of fluorescent probes and research chemicals.* 1992-1994.

Monsigny, M., Roche, A.C., Sene, C., Maget-Dana, R. and Delmotte, F, 1980, Sugar-lectin interactions : How does wheat-germ agglutinin bind sialoglycoconjugates ? European Journal of Biochemistry 104, 147-153.

Moss S. E. Ed. 1992. in The Annexins, Portland Press, Ltd, London.

Nagatomo, S., Nagai, M., Shibayama, N. and Kitagawa, T, 2002, Differences in changes of the alpha1-beta2 subunit contacts between ligand binding to the alpha and beta subunits of hemoglobin A: UV resonance raman analysis using Ni-Fe hybrid hemoglobin. Biochemistry. 41, 10010-10020.

Narasimhulu, S, 1988, Quenching of tryptophanyl fluorescence of bovine adrenal P-450_{c-21} and inhibition of substrate binding by acrylamide. Biochemistry. 27 : 1147-1153.

Neyroz, P., Menna, **C.,** Polverini, E. and Masotti, L. 1996, Intrinsic Fluorescence Properties and Structural Analysis of p13^{suc1} from *Schizosaccharomyces pombe* Journal of Biological Chemistry. 271, 27249-27258.

Nishimura, C., Riley, R., Eastman, P. and Fink, A. L, 2000, Fluorescence energy transfer indicates similar transient and equilibrium intermediates in staphylococcal nuclease folding. Journal of Molecular Biology 299, 1133-1146.

Ntziachristos, N. and Chance, B. 2001, Breast imaging technology: Probing physiology and molecular function using optical imaging- applications to breast cancer. Breast Cancer Research 3, 41-46

Nyitral, M., Hild, G., Lakos, Zs and Somogyi, B, 1998, Effect of Ca^{2+}-Mg^{2+} exchange on the flexibility and/or conformation of the small domain in monomeric actin. Biophysical Journal 74, 2474-2481.

O'Connell, M. A., Gerber, A., and Keller, W. 1997, Purification of human double-stranded RNA-specific editase 1 (hRED1) involved in editing of brain glutamate receptor B pre-mRNA. Journal of Biological Chemistry 272, 473-478.

OECD Draft Document, January 2000, OECD guideline for the testing of chemicals. Earthworm reproduction test (*Eisenia fetida/andrei*).

Ogul'chansky, T. Y., Losytskyy, M. Yu, Kovalska, V. B., Yashchuk, V. M. and Yarmoluk, S. M, 2001, Interactions of cyanine dyes with nucleic acids. XXIV. Aggregation of monomethine cyanine dyes in presence of DNA and its manifestation in absorption and fluorescence spectra. Spectrochimica Acta Part A: Molecular and Biomolecular Spectroscopy, 57, 1525 – 1532.

Ohki, S., Ikura, M. and Zhang, M., 1997, Identification of Mg^{2+}-binding sites and the role of Mg^{2+} on target recognition by calmodulin. Biochemistry 36, 4309-4316.

Oien, N. and Stenersen, J, 1984, Esterases of earthworms – III. Electrophoresis reveals that Eisenia fetida (Savigny) is two species. Comparative Biochemistry and Physiology 78C, 277-282.

Oton J, Franchi D, Steiner R. F., Martinez, C. F. and Bucci, E. 1984. Fluorescence studies of internal rotation in apohemoglobin alpha-chains. Archives of Biochemistry and Biophysics 228, 519-524.

Page, C. C., Moser, C. C., Chen, X. and Dutton, P. L, 1999, Natural engineering principles of electron tunneling in biological oxidation-reduction. Nature 402, 47 –52.

Parisini, E., F. Capozzi, P. Lubini, V. Lamzin, C. Luchinat, and G. M. Sheldrick. 1999. Ab *initio* solution and refinement of two high potential iron protein structures at atomic resolution. Acta Crystallography D. 55, 1773-1784.

Parikh, H. H., McElwain, K., Balasubramanian, V., Leung, W., Wong, D., Morris, M. E. and Ramanathan, M. 2000, A rapid spectrofluorimetric technique for determining drug-serum protein binding suitable for high-throughput screening. Pharmaceutical Research 17, 632-637.

Perotti, C., Fukuda, H., DiVenosa, G., MacRobert, A. J., Batlle, A. and Casas, A, 2004, Porphyrin synthesis from ALA derivatives for photodynamic therapy. In vitro and in vivo studies. British Journal of Cancer. 90, 1660-1665.

Perutz, M. F, 1970, Stereochemistry of cooperative effects in haemoglobin. Nature 228, 726-739.

Pervais, S. and Brew, K, 1985, Homology of beta-lactoglobulin, serum retinol-binding protein and protein HC. Science 228, 335- 337.

Piston, D.W., Masters, B. R. and Webb, W.W. 1995, Three-dimensionally resolved NAD(P)H cellular metabolic redox imaging of the in situ cornea with two-photon excitation laser scanning microscopy. Journal of Microscopy 178, 20-27.

Privat J. P, Wahl, P. and Auchet J.C, 1979, Rates of deactivation processes of indole derivatives in water-organic solvent mixtures--application to tryptophyl fluorescence of proteins. Biophysical Chemistry 9, 223-233.

Pjura, P. E, Grzeskowiak, K and Dickerson, R. E, 1987, Binding of Hoechst 33258 to the minor groove of B-DNA. Journal of Molecular Biology 197, 257-271.

Poveda, J. A., Prieto, M., Encinar, J. A., González-Ros, J. M. and Mateo, C. R, 2003, Intrinsic Tyrosine Fluorescence as a Tool To Study the Interaction of the Shaker B "Ball" Peptide with Anionic Membranes. Biochemistry, 42, 7124 -7132.

Prats, M, 1977, Complexes between L (+) lactate: cytochrome c oxidoreductases from the yeasts Saccharomyces cerevisiae or Hansenula anomala and horse heart cytochrome c. Biochimie 59, 621-626.

Rachofsky, E. L., Osman, R. and Ross, J. B, 2001, Probing structure and dynamics of DNA with 2-aminopurine: effects of local environment on fluorescence. Biochemistry 40, 946-956.

Ragone, R., Colonna, G., Balestrieri, C., Servillo, L. and Irace, G, 1984, Determination of tyrosine exposure in proteins by second-derivative spectroscopy. Biochemistry 23, 1871-1875.

Raja, S. M., Rawat, S. S., Chattopadhyay, A. and Lala, A. K. 1999. Localization and Environment of Tryptophans in Soluble and Membrane-Bound States of a Pore-Forming Toxin from *Staphylococcus aureus.* Biophysical Journal 76, 1469-1479.

Rattee, I. D. and Greur, M. M. 1974, in The Physical Chemistry of Dye absorption, Academic Press, New York.

Rao Ch. M., Rao, S.C. and Rao. P. B, 1989, Red edge excitation effect in intact eye lens. Photochemistry and photobiology 50, 399-402.

Ramasamy, S. M., Senthilnathan, V. P. and Hurtubise, R. J, 1986, Determination of room-temperature fluorescence and phosphorescence quantum yields for compounds adsorbed on solid surfaces. Analytical Chemistry 1986, 58, 612-616.

Ray, S., Mukherji, S. and Bhaduri, A. 1995, Two tryptophans at the active site of UDP-glucose 4-epimerase from Kluyveromyces fragilis J. Biol. Chem. 270, 11383-11390.

Renoir, J.-M., Mercier-Bodard, C., Hoffmann, K., LeBihan, S., Ning, Y.-M., Sanchez, E. R., Handschumacher, R. E. and Baulieu, E.-E., 1995, Cyclosporin A potentiates the dexamethasone-induced mouse mammary tumor virus-chloramphenicol acetyltransferase activity in LMCAT cells: a possible role for different heat shock protein-binding immunophilins in glucocorticosteroid receptor-mediated gene expression. Proceedings of Natural Academy of Sciences. U.S.A. 92, 4977-4981.

Rholam, M., Scarlata, S. and Weber, G, 1984, Frictional resistance to the local rotations of fluorophores in proteins. Biochemistry 23, 6793-6796.

Riccio, A., Vitagliano, L., di Prisco, G., Zagari, A., and Mazzarella, L, 2002, The crystal structure of a tetrameric hemoglobin in a partial hemichrome state. Proceedings of Natural Academy of Sciences. U.S.A. 99, 9801–9806.

Riordan, J.F., Wacker, W.E.C. & Vallee, B.L. 1966, Tetranitromethane. A reagent for the nitration of tyrosine and tyrosyl residues of proteins. Journal of The American Chemical Society. 88, 4104-4105.

Roch, P., Valembois, P. and Lassegues, M, 1980, Biochemical particulars of the antibacterial factor of the two subspecies *Eisenia fetida fetida* and *Eisenia fetida andrei*. American Society of Zoologists. 20, 794.

Rosenberg, R. D., Bauer, K. A. and Marcum, J. A, 1986, Antithrombin III, the heparin antithrombin system. Reviews in Hematology, 2, 351-416.

Rouvière, N., Vincent, M., Craescu, C. and Gallay, J, 1997, Immunosuppressor binding to the immunophilin FKBP59 affects the local structural dynamics of a surface beta-strand: time-resolved fluorescence study. Biochemistry 36, 7339-7352.

Royer, C.A., Tauc, P., Hervé, G. and Brochon, J-C, 1987, Ligand binding and protein dynamics: a fluorescence depolarization study of aspartate transcarbamylase from Escherichia coli. Biochemistry 26, 6472 – 6478.

Ruan, K.-C. and Weber, G, 1993, Physical heterogeneity of muscle glycogen phosphorylase revealed by hydrostatic pressure dissociation. Biochemistry 32, 6295-6301.

Ruan, K and Balny, C, 2002, High pressure static fluorescence to study macromolecular structure-function. Biochimica Biophysica Acta. Proteins and Proteomics. 1595, 94-102.

Rye, H. S., Yue, S., Wemmer, D. E., Quesada, M. A., Haugland, R. P., Mathies, R. A. and Glazer, A. N. 1992, Stable fluorescent complexes of double-stranded DNA with bis- intercalating asymmetric cyanine dyes: properties and applications. Nucleic Acids Research 20, 2803-2812.

Safo, M. K., Burnett, J. C., Musayev, F. N., Nokuri, S. and Abraham, D. J, 2002, Structure of human carbonmonoxyhemoglobin at 2.16 Å: a snapshot of the allosteric transition. Acta Crystallography D. 58, 2031-2037.

Sagan, S., Amiche, S., Delfour, A., Mor, A., Camus, A. and Nicolas, P. 1989, Molecular determinants of receptor affinity and selectivity of the natural delta-opioid agonist, dermenkephalin. Journal of Biological Chemistry 264, 17100-17160.

Sager, G. Nilsen, O.G., and Jackobsen, S, 1979. Variable binding of propranolol in human serum. Biochemical Pharmacology 28, 905 – 911.

Sagot, I., Bonneu, M., Balguerie, A. and Aigle, M. 1999, Imaging fluorescence resonance energy transfer between two green fluorescent proteins in living yeast. FEBS Letters 447, 53-57.

Sassaroli, M., Bucci, E., Liesegang, J., Fronticelli, C. and Steiner, R. F. 1984, Specialized functional domains in hemoglobin: dimensions in solution of the apohemoglobin dimer labeled with fluorescein iodoacetamide. Biochemistry 23, 2487 – 2491.

Sau, A. K., Chen, C. A., Cowan, J. A., Mazumdar, S. and Mitra, S. 2001, Steady-State and Time-Resolved Fluorescence Studies on Wild Type and Mutant *Chromatium vinosum* High Potential Iron Proteins: Holo- and Apo-Forms. Biophysical Journal 81, 2320-2330.

Sato, K., Nishina, Y. and Shiga, K., 2003, Purification of Electron-Transferring Flavoprotein from *Megasphaera elsdenii* and Binding of Additional FAD with an Unusual Absorption Spectrum. Journal of Biochemistry 134, 719-729.

Scarlata, S., Rholam, M. and Weber, G, 1984, Frictional resistance to local rotations in aromatic fluorophores in some small peptides. Biochemistry 23, 6789-6792.

Schaberle, F. A., Kuz'min, V. A. and Borissevitch, I. E, 2003, Spectroscopic studies of the interaction of bichromophoric cyanine dyes with DNA. Effect of ionic strength. Biochimica Biophysica Acta 1621, 183-191.

Sebban Pierre, Thesis, 1979, University of Paris 7. « Etude par fluorescence de molécule Hb^{desFe} ».

Shaanan B., Lis, H. and Sharon, N, 1991, Structure of a legume lectin with an ordered N- linked carbohydrate in complex with lactose. Science 254, 862-866.

Sharma, 0. P. 1976, Antioxidant activity of curcumin and related substances. Biochemical Pharmacology 25, 1811-1812.

Schmid, K. 1953, Preparation and properties of serum and plasma proteins. XXIX. Separation from human plasma of polysaccharides, peptides and proteins of low molecular weight. Crystallization of an acid lipoprotein. Journal of The American. Chemical. Society 75, 60-68.

Schmid K, Kaufmann H, Isemura S, Bauer F, Emura J, Motoyama T, Ishiguro M, Nanno S, 1973, Structure of α_1 -acid glycoprotein. The complete amino acid sequence, multiple amino acid substitutions, and homology with the immunoglobulins. Biochemistry 12, 2711-24.

Schmid, F. X., 1991, Catalysis and assistance of protein folding Curruent Opinion of Structural Biology 1, 36-41.

Shea, M.A., Verhoevenn, A. S. and Pedigo, S, 1996, Calcium-induced interactions of calmodulin domains revealed by quantitative thrombin footprinting of Arg37 and Arg106. Biochemistry 35, 2943-2957.

Siano, D. B., and Metzler, D. E., 1969, Band shapes of the electronic spectra of complex molecules. Journal of Chemical Physics 51, 1856-1861.

Sillen, A.,Verheyden, S., Delfosse, L., Braem, T., Robben. J., Volckaert, G. and Engelborghs, Y., 2003, Mechanism of fluorescence and conformational changes of the sarcoplasmic calcium binding protein of the sand worm Nereis diversicolor upon Ca2+ or Mg2+ binding. Biophysical Journal 85, 1882-1893.

Smith, K. D., Elliott, M.A., Elliott, H. G., McLaughlin, C. M., Wightman, P., Wood, G. C, 1994, Heterogeneity of α_1 -acid glycoprotein in rheumatoid arthritis. Journal of Chromatography B Biomedical Applications. 661, 7-14.

Soengas, M. S., Reyes Mateo, C., Salas, M., Ulises Acuña, A. and Gutiérrez, C. 1997, Structural Features of φ29 Single-stranded DNA-binding Protein. I. Environment of tyrosines in terms of complex formation with DNA. J. Biol. Chem. 272, 295-302.

Sopkova, J., Reneouard, M. and Lewit-Bently, A. 1993, The crystal structure of a new high-calcium form of annexin V. Journal of Molecular Biology 234, 816-825.

Sopkova, J., Gallay, J., Vincent, M., Pancoska, P. and Lewit-Bentley, A. 1994. The dynamic behavior of annexin V as a function of calcium ion binding : a circular dichroism, UV absorption, and steady-state and time-resolved fluorescence study. Biochemistry. 33, 4490-4499.

Spencer, R. D. PHD thesis (1970) University of Illinois at Urbana Champaign. *Fluorescence lifetimes : theory, instrumentation and application of nanosecond fluorometry.* Published by : University Microfilms International.

Spinelli, S., Vincent, F., Pelosi, P., Tegoni, M. and Cambillau, C, 2002, Boar salivary lipocalin. Three-dimensional X-ray structure and androsterol/androstenone docking simulations. European Journal of Biochemistry 269, 2449-2456.

Sridhar M.S, and Sharma S, 2003, Microsporidial keratoconjunctivitis in a HIV-seronegative patient treated with debridement and oral itraconazole. American Journal of Ophthalmology 136, 745-746.

Srinivasan, R. and Rose, G. D, 1994, The T-to-R Transformation in Hemoglobin: A Reevaluation. Proceedings of Natural Academy of Sciences USA. 91, 11113–11117.

Steiner, R. F. and Norris, L, 1987, Fluorescence dynamics studies of troponin c. Biopolymers 26, 1189-1204.

Stern, O and Volmer, M, 1919, Uber die Abklingungszeit der Fluoreszenz, Physikalische Zeitschrift. 20, 183-188.

Stolle-Smits, T., Gerard Beekhuizen, J., Kok, M. T. C., Pijnenburg, M., Jan Derksen, K. R. and Voragen, A. G. J, 1999, Changes in Cell Wall Polysaccharides of Green Bean Pods during Development. Plant Physiology 121, 363–372.

Storch, E. M., Grinstead, J. S., Campbell, A. P., Daggett, V. and Atkins, W. M. 1999, Engineering Out Motion: A Surface Disulfide Bond Alters the Mobility of Tryptophan 22 in Cytochrome b_5 As Probed by Time-Resolved Fluorescence and ^{1}H NMR Experiments. Biochemistry 38, 5065- 5075

Strickland, S., Palmer, G., and Massey, V, 1975, Determination of dissociation constants and specific rate constants of enzyme-substrate (or protein-ligand) interactions from rapid reaction kinetic data. Journal of Biological Chemistry 250, 4048-4052.

Struck, D. K., Hoekstra, D. and Pagano, R. E, 1981, Use of resonance energy transfer to monitor membrane fusion. Biochemistry 20, 4093-4099.

Stryer, L, 1965, The interaction of a naphthalene dye with apomyoglobin and apohemoglobin. A fluorescent probe of non-polar binding sites. Journal of Molecular Biology 13, 482 - 495.

Szabo, A. G., Krajcarski, D., Zuker, M. and Alpert, B, 1984, Conformational heterogeneity in hemoglobin as determined by picosecond fluorescence decay measurements of the tryptophan residues. Chemical Physical letters. 108, 145-149.

Swaminathan, R., Nath, U., Udgaonkar, J. B., Periasamy, N. and Krishnamoorthy, G. 1996, Motional Dynamics of a Buried Tryptophan Reveals the Presence of Partially Structured Forms during Denaturation of Barstar. Biochemistry 35, 9150 –9157.

Tadokoro, H. in *Structure of Crystalline Polymers,* John Wiley and Sons, New York, 1979.

Tahirov, T. H., Lu, T.H., Liaw, Y. C., Chen, Y. L. and Lin, J. Y, 1995, Crystal structure of abrin-a at 2.14 Å. Journal of Molecular Biology. 250, 354-367. Erratum in: Journal of Molecular Biology 1995, 252, 154.

Tai, P.-K. K., Albers, M. W., McDonnell, D. P., Chang, H., Schreiber, S. L. and Faber, L. E, 1994, Potentiation of progesterone receptor-mediated transcription by the immunosuppressant FK506. Biochemistry 33, 10666-10671.

Talbott L. D. and Ray, P. M, 1992, Molecular size and separability features of pea cell wall polysaccharides: implications for models of primary structure. Plant Physiology 98, 357–368

Tanford, C., Kawahara, K and Lapanje, S, 1967, Proteins as random coils. I. Intrinsic viscosities and sedimentation coefficients in concentrated guanidine hydrochloride Journal of The American Chemical Society 89, 729-736

Tegoni, M., White, S. A., Roussel, A., Mathews, F. S. and Cambillau, C, 1993, A hypothetical complex between crystalline flavocytochrome b_2 and cytochrome c. Proteins : Structure, Function and Genetics. 16, 408-422.

Teilum, K., Maki, K., Kragelund, B. B., Poulsen, F. M. And Roder, H. 2002, Early kinetic intermediate in the folding of acyl-CoA binding protein detected by fluorescence labeling and ultrarapid mixing. Proceedings of the Natural Academy of Sciences. U. S. A. 99, 9807-9812.

Thomas, M. A., Gervais, M., Favaudon, V. and Valat, P. 1983, Study of the Hansenula anomala yeast flavocytochrome-b2-cytochrome-c complex 2. Localization of the main association area. European Journal of Biochemistry 135, 577-581.

Tomlin A.D. and J.J. Miller (1989) Development and fecundity of the manure worm, *Eisenia fetida* (Annelida: Lumbricidae) under laboratory conditions. *In*: Dindal, D.L. (Ed.), Soil biology as related to land use practices. Proceedings of Seventh International Soil Zoology Colloquium, Syracuse, New York, 673-678.

Travis, J. and Salvesen, G. S, 1983, Human plasma proteinase inhibitors. Annual Review of Biochemistry 52, 655-709.

Truong, T., Bersohn, R., Brumer, P., Luk, C. K. and Tao, T. 1967, Effect of pH on the phosphorescence of tryptophan, tyrosine and proteins. Journal of Biological Chemistry 242, 2979-2985.

Truong, K. and Ikura, M. 2001, The use of FRET imaging microscopy to detect protein-protein interactions and protein conformational changes in vivo. Current Opinion of Structural Biology 11, 573-578.

Tsai, C. H., Simplaceanu, V., Ho, N. T., Shen, T. J., Wang, D., Spiro, and Ho, C, 2003, Site mutations disrupt inter-helical H-bonds (alpha14W-alpha67T and beta15W-beta72S) involved in kinetic steps in the hemoglobin R-->T transition without altering the free energies of oxygenation. Biophysical Chemistry. 100, 131-142.

Turoverov, K. K., Biktashev, A. G., Khaitlina, S. Y. and Kuznetsova, I. M, 1999, The Structure and Dynamics of Partially Folded Actin, Biochemistry 38, 6261 – 6269.

Urien S., Bree F., Testa. B., Tillement, J-P. 1991, pH-dependency of basic ligand binding to α_1-acid glycoprotein (orosomucoid). Biochemical. Journal 280, 277-280.

Urien, S., Giroud, Y., Tsai, R. S., Carrupt, P. A., Bree, F., Testa, B., Tillement J-P, 1995, Mechanism of ligand binding to α_1-acid glycoprotein (orosomucoid): correlated thermodynamic factors and molecular parameters of polarity. Biochemical Journal 306, 545-549.

Unzai, S., Eich, R., Shibayama, N., Olson, J. and Morimoto.H. J, 1998, Rate constants for O2 and CO binding to the alpha and beta subunits within the R and T states of human hemoglobin. Journal of Biological Chemistry 273, 23150 - 23159.

Valeur, B. and Weber, G. 1977, Resolution of the fluorescence excitation spectrum of indole into the ^1L_a and ^1L_b excitations bands. Photochem. Photobiol. 25, 441-444.

Valeur B. 2002, *Molecular Fluorescence. Principles and Applications.* Wiley-VCH.

Van Berkel, P. H. C., Van Veen, H. A., Geerts, M. E. J., De Boer, H. A.and Nuijens, J. H., 1996, Heterogeneity in utilization of N-glycosylation sites Asn[624] and Asn[138] in human lactoferrin: a study with glycosylation-site mutants. Biochemical Journal 319, 117–122.

Van Dijk W, Havenaar EC, Brinkman-van der Linden EC, 1995, α_1-acid glycoprotein (orosomucoid): pathophysiological changes in glycosylation in relation to its function. Glycoconjugate Journal 12, 227 - 233.

Varriale, L., Coppola, E., Quarto, M., Veneziani, B. M., and Palumbo, G. 2002, Molecular aspects of photodynamic therapy: low energy pre-sensitization of hypericin-loaded human endometrial carcinoma cells enhances photo-tolerance, alters gene expression and affects the cell cycle. FEBS Letters, 512, 287-290.

Vasquez, G. B., Ji, X., Fronticelli, C, and Gilliland, G. L, 1998, Human carboxyhemoglobin at 2.2 Å resolution: structure and solvent comparisons of R-state, R2-state and T-state hemoglobins. Acta Crystallography D Biological Crystallography 54, 355-66.

Verheyden, S., Sillen, A., Gils, A., Declerck, P. J. and Engelborghs, Y. 2003, Tryptophan properties in fluorescence and functional stability of plasminogen activator inhibitor 1. Biophysical Journal 85, 501-510.

Vidugiris, G. J. A. and Royer, C. A. 1998, Determination of the volume changes for pressure-induced transitions of apomyoglobin between the native, molten globule, and unfolded states. Biophysical Journal 75, 463-470.

Villette J. R, Helbecque, N., Albani, J. R. , Sicard, P. J. and Bouquelet, S. J. 1993, Cyclomaltodextrin glucanotransferase from Bacillus circulans E 192: nitration with tetranitromethane. Biotechnology and Applied Biochemistry. 17, 205-216.

Vincent, M., Deveer, A. M., De Hass, G. H., Verhu, V. H. and Gallay, J, 1993, Stereospecificity of the interaction of porcine pancreatic phospholipase A_2 with micellar and monomeric inhibitors. A time-resolved fluorescence study of the tryptophan residue. European Journal of Biochemistry 215, 531-539.

Wahl, P. and Weber, G, 1967, Fluorescence depolarization of rabbit gamma globulin conjugates. Journal of Molecular Biology 30, 371 – 382.

Ward, D. C., Reich, E., and Stryer, L, 1969, Fluorescence studies of nucleotides and polynucleotides. I. Formycin, 2-aminopurine riboside, 2,6-diaminopurine riboside, and their derivatives. Journal of Biological Chemistry 244, 1228-1237

Webb, L. E. and Fleischer, E. B. 1965. Crystal structure of porphine. J. Chem. Phys. 43, 3100-3111.

Weber, G, 1950, Fluorescence of riboflavin and flavin-adenine dinucleotide. Biochemical Journal 47, 114-121.

Weber, G, 1952, Polarization of the fluorescence of macormolecules. I. Theory and experimental method. Biochemical Journal 51, 145-155.

Weber, G. and Young, L. B, 1964, Fragmentation of bovine serum albumin by pepsin. I. The origin of the acid expansion of the albumin molecule. Journal of Biological Chemistry 239, 1415-1423.

Weber, G. and Farris, F. G, 1979, Synthesis and spectral properties of a hydrophobic fluorescent probe : 6-propionyl-2-(dimethylamino)naphthalene. Biochemistry 18, 3075-3078.

Weber, G, 1984, Order of free energy couplings between ligand binding and protein subunit association in hemoglobin. Proceedings of The Natural Academy of Sciences. U.S.A. 81, 7098 - 7102.

Weber, G., Scarlata, S. and Rholam, M, 1984, Thermal coefficient of the frictional resistance to rotation in simple fluorophores determined by fluorescence polarization. Biochemistry 23, 6785-6788.

Wetlaufer, D. B., 1962, Ultraviolet absorption spectra of proteins and amino acids, Advan. Protein Chem. 17, 303-390.

Wrighton, M. S., Ginley, D. S. and Morse, D. L, 1974, A technique for the determination of absolute emission quantum yields of powdered samples. Journal of Physical Chemistry 78, 2229-2233.

Wyss, D.F. and Wagner, G, 1996, The structural role of sugars in glycoproteins. Current Opinions in Biotechnology 7, 409-416.

Xia, Z.-X. and Mathews, F. S, 1990, Molecular structure of flavocytochrome b_2 at 2.4 Å resolution. Journal of Molecular Biology 212, 837-863.

Yang, C.-C., Sklar, P., Axel, R. and Maniatis, A, 1997, Purification and characterization of a human RNA adenosine deaminase for glutamate receptor B pre-mRNA editing. Proceedings of The Natural Academy of Sciences. U.S.A. 94, 4354-4359.

Yi-Brunozzi, H. Y., Stephens, O. M. and Beal, P.A, 2001, Conformational changes that occur during an RNA-editing adenosine deami- nation reaction. Journal of Biological Chemistry 276, 37827- 37833.

York, G. M. and Walker, G. C, 1998, The *Rhizobium meliloti* ExoK and ExsH glycanases specifically depolymerize nascent succinoglycan chains. Proceedings of The Natural Academy of Sciences. U.S.A. 95, 4912-4917.

Young, R. C., Patel, A., Meyers, B. S., Kakuma, T. and Alexopoulos, G. S, 1999, $Alpha_1$-Acid Glycoprotein, Age, and Sex in Mood Disorders. American Journal of Geriatric Psychiatry, 7, 331-334.

Yu, P. and Pettigrew, D. W, 2003, Linkage between fructose 1,6-bisphosphate binding and the dimer-tetramer equilibrium of Escherichia coli glycerol kinase: critical behavior arising from change of ligand stoichiometry. Biochemistry 42, 4243-4252.

Zandomeneghi, M, 1999, Fluorescence of Cereal Flours. Journal of Agricultural and Food Chemistry 47, 878-882.

Zandomeneghi M, Carbonaro L, Calucci L, Pinzino C, Galleschi L, Ghiringhelli S. 2003, Direct fluorometric determination of fluorescent substances in powders: the case of riboflavin in cereal flours. Journal of Agricultural and Food Chemistry 51, 2888-2895.

Zargarian, L., Le Tilly, V., Jamin, N., Chaffotte, A., Gabrielsen, O. S., Toma, F. and Alpert, B. 1999, Myb-DNA recognition: role of tryptophan residues and structural changes of the minimal DNA binding domain of c-Myb. Biochemistry, 38, 1921-1929.

Zentz, C, Glandières J. M., El Moshni S. and Alpert, B. 2003, Protein matrix elasticity determined by fluorescence anisotropy of its tryptophan residues. Photochemistry and Photobiology 78, 98-102.

Zhang, M., Tanaka, T. and Ikura, M., 1995, Calcium-induced conformational transition revealed by the solution structure of apo calmodulin. Nature Structural Biology 2, 758-767.

Index